Habsburg Wings 1915

Austro-Hungarian Aviation in the 1915 Campaign Over Galicia and the Balkans

Andrzej Olejko

Translated by Aleksandra Kalbarczyk-Biela

To Dr Gerhard Artl, Gerald Penz, and Boris Ciglić - the best researchers in the aviation history of Austria-Hungary and Serbia that I have ever met.

Acknowledgements

For our aviation books, please see our website at: www.aeronautbooks.com.

I am looking for photographs of the less well-known German aircraft of WWI to complete this series. For questions or to help with photographs you may contact me at jherris@me.com.

Interested in WWI aviation? Join The League of WWI Aviation Historians (**www.overthefront.com**), Cross & Cockade International (**www.crossandcockade.com**), and Das Propellerblatt (**www.propellerblatt.de**).

ISBN: 978-1-953201-56-0

Translation: Aleksandra Kalbarczyk-Biela
Design and layout: Jack Herris
Cover design: Aaron Weaver
Digital photo editing: Jack Herris

Aeronaut Books

Books for Enthusiasts by Enthusiasts

www.aeronautbooks.com

Table of Contents

Introduction 3

Chapter I. 1914 in the Air. The First Use of Aviation in the Decisive Battles of the War 10

Chapter II. The first aviation winter at the Nord-Ostfront or… Ohne Flieger kein Sieg! 104

Chapter III. Gorlice Breakthrough. Aviation in the Tarnów-Gorlice Operation in 1915 174

Chapter IV. Punitive Expedition Against Serbia … Aviation Activities in the Balkan Campaign 1915–1916 211

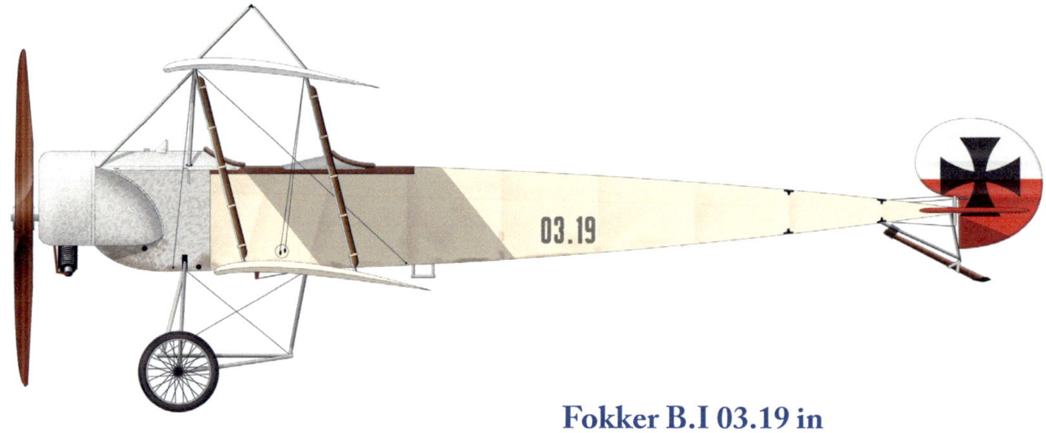

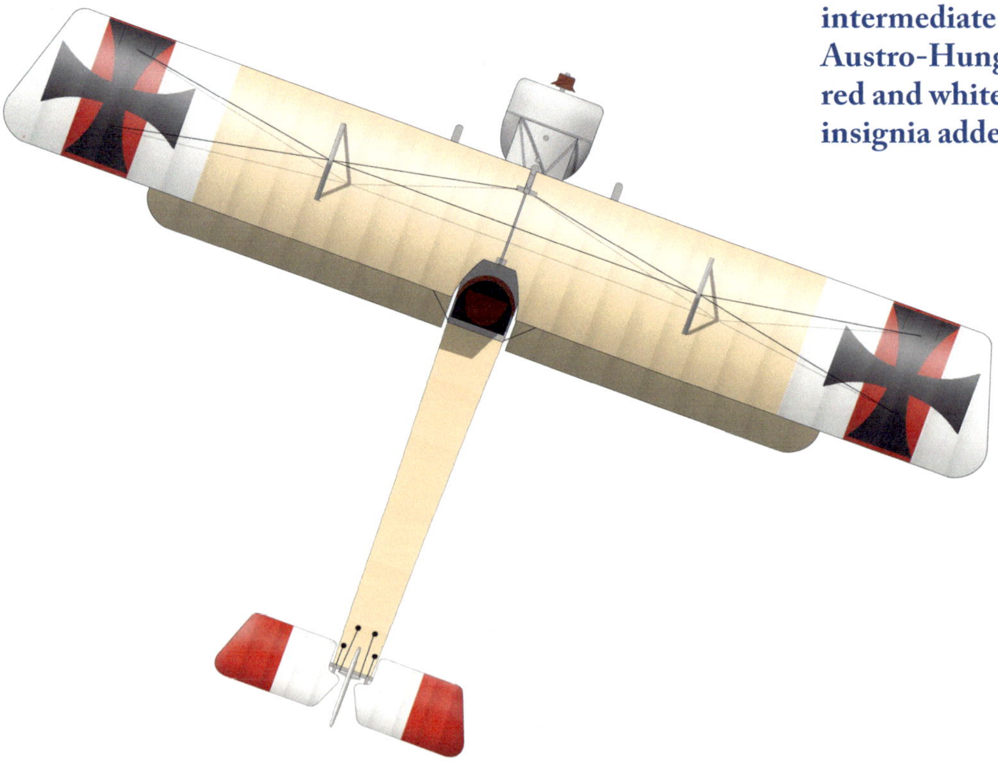

Fokker B.I 03.19 in intermediate markings of Austro-Hungarian colors of red and white with iron cross insignia added.

Introduction

April 29th, 1915.
For three days we have been observing [in Biecz] *areoplanes* [areoplanes - airplanes] *flying above us. Sałdaty* [spelling consistent with the original in the entire quoted text - soldiers of the tsarist army] *shoot with handguns and polymaths* [heavy machine guns] *because they are Austrian* [Austro-Hungarian] *(*Kaiserliche und Königische - KuK Luftschifferabteilung / Luftfahrtrupkiepen[1], *and Teuton* [German]. *Sensation! Areoplanes* [aeroplanes] *throw bombs. In Górne Przedmieście* [in Biecz] *one fell and plunged into the ground. This displeased the soldiers, and they ran away with the horses. But we are watching aeroplanes* [aeroplanes] *with concern, as well*[2].

The quoted entry from the Galician "Diaries of war events in Biecz" by Maria Kromkayowa presented the "fifth weapon"[3], i.e., aviation, seen over the Low Beskid, on the Eastern Front in the spring of 1915. This event took place just before the Gorlice Breakthrough (German: Die Durchbruchsschlacht von Gorlice -Tarnów 1915, Durchbruchsschlacht bei Gorlice-Tarnów, Die Durchbruchsschlacht bei Gorlice, Durchbruch bei Tarnow-Gorlice 1915; (Russian: Горлицкий Прорыв) from May 2nd 1915, during preparations to which from April 18th 1915, the commands of Austro-Hungarian (KuK Armeeoberkommando - AOK) and German (Oberste Heeresleitung - OHL) armies, for the first time used the possibilities of photo recognition on a large scale[4].

The warfare in Europe, which started in 1914, resulted in the establishment of several theaters of warfare (tw) in the European theater of war (etw), covering not only the land area, but also the sea and the newest one – the air, where the youngest, fifth weapon- aviation, dominated. Father Witalis Kapuśnik from the monastery in Biecz has described the September days of 1914 of the retreat of I.R.(Imperial-Royal) army from Galicia:

September 18th [1914]. *A lot arrived today from the site - 17 horses, with us 5 officers, are to stay until tomorrow. Constant movement. You can see that there is something going on in Przemyśl since such a movement is taking place here ... Site [camp] and the officers left on September 19th* [1914] *in the evening. The captain and the doctor stayed overnight in the barn and did not leave until September 20th [*1914]. *September 19th [1914] strangely calm. Three trains carried bridge-making boats. Towards Jasło - larger ones, and from Jasło a train with small boats. There is going to be a battle near Sanok - so the rumor says - we will see ... The skirmishes near Lviv are clearly going* on [there were two battles near Lviv in 1914], *people are talking a lot, and mostly those who escaped first and haven't seen anything, but heard from reliable people, and those again from others. The rail traffic is getting bigger and bigger, the transport of soldiers and various war props goes one way and the wounded the other way, day after day. The clergy began studying indoors, as it was possible with few in the cell, lectures in the Refectory, but there* areoplanes [airplanes] *began to fly over the Monastery, and more and more troops in the monastery, on September 20th [1914 r] I sent some to Zakliczyn by rail - evacuation train*[5]

Above: Viribus Unitis ... The beginnings of Austro-Hungarian aviation. – author's collection…

The October counter-offensive of the AOK, launched from the area of the Dukla Pass towards the Przemyśl Fortress, which was built as a 1st class fortress in the so-called The Przemyśl Gate, was to cover the main Hungarian direction, becoming for the AOK a cover for the Budapest direction, led to its unblocking on 10th October 1914 (the end of the first siege), but soon the November retreat of the I.R. army from the San River to the west began, as a result from another, this time decisive, counterattack by Stawka. It had a bad influence on the morale of the Galician society, which was afraid of the arrival of the Russian army - "Muscovites" or "Mochs" - M. Kromkayowa wrote in her diary as follows:

Above: Forgotten symbolism ... The seal of K.u.K. Luftschifferabteilung from the early days of Austro-Hungarian aviation. Fotoarchiv SAW / KA Wien collection

November 1st [1914]. *Airplanes above us [in Biecz]. God! What does it mean? We can't do anything. We grab newspapers ... Full of Army. Unfortunately, our* [I.R.] *army is retreating*[6]

Looking at the engulfed in war "old continent", where the longest front of the Great War was established in the summer of 1914 - the Eastern Front[7], its southern part naturally became a Galician tw, and the first two months of fighting on the border of the Russian and Austro-Hungarian partitions and on the Galician tw showed that this war, according to theoreticians, was supposed to be short-lived, but it would not be. November 1914 was the month of peak successes of the Russian Supreme Command (Stawka) on the Galician section of the Eastern Front. The corps of the Russian field armies that made up the Southwestern Front of the tsarist army regrouped and Stawka pushed the "Russian steam roller" ("rusissche Dampfwalze") with great force to the west, towards the "city of kings" - Krakow, which was a 1st Class Fortress, and to the south, in the side of the Carpathians blocking the road to Upper Hungary - today the Slovak Republic. The success of the Russian field armies achieved by the North-Western Front in the territory of the Kingdom of Poland ("Kongresówka") in October 1914, and especially the chase of the Russian 1st and 9th Armies for the beaten near Dęblin 1st I.R. filed Army, led towards Kraków, additionally worsened the position of other Imperial-Royal field armies in Galicia. As a result of the rapid change of the situation on the Galician front - in the AOK Nord-Ost Front nomenclature - and the necessity of the mentioned retreat of the 1st I.R. Army, the planned creation of a front on the San river turned out to be impossible for AOK, and thus the general retreat of the Imperial-Royal army in Galicia began. It happened very quickly but with less losses than during the previous retreat in September 1914. Royal-imperial 2nd and 3rd field armies were retreated under pressure by the Russian 8th field army towards the main ridge of the Carpathians - towards the Beskids- while the field forces of the 3rd Russian army, after crossing the San river, managed to break the position of the Imperial-Royal 4th field army near Nisko and force it to retreat towards Krakow. In this situation, the second siege of the Przemyśl Fortress took place. The location of the Przemyśl Fortress, from then on, attracted special attention of the AOK and Stawka, when it was cut off by the Russian Siege Army (later the 11th Army), becoming increasingly heavy, and its connection with the outside world was only ensured by ... aviation. *In modern warfare, aviators, apart from their professional duties, fulfill a grateful task for the families of fighting soldiers. Especially in our fortress* [Przemyśl], *to which communication* [spelling consistent with the original throughout the quoted text] *is suspended, the benefit of carrying letters from soldiers is even more appreciated ... In Vienna, they await the news about Przemyśl with bated breath ... Five soldiers can send one sheet of paper, since aviators can only carry additional loads, such as bombs or packages, up to 130 kg*[8].

In November 14th1914, the Stawka offensive in Galicia directed its main offensive efforts towards Silesia and Wrocław. The Imperial-Royal 2nd and 3rd field armies withdrew to the line of the Carpathians between Galicia (Przedlitawia) and the Kingdom of Hungary (Zalitavia), while the Imperial-Royal 4th field army withdrew to the Dunajec line having behind its backs, another 1st class fortress, the Kraków Fortress which was protecting the so-called The Moravian Gate and the Vienna direction[9]. "Stawka" hoped that an attack on German Silesia, breaking through to Moravia and the Czech Republic, which was heavily saturated with pan-Slavic, pro-Russian propaganda, could cause the final defeat of the Austro-Hungarian monarchy. The launch of the counteroffensive planned by AOK in the south depended primarily on how quickly the Imperial-Royal 2nd field army, transferred to Silesia, would be able to group together and how quickly it is possible to replenish the army ranks, thinned by losses in the fights so far in Galicia, to at least the average level. If the Russian offensive would become successful, it could be a cut off the field armies of the Central States from each other, to take control of industrial Silesia and to move closer to Czech Republic, a stronghold of Pan-Slavism. The intercepted messages (sparks messages), thanks to the Russian key getting into the AOK's hands, perfectly informed them and the OHL about the enemy's intentions, and yet it was not possible to forestall the Russian offensive, which was aimed at driving a wedge between the Royal-Imperial 1st and 4th field Armies in the north of Kraków (the command of the Kraków Fortress at that time was in the hands of Lieutenant General [FML][10] Karl Kuk).

The mentioned Imperial-Royal 4th Army was fully transferred from the right to left bank of the Vistula in the north of Krakow, and at the cost of emptying a large part of the Nord-Ostfront between the Vistula and the Beskids, where only the Lviv Imperial-Royal XI Corps was left (2 ½ Infantry Divisions - DP) and between the Vistula and the Warta, 30 Austro-Hungarian and German Infantry Battalions (approx. 360 infantry battalions) were gathered, outnumbered by the 21 ½ Infantry Divisions of the Russian 4th and 9th Army, numbering 340 infantry battalions. During 16-25 November 1914, the first battle of Krakow took place, and the Russian army was approaching from Gdów and Wieliczka not only to Krakow and Silesia, but also to the Beskidy belt, the forces of the Russian 8th field Army approached Nowy Sącz, hitting towards Limanowa. Stawka's offensive met with a relatively well-prepared counteroffensive by the Imperial-Royal Of the Army - counterattack of the Imperial-Royal 4th field Army in November 17th 1914, on north of Krakow, took on a wider scale, and in the same day the Imperial-Royal 1st field Army started launched offensive actions. The field armies subordinate to AOK stopped the Russian "Silesian offensive", which resulted in the need for Stawka to quickly send reinforcements from across the Vistula from the Russian 3rd Army, whose commander directed one corps to the south with the help of the fighters in the Krakow area, with the rest of the forces, continuing the strong offensive between the Vistula and the Carpathians, which was covered by only one royal-imperial XI Corps. This corps, attacked by three Russian corps of the 3rd and 8th field Armies fighting against it, was forced to retreat behind Bochnia, on the Wieliczka-Dobczyce line the way that from the south-eastern side the Russian 8th filed Army units approached Krakow. The situation was even more serious because the units also threatened the Moravian Gate[11].

The December breakthrough of 1914 was brought by the decisive battle of this campaign on the southern section of the Eastern Front, fought near Limanowa as the key clash of the Łapanów-Limanowa operation, also called, with some exaggeration, the "Poor Eastern Front"[12]. Mainly based on the Hungarian sources, the *Hungarian command, and the Hungarian heroism*[13] of soldiers going to battle not only in "blau"-colored uniforms but also in ... red cavalry pants, often with a sapper shovel in their hands triumphed in the Limanowa battle. Despite the terrible weather conditions, the air reconnaissance gave invaluable services to AOK at that time, both in the Second Battle of Krakow and in the above-mentioned breakthrough operation in 1914. During the ongoing battle at Limanowa, the following was written in the "Diariusz of the College of the Society of Jesus in Nowy Sącz":

12. [XII] 1914 All night and from the morning retreat in the panic of the Russian troops towards Zabełcze ... Aeroplanes [planes] over Sącz - Russian troops are constantly pushing towards Zabełcze ... Before five, huge traffic in the city and a scream - what is going on? Various Austrian patrols [c.k. army] entered the city. Then, the army entered, amidst

Above: War fear of the tsarist power ... A caricature from the epoch presenting in 1914 the advent of the Russian army. the collection of R. Bańkosz

the singing mostly Hungarian: it is therefore Austrian, Hungarian and Polish singing language[14] and in the chronicle of the Łącko Parish, it was noted:

On December 12 [1914] the Muscovites were closely following the movements of the Austrian army [c.k. armii - spelling consistent with the original throughout the quoted text] by aeroplanes rumbling over our mountains. Seeing the danger, on December 14 [1914] they began to quickly flee from the Limanowski and Sądecki districts. They did not have time to take from [Nowy] Sącz the provisions from the robbery of the entire district - supplies for the winter[15].

"Singing and colorful war" quickly turned into a nightmare full of completely different images, although with the retreat of the Russian troops in December 1914, on most of Western Galicia" returned "c.k. reality". Over the days, the fights in the Carpathian foothills and in the Carpathians turned into a typical trench warfare, and in close-range clashes, where the attackers alternately found themselves either in mud or in snow, bayonets, butts, and most often shovels dominated. Fogs and the abundance of rain and snowfall made it difficult

Above: The power of the Russian sword... A caricature from the epoch depicting beheading a multi-headed serpent symbolizing the Central Powers. the collection of R. Bańkosz

not only to maneuver the units, but above all to conduct reconnaissance from the air. Due to the disastrous weather conditions, the air reconnaissance, which provided such invaluable services to the AOK in the Battle of Krakow and in the Lapland-Limanowa operation in 1914, became illusory during the "Battle of Christmas" at the end of December 1914. Despite the exceptional dedication of the Fliegerkompanien crews (eight pilots and six planes each), K.u.K. Luftschifferabteilung, they were, like the Russian Korpuśne Awioatriady (KAO) literally "chained to the ground", at their field airfields in Przedlitavia and Zalitavia.[16] After the success of the AOK in the Łapanów-Limanów operation in December 1914, after a series of disasters and successes on both sides, the Nord-Ostfront line, based on the main Carpathian range, froze for the winter 1914-spring 1915 on the Biała- After the success of the AOK in the Łapanów-Limanów operation in December 1914, after a series of disasters and successes on both sides, the Nord-Ostfront line, based on the main Carpathian range, froze for the winter 1914-spring 1915 on the Biała-Dunajec-Karpaty rivier line. The military operations on the Galician tw ended together with the shift of the Eastern Front line to the east in May 1915, because of the (Durchbruchschlacht Gorlice) Gorlice Breakthrough by the field armies of the Austro-Hungarian-German, which took place as part of the Tarnów-Gorlice operation (for AOK and in Austrian historiography this is the wording of this operation and for OHL and in German historiography it is the Gorlice-Tarnów operation)[17]. The Balkan war failures of the Imperial-Royal field armies in the Serbian campaign of 1914 on the local front (in AOK Süd-Ostfront terminology - the Cerska Battle and the Battle of Kolubara) and the total operational defeat of the AOK (mistakes made in the "arrogant" planning of operations against the weaker Serbian army; errors in the command of Gen. Oskar Potiorek with the field forces of the 5th and 6th field armies; failure to use numerical and technical advantage; insufficient supply of the Imperial-Royal field armies fighting in difficult mountain terrain; underestimation of an opponent fighting with strong ideological support from the civilian population) became the so-called " Serbian's cheek" and caused the collapse of the prestige of the Imperial-Royal the Habsburg Monarchy in Europe, especially in the Balkans. For the Serb side, the 1914 campaign ended with practically no territorial changes and proved that in the clash between "Serbian David" and "c.k. Goliath ", the weaker one, fighting in his own territory and having great moral support, was the winner. The Serbian army, having no advantage either on land or in the air, fighting on its territory, not only defended the integrity of the borders of the Kingdom of Serbia, but from the military point of view it displaced after five months of fighting the overwhelming forces of the c.k 5. and 6. field armies to their starting positions in summer 1914![18] The Balkan campaign of 1914 ended with losses unheard of for both sides in the region (and not only), which were a harbinger of the terrible balance sheet of the Great War: the Serbian army lost 163,557 soldiers (including 2,110 officers) killed, wounded and missing; field c.k. 5. and. 6. Armies lost 273,804 soldiers (including 7,592 officers) killed, wounded and missing.[19] As a result of the failures and the policy of personnel changes, during and after this campaign, General O. Potiorek was removed from command in the Balkans, replaced on October 28, 1914 by General Moritz Freiherr Auffenberg von Komarow, and finally the commander-in-chief in the Balkans tw - Süd-Ostfront - became Archduke Eugeniusz F. "Erzi" Habsburg[20]. As a result of this chain of events, AOK suspended combat operations against the Kingdom of Serbia, which led to an almost nine-month break in fighting in the Balkan region and the introduction of a typical trench warfare instead of maneuvering. Although the losses significantly weakened the Serbian army, the 1914 campaign was a great success for it and for the Serb people. The Serbian army managed to repel three offensives of a much stronger and better-armed enemy, but the price-they paid for it was exceptionally high. Thus, there appear question, whether in the case of the next AOK offensive, which was only a matter of time, the strongly weakened Serbian army would be able to stop the attack from Imperial-Royal field armies once again. On December 24th 1914, the famous journalist Ferenc Molnár (Ferenc Neumann) in an interview published in the Austro-Hungarian press after an interview with the commander-in-chief of the Imperial-Royal in the matter of the Balkan campaign, Archduke Frederick Habsburg noted: We did not achieve great victories ... There was a catastrophe in Serbia[21]. Unfortunately, the passing time and the alliance of the Central Powers with the natural and long-established enemies of the Kingdom of Serbia in the Balkans (Bulgarian Empire, the Ottoman Empire) acted against him during the 1915 campaign.

The study handed over to you is an attempt to show the unique role of aviation in the military operations on the Galician section of the Eastern Front and on the Serbian Front in 1915, where at the turn of 1914/1915 and in 1915 alone, so many events, unknown to a wider audience, took place, based on an extensive archival query and multilingual literature on the subject[22]. I would like to thank the wide group of people who helped me in collecting the source material for the above study, hoping that the issues presented in this study will enrich, Dear Reader, the current state of your knowledge about those, for most of the now forgotten days of the Great War.

Andrzej Olejko

Endnotes

1 The original name of the Imperial-Royal (I.R.) aviation used during the fighting in 1914 was K.u.K. Luftschifferabteilung and from June 8, 1915 it was changed to K.u.K. Luftfahrtruppen. In August 1914, the Imperial-Royal's aviation had 85 pilots and 12 observers, and at the end of the Great War (World War I), about 36 thousand people served in the Austro-Hungarian air force, including over 1 thousand. officers, while the flying staff consisted of 964 officers and 1,854 non-commissioned officers. At the beginning of the fighting in 1914 there were 39 planes, and during the Great War, the aviation industry of the Imperial-Royal's Austria-Hungary delivered 4,356 aircraft engines and 5,431 aircraft to combat units (Fliegerkompanien). Losses incurred by the I.R. aviation during the fights until 1918 amounted to approx. 2,800 officers and non-commissioned officers, of whom, among the flying personnel, approx. 1,700 people, including 650 dead and missing - Lieutenant J. Rajlich," *Major polni pilot Otto Jindr"a* (1886-1932), "*Stories a Vojenstvi* "(HAV) 2009, No. 2, p. 8–11.

2 M. Kromkayowa, "*Diaries of war events in Biecz*", a copy in the private archive of Damian Nowak; http://bitwa.lobiecz.pl/kromkayowa.htm - entry from June 28, 2019.

3 "The fifth weapon" after the land forces - infantry, cavalry, artillery - and naval forces - after the Navy - see Der Luftkrieg, Berlin-Leipzig 1915, p. 9

4 H. Jäger, Erkundung mit der Kamera. Die Entwicklung der Photographie zur Waffe und ihr Einsatz im 1. Weltkrieg, München 2014, p. 120; M. Rauchensteiner, *Der Erste Weltkrieg und das Ende der Habsburgsmonarchie 1914–1918*, Wien-Köln-Weimar 2013, p. 212.

5 Archives of the Province of the Franciscan Fathers Minoris of the City of Krakow (APOFMMK), Chronicle of the monastery in Biecz, p.176 via D. Nowak.

6 M. Kromkayow, op. cit., p. 1.

7 The Russian Western Front, i.e., in Western historiography, the Eastern Front was 1770 km long, not including the Caucasus region. For comparison, the German Western Front in France was only 643 km long! In the armies of the countries fighting on both fronts, field armies were the basic operational units, divided into corps, divisions and smaller tactical units. In the case of the Russian army, the largest operational group was the Front, consisting of field armies, corps, divisions and smaller tactical units. During 1914–1918, there was a division of the Russian Western Front into operational fronts: The North-Western Front, the South-Western Front, the Caucasian Front and, from 1916, the Romanian Front

8 "The Was news" from October 10th 1914. Details of the air operations over the Przemyśl Fortress and over Western Galicia were presented by the author, among others, in the study *Karpacka war of three emperors* (Rzeszów 2013), *Wings of the Habsburgs*, Warsaw 2015, *The habsburgs's wings 1914. From the Balkans and the Adriatic to Galicia. Austro-hungarian aviation in the first campaignsof the Great war*, T. I, Lublin 2018, and in the periodical "Rocznik Przemyski - Historia Wojskowości" – *Z walk lotniczych K.u.K Luftschiggerabteilung over Galicia in autumn 1914 and during the first and the beginning of the second siege of the Przemyśl Fortress in the light of the Viennese Staatsarchiv-Kriegsarchiv archives and memoirs* (2013, Vol. 49, issue 1, p. 67–121).

9 *The Great War of 1914-1918 based on recent sources. Compiled and collected by Dr. Jan Dąbrowski, professor of the Jagiellonian University*, part I, vol. VII, Kraków 1937, pp. 243, 245, 247, 250–251.

10 FML - the spelling of military ranks from the period was used in the study.

11 A. Olejko, *Wings of the Habsburgs* 1914, Warsaw 2015 - passim; idem - *The habsburgs wings 1914*..., op. cit. passim. In mid-November 1914, the AOK and OHL had around 60 DPs across the Eastern Front (including the crew of the besieged Przemyśl Fortress), while Stawka had ten field armies, four of which fought in Western Galicia - see Z. Palski *Activities in the south-eastern territories during the First World War*, [in:] *Military activities in south-eastern Poland*, ed. W. Wróblewski, Warsaw 2000, p. 241, 244.

12 J. Roth von Limanowa-Lapanów, *Die Schlacht von Limanowa-Lapanów: Dezember 1914*, Insbruck 1928 - passim; M. Rauchensteiner, op. cit., p. 278; H. Münkler,

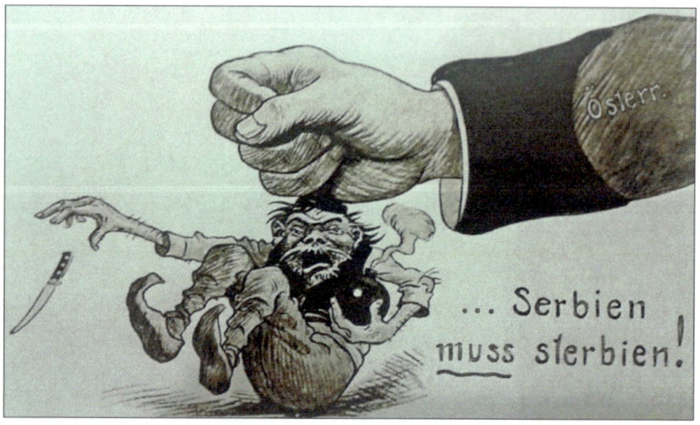

Above: A caricature from the era showing the line of Austro-Hungarian Balkan policy in 1914 - source: B. Hamann, Der Erste Weltkrieg. Wahrheit und Lüge in Bildern und Texten, München-Zurich 2014

Der Grosse Krieg, Berlin 2014, p. 311; Ł. Poremba, *World War I in the Nowy Sącz poviat - outline of fights, witness accounts, traces*, [in:] *Nowy Sącz and Sądecczyzna during World War I*, Nowy Sącz 2015, pp. 59-80; F. Julier, *Bitwa pod Limanowa* 1914, Oświęcim 2015 [A magyar katona csatadöntö küzdelmeinek okmányok és szemtanuk napoi vagy adatai alapján megirt részletes története, Budapest 1937] - passim; W. Polakiewicz, Limanowa 1914, Warsaw 2014 - passim. The authors of the introduction to the Polish-language version of the memoirs of Gen. J. Freiherr Roth von Limanowa-Łapanów (Josef Roth's baron von Limanowa-Łapanów, Bitwa pod Limanowa-Łapanowem December 1914, Limanowa 2018) as well as a critical commentary to them, W. Szczepanik and T. The janitor, while writing them, showed a surprising lack of full knowledge of the subject and ignorance of the full literature of the subject, which proves a "workshop" straight from internet forums. Among the decisive clashes of the 1914 campaign on the Eastern Front is also the largest maneuvering battle of this campaign at the turn of November and December 1914 - the Battle of Lodz - in which the German air force played a role similar to that in the Łódzkie-Limanów operation. It was the flight report of the Feldfliegerabteilung (FFA) 15 crew that decided about another German success near Łódź - the attack of the XX Corps in the area of Brzeziny on 23-24 November 1914! - *Lieutenant Der Weltkrieg. Die Schlacht von Lodz, [in:] Litzmanstadt, Geschichte und Entwicklung einer deutschen Industriestadt*, Poland 1941, pp. 96-116; A. Olejko, *A little about the history of the underestimated fifth weapon on the forgotten front of the Great War of 1914-1915*, [in:] *Aviation in the Gorlice Breakthrough 1915. A look after 100 years, ed. A. Olejko, A. Kharuk - passim.*

13 F. Julier, op. cit., pp. 9.

14 Ł. Poremba, *World War I in the Nowy Sącz poviat - outline of fights, witness accounts, traces*, [in:] *Nowy Sącz and Sądecczyzna during World War I*, Nowy Sącz 2015, pp. 83–84

15 Ł. Poremba, op. cit., pp. 110.

16 G. A. Tunstall, *Blood on the Snow: The Carpathian Winter War of 1915 (Modern War Studies)*, University Press of Kansas 2010, pp. 160.

17 A. Olejko, *Nieco...*, op. cit., pp. 18–44.

18 A. Olejko, *The habsburgs Wings 1914...*, op. cit .; *The Great War 1914-1918 ...*, op. cit., p. 233; A. Krzak, *The Great War in the Balkans. Military and political activities during World War I 1914–1918*, Czestochowa 2018, passim.

19 M. Радојевић, Љ. Димић, Србија у великом рату *1914–1918*, Београд 2014, s. 135; M. Rauchensteiner, *Der Tod des Doppeladlers. Őesterreich-ungarn und der erste Weltkrieg*, Graz 1997, pp. 153. A. Szelągowski (*Historia Powszechna*, T. II, Warsaw 1938, p. 746) states that according to the report of the c.k Minister of War presented on September 16, 1914 at the council of ministers of the Imperial-Royal army from the original state of 800-900 thousand soldiers fell to 130 thousand. C. Falls (*The Great War*, London 1960, p. 54), reports that the Serbian army lost 170,000. the fallen, wounded, captured and missing, and the c.k. the army of more than half of the states - approx. 215 thousand soldiers. J. Piekałkiewicz (*Calendar of World War I*, Warsaw 1998, p. 115) states that for the activities in the Balkans in the campaign of 1914, the army used 462 thousand. soldiers, losing 274,000 in combat (killed, wounded, missing and prisoners), and the Serbian army 132,000. soldiers. D. Jordan (The Balkans, Italy & Africa 1914–1918: *From Sarajevo to the Piave and Lake Tanganyika*, London 2008 - Polish edition - Balkans, Italy and Africa 1914-1918, From Sarajevo to Piawy and Lake Tanganyika, Poznań 2011, pp. 42) states that ck the army lost 227 thousand. soldiers, and H. Stratchan (*The First Word War*, London-New York-Sydney-Tokyo-Singapore-Toronto-Dublin 2003, pp. 30–31) reports that during four months of fighting on the Nord-Ostfront and Süd-Ostfront ck the army lost a total of 957 thousand soldiers. In the winter of 1914/1915, an epidemic of typhoid fever broke out in the Kingdom of Serbia, which claimed thousands of lives.

20 M. Rauchensteiner, op. cit., p. 188.

21 D. Jordan, *Balkans, Italy and Africa 1914–1918, From Sarajevo to Piawa and Lake Tanganyika, Poznań* 2011, p. 43; J. Piekałkiewicz, op. cit., p. 115; M. Rauchensteiner, op. cit., p. 153. H. Münkler (op. cit., p. 185) states that until the end of 1914, the army lost 450,000 in the Balkans soldiers of 30 thousand killed, 173 thousand. injured and 70 thousand. captives. On December 7, 1914, the Serbian government, deliberating after it was taken over by the Imperial-Royal Belgrade's army in Niš, he announced a declaration of the goals of the war, which included a record that the Serbs were waging *a*

war for the liberation and unification of all our unbroken brothers, Serbs, Croats, and Slovenes - see M. Tanta, *Balkans in the 20th century. History of Politics*, Warsaw 2003, p. 121

22 Bringing a new perspective to science requires the researchers of this era to use the widest selection of sources in order to obtain the most objective picture of the described events. The comprehensive issue of the participation of aviation in the fighting on the Eastern Front at the turn of 1914/1915 is presented in the study edited by A. Olejko and A. Kharuk (*Aviation in Przełamanie Gorlickie 1915*, op. Cit.), In which an international team of authors presented a previously unknown contribution aviation success achieved by the field armies of the Central Powers in Durchbruchschlacht Gorlice. It is hard to find such a point of view in the thematically unsuccessful and compilation studies by J. Cent (*Korpus Gwardii w bitwie pod Gorlicami,* Warsaw 2015) and edited by J. Centek, S. Kułacz and K. Ruszała (*Bitwa pod Gorlicami.* Warsaw 2015). However, this trend is alien to T. Woźny (*Operat 3. Rakusko uhorskej armady v priebehu bitky pod Gorliciami* (May 2–8, 1915), [in:] *Karpatský front 1914/1915,* Bratislava 2016). These authors, disregarding the Russian archival sources, did not contribute anything new to the study of the subject of the Great War in the Carpathians.

Above: Black Peter ... Political deck of cards 1914. In the photo King of Serbia Piotr, I Karajordziewič with a bloodied knife suggesting responsibility for the attack in Sarajevo. source: B. Hamann, *Der Erste Weltkrieg.* Wahrheit und Lüge in Bildern und Texten, München-Zurich 2014.

Donnet-Leveque 11

Chapter I: 1914 in the Air. The First Use of Aviation in the Decisive Battles of the War

After 1903, an airplane appeared in the arsenals of many countries around the world, with an engine and therefore able to compete with the wind[1], and when W. Wright in 1908 brought his plane to Europe (he founded an aviation school in the Republic of France)[2], it became quite clear that it was the Americans in the development of aviation – ahead the European pioneers of aviation[3]. The progress in aviation meant that even before the outbreak of the Great War, the position of the air leader in the number of fifth weapon[4], was given to republican France, and in the United Kingdom of Great Britain and Ireland, in the spring of 1912, the Royal Flying Corps (RFC) was established. A similar involvement took place in the Russian Empire[5] and in the Kingdom of Italy. Year of 1910 became a landmark in the history of the air force of the dualist, imperial-royal Austro-Hungarian monarchy[6]. On January 1st 1910, the Aero Club of Austria was established, based on the Vienna Aero Club[7] founded in 1901 by Victor Silberer, headed by Archduke Franz Ferdinand d'Este Habsburg[8]. And now…

On June 30, 1914 we were instructed to equip the Fischamend airport, for this purpose we were on our feet already at 6 am ready to leave Vienna (Wiener Neustadt). What equipment we were supposed to get was still a mystery. Then Uz (the nickname of Obstlt. [Lt. Col.] Milan Emil Uzelac - commander of the KuK Luftschifferabteilung [CK Aviation]) appeared and led us to the old balloon hall, where the disassembled C-Apparate [Lohner C Gebirgspfeillflieger planes] stood and were nested friendly and in peaceful condition robins. Suddenly I turned to the general [then Obstlt. M. E.] Uzelac: "For God's sake, are we supposed to fly with such coffins ?!". Uz quickly replied: "In the year [18] '66 [July 20] our brave and bold navy [- K.u.K. Kriegsmarine] drilled steel [Italian] colossi in wooden shells and thus won [with the Italian Navy (owned by Regia Marina) in the naval battle of Lissa - today Vis][9]. So, you probably won't want to tell me that you are like that, otherwise I don't know you!". He turned on his heel and walked away. And we went to the front with those coffins[10].

In 1909, the French Ministry of War purchased the first ... seven aircrafts, a year later ... 13 aircrafts and four airships were used during maneuvers in Picardy, and in 1913, the production of aircraft in French factories was 96 each month, and in 1914: in August 50, 62 in September and 100 in October[11]. In 1913, 115 aircrafts and 49 aircrafts for the Royal Navy Air Service (RNAS) were ordered for 1914

Above: Year 1914 – the era of the "fifth weapon" ... The photo shows an engraving from the era of air combat from 1914. source: B. Hamann, *Der Erste Weltkrieg.* Wahrheit und Lüge in Bildern und Texten, München-Zurich 2014.

at the British aviation plants for the RFC, which "father" is to believe Sir Hugh Trenchard, and the presentation of the RFC force took place on June 29th 1914 in Netheravon. On the other hand, analyzing the global development of the rivalry between the aerostat-airship and the plane (at the end of 1908, one of the "fathers of German aeronautics", specializing in the construction of semi-rigid Parseval airships [22 were built] - August von Parseval said that despite the lower engine power, the Wright's apparatus always stay ahead of the airship[12]) it is worth to mention that the state of this rivalry in the following years was to the decisive advantage of the second:

- in 1910 - Kaiser Germany: planes - nine, airships - five; Great Britain: planes - two, airships - four, republican France: planes - three, airships - 36, royal Italy: planes - two. airships - three, c.k. Austria-Hungary: planes - two, airships - three, imperial Japan: planes - two, airships - one, imperial Russia: planes - three, airships - three, royal Belgium: planes - two, airships - two, United States of America (USA): planes - two, airships - two.

In 1911, the number of airplanes in the world aviation was as follows:

	Army	Fleet	Private
United Kingdom of Great Britain and Ireland	57	31	167
Republic of China	1	0	1
Republics of France	161	10	400
Empire of Germany	50	2	101
Kingdom of Italy	31	4	36
Empire of Japan	4	4	3
Empire of Russia	23	5	41
USA	11	3	301!

In 1912, the number of aircraft in the above-mentioned countries were as follows:

	Army	Fleet	Private
United Kingdom of Great Britain and Ireland	23	6	130
The Republic of China	1	0	2
Republic of France	259	1	422
Empire of Germany	46	2	100
Kingdom of Italy	22	4	50
Empire of Japan	10	4	2
Empire of Russia	99	1	50
USA	3	2	750[13]

According to German sources, aviation spending in the run-up to the Great War in the German Empire, the Republic of France and the United Kingdom of Great Britain and Ireland were as follows:

1910

3,918,400 Marks - 5,212,350 Francs - £ 160,000
for flying equipment
307,500 Mark - 2,600,000 francs

1911

4,653,500 Marks - 15,279,800 Francs - £ 131,000
for flying equipment
2 359 100 Marks - 9 770 000 francs

1912

10,098,550 Marks - 32,528,000 francs - 232,000 pounds
for flying equipment
3,959,150 Marks - 23,464 230 Francs –

Above: The Kingdom of Montenegro was founded based on the declaration of independence of August 28, 1910, when Fr. Nikola I Petrović-Njegoš, during the celebration of the 50th anniversary of his reign, took the royal title and transformed the state into an "exotic kingdom" with himself as king Nikola I (Nicholas I Petrović) on the throne ... In the photo from the left are princesses Xenia and Militza, prime minister of the Montenegrin government, heir to the throne prince Daniło, below is the reigning royal couple - the king and his wife. source: "*Das interessante Blatt*" 1916

1913

39,985,650 Marks - 41,211,000 Francs - 526,000 Pounds
for flying equipment
17 475 100 Mark - 29 767 800 Francs –

1914

62,759,350 Mark - 46,733,900 Francs - £ 1,000,000
for flying equipment
27,029,600 Mark - 29,256,900 Francs[14]

The plane goes to war ...

In August 1914, so, in the first month of the fighting of the Great War, the front line of the air forces of the fighting parties - the Central Powers and the Entente - were as follows:

Above: King of Serbia, Piotr I Karajordziewič, during the campaigns of 1914 and 1915, was with his nation "for better and for worse"... In the photo, the Serbian monarch at the front. source: *"Das interessante Blatt"* 1916

German aviation: Western Front - 190 aircraft, five airships and 16 observation balloons; Eastern Front - 42 planes, three airships and six observation balloons
Austro-Hungarian aviation: Eastern Front - 48 aircraft, one airship and 12 observation balloons
French aviation: Western Front - 165 aircraft, 10 airships and 10 observation balloons
Russian aviation: Eastern Front - 263 planes, four airships and 46 observation balloons.
British Aviation: Western Front - 63 planes, two airships and two observation balloons.
Belgian Aviation: Western Front - 16 aircraft, two airships and two observation balloons[15]

Until the end of hostilities in 1918, the thriving German aviation industry produced 47,831, while the industry of allied Austro-Hungary only or as many as 5,391 aircrafts. Thus, the air force of the two major powers of the Central Powers bloc in the air war which it waged against the Entente bloc since 1914 used 53,222 aircraft in total. In the hostile Ententa bloc, until the fall of 1918, only British industry produced 55,061 aircrafts, French industry 52,146, American industry 13,840, Italian industry 12,031 and Russian industry only 5,607 aircrafts, for a total of 138,685 aircrafts for the Ententa air force. giving them 2 and ½ air advantage! By analyzing the production of aircrafts during the Great War in individual countries involved in it, we will obtain the following statistics:

The year 1914

Central States: German industry - 1,348 aircraft, Austro-Hungarian industry - 70 aircraft - a total of 1,418 aircraft
Ententa: British industry - 245 aircraft, French industry - 541 aircraft, American and Italian industry - 0 aircraft, Russian industry - 535 aircraft - a total of 1,321 aircraft

The year 1915

Central States: German industry - 4,532 aircraft, Austro-Hungarian industry - 238 aircraft - totaling 4,770 aircraft
Ententa: British industry - 1933 aircraft, French industry - 4489 aircraft, American industry - 0 aircraft, Italian industry - 382 aircraft, Russian industry - 1305 aircraft - 8109 aircraft in total

The year 1916

Central States: German industry - 8,182 aircraft, Austro-Hungarian industry - 931 aircraft - 9,113 aircraft in total
Ententa: British industry - 6,099 aircraft, French industry - 7,549 aircraft, American industry - 83 aircraft, Italian industry - 1,255 aircraft, Russian industry - 1,870 aircraft - 16,856 aircraft in total

The year 1917

Central Powers: German industry - 19,646 aircraft, Austro-Hungarian industry - 1,714 aircraft - 21,360 aircraft in total
Ententa: British industry - 14,748 aircraft, French industry - 14,915 aircraft, American industry - 1,807 aircraft, Italian industry - 3,871 aircraft, Russian industry - 1,897 aircraft - 37,238 aircraft in total

The year 1918

Central Powers: German industry - 14,123 aircraft, Austro-Hungarian industry - 2,438 aircraft - total 16,561 aircraft
Ententa: British industry - 32,036 aircraft, French industry - 24,652 aircraft, American industry - 11,950 aircraft, Italian industry - 6,532 aircraft, Russian industry - 0 aircraft - 75,170 in total[16]

With the passage of time and the development of aviation, important events from 1914 were gradually forgotten: on August 16-21, 1914 it was the German air force that recognized the British land forces on the Western Front at Le Cateau and during the subsequent battles at Mons; On August 23, 1914, the crew of the German airship

Above: On the route from the train station to the city hall, the first car was driven by the mayor of Sarajevo, Fehim Efendi Curčic and the police chief, and the archducal couple drove a Gräf und Stift sports car, lent by hr. Franz Harrach, who sat down next to the driver, and the governor of Bosnia and Herzegovina, General Oskar Potiorek, sat opposite her ... In the photo Wien, June 28, 2014, 100 years later. In the Heeresgeschichtliches Museum, editors Jerzy Pasierb and Andrzej Olejko are standing next to the historic car. source: author's collection.

SL II "Liegnitz" on the route Kraśnik-Chełm-Lublin-Kraśnik made the first reconnaissance on the Eastern Front for the benefit of Imperial-Royal Armeeoberkommando (AOK); in the period from August 15th to September 9th 1914 preceding the Battle of the Marne and during its duration, with the exception of two rainy days, every day the command of both warring parties needed air reports[17] In 1918, the above-mentioned events, such as the so-called "The Fokker Scourge" and the air dominance in the fall of 1916 of Albatros fighter planes were just ... a breath of the past in the face of the massive use of the "fifth weapon".[18] The year 1918 before the end of the world conflict, which on the Western Front was marked by "bloody" butchery related to the last OHL offensives, which were the last hopes for victory. If you were to compile a factual statement in the style of do you know that concerning the Air Force of the Great War of 1914–1915, then in the light of the German-language literature on the subject, the following data should be included:

In the period of August–October 1914, the German air force lost as many as 100 planes and 52 pilots, while the air force of the Russian Southwestern Front lost from 99 planes in August 1914 to December 93![19]

In 1914, the price of the Rumpler "Taube" was 25,000 marks - the first-class C planes already cost 35,000. marks, and at the turn of 1917/1918, a class C plane cost 40-60 thousand marks! In the final period of hostilities, the production costs of AEG, Gotha or Staaken planes as class G planes - Grossenflugzeug - amounted to 500,000 marks, and the Siemens Schuckert type with six Type VIII engines - 750,000 marks.[20]

On October 25th, 1914, a radio was used for the first time over the Western Front to coordinate artillery fire with the participation of an air observer.[21]

Left: Serbian manual car wash ... A caricature from the era depicting the washing of hands by the Serb side from the Balkan tensions in 1914 – source: B. Hamann, *Der Erste Weltkrieg. Wahrheit und Lüge in Bildern und Texten*, München-Zurich 2014.

On April 18th, 1915, a French Morane Saulnier L "Parasol" fighter piloted by Rolland Garros was shot down over the Western Front by the German anti-aircraft defense (opl.), Landing on the German side of the front line, and then copying the km-u synchronization system used in it, caused it to be installed in a Fokker EI aircraft on which already on August 1, 1915, Oblt. Max Immelmann ("Eagle of Lille") achieved his first aerial victory[22]

On September 29th , 1914, two squadrons of Voisin planes created the first bomber squadron in the French aviation, which from November 13th , 1914 was in action at the front.[23]

The well-known German bomber aviation unit Briefabteilung Ostende (BAO) in 1915 had 36 aircraft grouped in six squadrons[24]

In December 1914, the first experiments were carried out on the K.u.K. Luftschifferabteilung of radio apparatus then Oblt. Ludwig Dumbacher, Hptm. Viktor Schüntzel and Oblt. Stefan Horwath together with Zgsf. Max Libano and Lt. Benno Fiala Ritter von Fernbrugg performed radio flights in the area of Zakliczyn, Gorlice and Bobowo for the benefit of the Imperial-Royal Csurilo Artillery Group[25]

By 1913, the German Fleet (Kaiserliche Marine) received 11 airships, and when the air war over the British Isles started in 1915 with their participation, the famous erudite and creator of Scherlock Holmes, Arthur Conan Doyle and 200 other personalities the world of culture protested against the air war conducted in this way, emphasizing that in a civilized world it is unacceptable[26] (the first airship raid on Great Britain on January 19th , 1915 was carried out by three Fleet airships, initiating the implementation of the propaganda slogan "England durch feuer zu zerstören" and the first victims of German air bombing in the British Isles were 72-year-old Martha Taylot and 53-year-old Samuel Smith)[27]

On May 31, 1915, the German airships of the Fleet, following the propaganda slogan "Gott strafe England", bombed London for the first time. In the first phase of airship raids on the British Isles, in the period of January-November 1915, 50 airships were made, and in 23 raids 35 tons of bombs were dropped (308 British were killed and 531 were wounded). In the second phase of the airship attacks on the British Isles, in the period of February-November 1916, 22 airstrikes (123 airships) were carried out and 120 tons of bombs were dropped on British ground targets (193 British were killed and 692 were wounded), however, losing as many as 13 airships, including five shot down by RFC fighter planes[28]

The specialization of individual types of aircraft was created extremely quickly during the fights of the Great War, which was forced by the requirements of the land battlefield and technical progress in aviation. The following data addresses this issue in the development of the British RFC in 1914–1916:

In 1914, the RFC had 100% of the reconnaissance aircraft in its composition

Above: Aviation Fantasy… Oblt. Hassan Riza Effendi Pieler, pilot of Fliegerkompanie (Flik) 4 in the cabin of a fighter plane type Fokker A. III (E. III). On September 13, 1915, during a combat flight on the Italian Front, he flew an Albatros B.I aircraft from Flik 4 whose Observer Oblt. Roman Florer fought an air battle with an Italian plane, shooting it down. It was the first ever shot down on the Italian Front, how important it was because it was a Pole! - source: Fotoarchiv SAW-KA Wien

By mid-1915, 90% of the RFC state were bombers and 10% fighter planes
In mid-1916, 68% of the RFC state were bombers, 32% fighter planes and aircraft of direct support of the battlefield[29]

In the light of the German archives, the list of losses of enemy planes shot down by the German air force on the Western and Eastern Fronts of the Great War in the period from September 1915 (the beginning of the "bloody war in the air") until the end of this year were as follows: September 1915 - Western Front - 30, Eastern Front - 0 - October 1915 - Western Front - 17, Eastern Front - four - November 1915 - Western Front - 13, Eastern Front - one - December 1915 - Western Front - 17, Eastern Front - one[30]

According to British archives, in the years 1915–1916 the German air force suffered the following losses:

Above: From infantry to aviation ... With the creation of the KuK Luftschifferabteilung, many of the army's officers began to think about changing their service assignment, ceasing to be a cadre of infantry, artillery, or cavalry. An example of a typical "aviation career" in the KuK Luftschifferabteilung / Luftfahrtruppen during the Great War is Hptm. Roman Florer - in the photo - which of the Imperial-Royal 45th Infantry Regiment"Erzherzog Joseph Ferdinand"from Przemyśl / Sanok went to the "fifth weapon". source: collection of the Polish Aviation Museum in Krakow

Above: In Kotor the air is fragrant, and the light is gray, gray like cliffs, when the clouds cover the peak of Lovćen, and when it rains ... The photo shows the beauty of Boka Kotorska, the port in Kotor, the anchorage of KuK Kriegsmarine and the majesty of the Lovćen massif, where the national park is now and in 1914, dominating the Bay of Kotor, it was within the borders of the Kingdom of Montenegro, the so-called by the Lovćen battery. source: "*Das interessante Blatt*" 1916.

1915 - air combat - 19, missing - 6, shot down - two planes - 27 planes in total
1916 - air combat - 44, missing - 19, shot down - six aircraft - total - 69 aircraft
At the same time, British aviation losses were as follows:
1915 - air combat - 35, shot down - eight planes, landings for various reasons, including forced landings - seven - 50 planes in total
1916 - air combat - 260, shot down - 80 aircraft, landings for various reasons, including forced landings - 17, in total - 357 aircraft[31]

In September 1914, the first ever air battering ram carried out by a tsarist aviation airman took place over Galicia[32], which until now has been little talked about in the western part of the continent[33], although the unwritten rule of the third dimension on all fronts of the Great War was: I let you live, and you let me[34]

The above-mentioned aviation episodes of the Great War are virtually unknown to the inhabitants of the Western world and not only. They show a different face of the fighting on the Eastern Front, for many military historians constantly overshadowed by the legend of the red baron[35] and so popularized on various levels of research in the western part of the Great War[36]

But also, in the biography of this most famous aviator of the Great War there are white fields in the minds of the inhabitants of the Western world, because it was on the Eastern Front in June 1915 that he made his first combat flights as an observer in FFA 69. In his memoirs he wrote:

In June, July, and August 1915, I was in the air unit that accompanied the entire Mackensen's [commander of the German field 11th Army - ed. ed.] offensive [gen. August von] from Gorlice to Brest [on the Bug River]. I came there as a young observer and had no clue ... It is important for the observer to find a diligent pilot. And then one fine day the news broke that "[Oblt.] Count [Erich von] Holck is coming towards us." A thought immediately occurred to me - this is the man you need ... After a long journey by train, he finally came to Jarosław. There he got off the train, as it was again one of those endless long stops, and he ordered his orderly to continue the train with the luggage. He himself decided to go on foot ... After walking fifty kilometers, he finally reached his destination in Rawa Ruska, and the orderly with the luggage did not appear there until twenty-four hours ... Count [E. von] Holck was not only a sportsman who fished out on the green grass; he, apparently, also enjoyed air sport. He was a pilot with extremely rare abilities and, most importantly, he was at least a class better than the enemy. With him I made many beautiful - who knows how far - reconnaissance flights over Russia. With this young pilot, I

Above: Machina demonica ... This is how the artist from the era presented in a watercolor the first ever aerial bombardment of Turkish troops in the oasis of Tagiura and Ain Zara by the Italian tenete pilot (second lieutenant) G. Gavotti, piloting a German-made Etrich "*Taube*". source: Domena Public.

never felt uncertain, and what's more, in critical situations he was always my support ... We were just changing the landing site to a more forward one, and we didn't know exactly which one was the right one. In order not to unnecessarily endanger our box when landing, we flew towards Brest [on the Bug river]. The Russians were in complete retreat and everything was on fire - it was a terrifying and beautiful at the same time. We wanted to find hostile columns of troops and we had to fly over the burning city of Wisznice [Lublin Province - 90% of the town's buildings were burned down during the Great War]. A huge cloud of smoke, perhaps even two thousand meters, made it difficult for us to continue our flight, because for better visibility we flew at an altitude of fifteen hundred meters. [E. von] Holck thought for a moment what to do, and I advised him to avoid the cloud, which would extend the flight perhaps five minutes. But he did not take it into account at all. On the contrary. The greater the danger, the more it attracted him. So, we're flying right through the middle of the smoke cloud! I also enjoyed being able to accompany such a perky guest. Our carelessness almost turned out to be too costly for us. As soon as the tail of our machine disappeared into the cloud of smoke, I immediately noticed the swing of the plane. I couldn't see anything; the smoke stung my eyes and the air was very hot. Only the sea of fire was visible below. Suddenly the plane lost its balance and began to tumble downward. I only managed to grab the strut, or I would have fallen out of the plane. The first thing I did was look at [E. von] Holck, whose face showed unwavering confidence, and that gave me immediate courage. All that crossed my mind was that it would be stupid to die a heroic death like this[37]. This combat flight ended in the fire of the Russian infantry with the shooting down of the German crew due to engine damage and its forced landing. Both pilots, however, were very lucky not to be captured. Late in the evening we finally reached our airport on a cart. Damn the machine[38].

In the years 1914–1915 reconnaissance was the main aviation

Above: Italian tenete pilot (second lieutenant) G. Gavotti. source: Domena Public.

Above: On October 15, 1910, a military airport was opened in Wiener Neustadt. The inaugural flight was carried out with the participation of the "Polish duo" ... Airplane designed by Ing. Adolf Warchałowski (Zivil Flieger Adolf Warchalowski, on April 22, 1910, received pilot diploma No. 1 in the Austro-Hungarian Empire) piloted by Oblt Mieczysław (Miecislaus) Miller, visible in the photo (on June 22, 1910, he received a pilot diploma No. 5) standing next to the Etrich *Taube*. source: collection of the Polish Aviation Museum in Krakow.

contribution to the war. As defined by one French aviation theorist in the early 20th century, the work of observation aviation was essentially to perform aerial reconnaissance for the army and corps, obstructed by bad weather; establishing leadership positions for use by infantry division commanders; supplying ground troops[39]. From 1915, the Great Air War became a field of fighter aviation.[40]

However, in 1914, during the first campaign of the Great War, the sight of an airplane on the Western, Eastern and Serbian Fronts over the Balkans evoked similar emotions as over the above-mentioned Biecz in the spring of 1915? The achievements of the late nineteenth and early twentieth century technology, such as the telephone and radio, often failed, and the warring parties referred to the nineteenth century, old and proven means, such as trumpets and signal flags, which were supplemented by the fighting spirit at the time when, in the initial period of the Great War, units the infantry of all fighting armies attacked the enemy positions traditionally in close ranks, going straight to the cannons and machine gun positions of the enemy, incurring unprecedented losses[41]. As soon as the "time of frontal slaughter" came the time of the above-mentioned and other "time of war" inventions that dominated all three "dimensions of war" - land, water, and air! On August 26-30, 1914, the German OHL managed to stop the attack of the Russian 2nd Army (the offensive actions of the Russian field armies on the Eastern Front were called the Russian "steam roller" - "Rusische Dampf Walze") in the battle of Tannenberg for the Eastern Front (today Stębark), after which General Paul von Hindenburg, commanding the victorious 8th Field Army, was to say:

The most wonderful people are my flyers. I cannot express what extraordinary feats they performed during their flights. After mobilization in 1914, the German air force - FFA / FeFA - was subordinate to the commands of the German field armies as well as the Army Corps and Cavalry Corps, and some eight Etapen-Flugzeugparks were quickly renamed the Armee Flugzeugparks constituting the technical support of eight German field armies. From the FFA group, four - FFA 14th, 15th, 16th and 17th - came from the fortresses Posen (today Poznań), Graudenz (today Grudziądz) and Königsberg (today Kaliningrad) with an assignment to the Eastern Front but along with the remaining 29 FFAs they were directed to the Western Front.[42] During the Battle of Tannenberg, the German air reconnaissance played a decisive role, providing the command of the above-mentioned field army with precise information about the enemy's movements[43], so that after its completion the words of the commander of the above-mentioned 8th Army- Without an airman, there would be no Tannenberg (German: Ohne Flieger kein Tannenberg), were the best assessment of the German aviation crews (Die Fliegertruppen des deutschen Kaiserreiches)[44]. In none of the groundbreaking battles of the first campaign of the Great War of 1914 on the Serbian, Western and Eastern Fronts,

Above: So they fly on this ... This is how you can read the speech of curious glances casted by German army soldiers towards a foyer transported somewhere in Eastern Galicia in 1915 in the hall behind ... a cart of an Aviatik B.I plane in the colors of the KuK Luftschifferabteilung. source: collections Polish Aviation Museum in Krakow.

aviation was used in a spectacular manner, reminiscent of the great operations of the Second World War with the participation of large air force formations. In fact, these were flights of single pilots or two-man crews, but the intelligence collected on the scale of three fronts by these few airmen was so exceptional that it decided on the highest-level decisions taken by the commanders. It should be emphasized that in 1914, for the general "chiefs of the older generation", aviation was an unreliable partner in providing information about the enemy in the confrontation with his "competition", which was the cavalry that was trustworthy and so far, reliable in the quality of reconnaissance[45]. It was the information from the "third dimension" provided by the few flying personnel of the "fifth weapon" that allowed the commanding Serbian, British and German armies in this order, who stood above the above-mentioned "staff prejudices", risking, however, believing in the information provided by several aviators to make groundbreaking decisions in the battles of 1914 fought at Mount Cer, at Mons, at Tannenberg and then at the Marne. It is the last of the above-mentioned battles, included in the group of 18 battles deciding the fate of the world, most often considered to be a fight where for the first time the use of the "fifth weapon" determined ground success, but was it so?

Until today, the symbol of the "miracle on the Marne", the battle fought on September 5-9, 1914 on the Western Front, is a Paris taxi, because on the breakthrough day of fighting units of the French 6th Army and the German 1st Army, which was on September 7th , 1914 ., about 600 Parisian taxis (mainly Renault cars), transported, on the order of the commander of the French field army, half of the French DP (about 6,000 soldiers) to the Marne, covering a distance of 74 km twice[46]. And yet, for the French command, the key to the above-mentioned breakthrough victory in the 1914 campaign on the Western Front were the air reports provided by the pilot L'aviation Militaire[47] about the recognition of the location of the German 1st Army's forces and the created frontal gap into which the counterattack of the French field has directed 6th Army. The battle itself showed that it was aerial reconnaissance that became an invaluable help in planning the French counterattack and the key to victory![48] The use of aviation in the Battle of the Marne, however, was the second incident in the history of the Great War on the Western Front, as for the first-time information from the "third dimension" determined the tactical British success in the August Battle of Mons! In the case of the Battle of Tannenberg, a landmark on the Eastern Front of the abovementioned campaign, it is surprising that so far, no comprehensive historical study has been published confirming the above words of General P. von Hindenburg. This gives food for thought, because the many times cited in the world literature on the subject have become a phrase summarizing the role of aviation in this breakthrough battle in East Prussia, which is not confirmed by the authors dealing with the subject of fights on it. For example, among German or Polish historians, the participation of aviation in the Battle of Tannenberg[49], so appreciated by the commander of the field of the 8th Army in 1914, was also not appreciated, who, in the light of the above-mentioned words, did not analyze the air operations over Tannenberg![50] For the first time, this issue was widely presented only in 2015.

However, it has not been noticed or forgotten in world historiography that aerial reconnaissance influenced the overall success of the land army during the Great War not only on the Western or Eastern Fronts, but also during the fights on the Serbian Front! In the battle of Mount Cer, which was decisive for this front, fought on August 15 (28) -24 (September 6), 1914[51], about 100 km from Belgrade, which ended in a painful defeat for the Austro-Hungarian army and was the first major success of the troops Ententas over the armies of the Central Powers in the Great War, it was a few air reports that determined the success of the Serbian army!

Endnotes

1 R. Witkowski, Bracia Wright, Warszawa 2003, s. 72-76, 96-97.

2 A. Nahum, Flying machines, Warsaw 1991, p. 14. The first military airman in history Lt. Thomas Selfridge of the US Army, died on September 17, 1908 in an air accident in Fort Myer, Virginia - see Lee Kennett, The First Air War 1914-1918, London 1991, p. 117

3 Der Luftkrieg 1914-1915, Leipzig 1915, p. 9. Piąta broń po piechocie, artylerii, kawalerii i marynarce wojennej..

4 H. de Graffigny, L'Aéronautique et l'aviation en 1909, Paris 1910 r. passim.

5 At the end of August 1911, the planes were used for the first time for reconnaissance purposes in army maneuvers in the Warsaw Military District (the equivalent of the Corps District in the organization of the Polish Army in 1918-1939), five Farman planes, and one airship was assigned to the exercising Northern side, while another five Bleriot aircraft of the South Group. On February 16 (1 March), 1913, a plan for the organization of the Russian aviation was drawn up, which assumed the allocation of aviation units - Aviaotriads (AO - each AO was to consist of four officers and two pilot soldiers, six aircraft and 34 service soldiers - three each for service on each aircraft) one for each corps, the formation of an AO for the needs of large cavalry units, and the organization of an AO for reconnaissance and combat of enemy aviation. In total, it was planned to create 63 AOs, including 37 AOs, 10 army AOs, three special-purpose AOs and eight AOs installed in fortresses. Eight Rotary Rotations were planned to be used for their service - see В. М. Ткачев, Крылья России. Воспоминая о прошлом русской военной авиации 1910-1917 гг, Санкт Петербург 2007, p. 43, 151–152.

6 R. Desoye, Die k.u.k. Luftfahrtruppe. Die Entstehung, der Aufbau und die Organisation der österreichisch-ungarischen Heeresluftwaffe, Wien 1999, s. 31.

7 Österreichische Staatsarchiv-Kriegsarchiv Wien (SAW-KA), August Raft-Marwil, Die Erste Luftverkehrslinie der Welt, s. 3-4; F. Hinterstoisser, Aus meinem Luftschiffertagebuch, Salzburg 2008 (org. Rzeszów 1904), p. 95–113.

8 J. Rajlich, On the Wings of the World Wolf, Cheb 2014, p. 42; M. Plavec, Archduke Leopold Salvator and his balloon Meteor, [w:] From balloon to transatlantic aircraft, ed. M. Plavec, Prague 2014, p. 14-15. 1 In 1902, the Aeroclub Węgier was built - Lt. F. Hinterstoisser, op. cit., p. 80–81.

9 The principle of placing place names in the languages of the majority of the inhabitants of the described regions of Europe was adopted throughout the study.

10 SAW-KA, General Uzelac, herausgegeben v. Jaromir Diakow, p. 54 - memory of major pil. Hugo Schwaba. S. Rosenboom (Im Einsatz über der "vergessenen Front". Der Luftkrieg an der Ostfront im Ersten Weltkrieg, Potsdam 2013, p. 23-24) reports that in 1910, planes were introduced to the c.k. of the army (the pilot's license in the K.u.K. Luftschifferabteilung was obtained after 60 flights - according to plans, until 1914 it was assumed to have 200 planes and 400 pilots) and the accident of the airship M III of June 20, 1914, closed the airship era in the army. During the mobilization period, 85 aircraft were in Fliegerkompanien (Flik) Along with 120 pilots (Feldpiloten - Fp.).

11 O. Groehler, Geschichte des Luftkriegs, Berlin 1975, p. 13.

12 Ibidem, p. 14.

13 Ibidem, p. 15–16.

14 Ibidem, p. 19.

15 Ibidem, p. 7, 15, 20, 24. In the period from mobilization to the beginning of 1915, the German aviation included five Fliegerbataillonen, 34 FFAs each with six airplanes and seven Fetsungsfliegerabteilungen (FeFA) each with four airplanes equipped with a total of 232 planes and five Fliegerparks. In 1914, Z IV (airship hall in Königsberg - now Kaliningrad, Z V (airship hall in Posen - now Poznań) and SL II "Liegnitz" (airship hall in Liegnitz - now Legnica) stationed from among the German airships on the Eastern Front. GP Neumann (Die deutschen Luftstreitkräfte im Weltkriege, Berlin 1920, p. 62, 64–65) states that as of August 2, 1914, the German air force consisted of 254 pilots and 271

Above: In the group of Serbian airmen of the Coastal Aviation Division in 1913 there was, among others while piloting a two-seater Bleriot XI-2 aircraft, Sgt. Miodar Tomič, holder of a record flight at an altitude of 1500 m. source: Б. Циглић, Крила Србије. *Ваздухопловна команда и авијатика српске војске 1912–1920*, Београд 2009

Above: Bleriot XI Serbian Air Force. source: Б. Циглић, Крила Србије. *Ваздухопловна команда и авијатика српске војске 1912–1920*, Београд 2009

Above: The basic flying equipment of Serbian aviation in the years 1913/1914 was Bleriot XI - in the photos during preparations for flights. source: Б. Циглић, Крила Србије. *Ваздухопловна команда и авијатика српске војске 1912–1920*, Београд 2009

observers serving in 34 FFA, seven FeFA, eight Etappenflugzeuparks and five Flieger-Eersatzabteilungen. The odeB of German aviation in May 1915 was as follows: 72 FFA, two FEFA, one Fliegerkorps with 36 aircraft, 18 Armeeflugparks and 11 Fliegerersatzabteilungen; The odeB of April 1, 1916 was as follows: 17 staff-aviators with their staffs, 81 FFA, 27 Artilerie-Fliegerabteilungen (A) FA, five Kampf (Bomben) Geschwader, six Staffeln each, six planes each, six independent Kampf (Bomben) Staffeln, 17 Armeeflugzeugparks, two Riesenflugzeugabteilungen (RFA), two Versuchs and Übungsparks (including one in Warsaw), Fliegerabteilung (FA) No. 300 "Pascha" composed of two FFAs, Fliegerkommando Sofia composed of one FFA, three FFAs in the composition of the air force at Militärmission Türkei. W. Murray, J. Keegan (*War in the Air 1914–45*, London 1970, p. 10), report that the air force of the powers that were in a world conflict in August 1914 was as follows: Great Britain - 113, Republic of France - 138, German Empire - 384, Russian Empire - 45, Austro-Hungarian Empire - 36 military aircraft. In the Western literature on the subject, there are a number of errors not always corrected even by recognized authors, such as the above-mentioned authors dealing with the history of aviation (op. Cit.). The aforementioned authors lowered the condition of the Russian air force five times compared to the actual state (!), and in the case of the air force of the Imperial-Royal Austria-Hungary more than twice! In order to source the above-mentioned words, it is worth presenting here the data provided by the Russian pilot and historian BM Tkachev (op. Cit., p. 259-260), according to his study the state of Tsarist Russia's aviation after mobilization in 1914 was 39 AO, including 25 KAO, one guard, one grenadier AO, three field AO, 1., 4., 5. Siberian AO and eight fortress AO in Brest-Litovsk, Kaunas, Grodno, Osowiec, Modlin, Sevastopol, Kars and Vladivostok, in the number of six aircraft each. According to the Stawka's staff assumptions in the Russian air force there were supposed to be 234 airmen and about as many planes, but in practice there were 213 airmen and 213 planes commanded by the so-called Awiakancę (Russian: Авиаканца) that is the Air and Ballooning Command Chancellery in the army. (S. Rosenboom, op. Cit., P. 26) states that the first maneuvers with the participation of aviation were carried out by the tsarist army in the fall of 1911, although the Minister of War, General Vladimir Sukhomlinov, had a very critical approach to aviation, and yet until 1914 230 planes, 12 airships and 46 tethered observation balloons were mobilized, and 40 Aviation Units with 16 types of aircraft, mainly of French production, were formed in the naval aviation. There is also no agreement among other authors dealing with aviation issues during the Great War as to the number of air forces taking part in it. According to Der Luftkrieg 1914–1915 (Leipzig 1915, p. 31–32) there were 13 airships in the Russian aviation and 30 seaplanes

Above: Known as the Pearl of Montenegro, the Bay of Kotor (Boka Kotorska). In 1914, the main base of the Imperial-Royal Naval Aviation on the southern Adriatic coast became Kumbor, and the seaplane water field was created in nearby Tivat. source: *"Das interessante Blatt"* 1916

stationed in the bases in St. Petersburg (Russian: Petrograd - today Petersburg), Rewel, Libau (today Libawa) and Sevastopol. BM Ткачев states (op.cit., P. 272) that in 1914 the French air force counted 25 Squadrons (French Escadrille) and 370 pilots, while the German air force consisted of 41 air units of which 34 FFA were assigned to the corps commands, and seven FEFA to the fortress commands. S. Rosenboom (op.cit., P. 20) states that in August 1914, in the German aviation, along with tethered observation balloons and airships, there were 240 aircraft with 254 pilots and 271 observers - the air force was assigned to 25 corps and eight army commands, and the British air force counted 70 aircraft grouped in squadrons. H. Potempa (The Royal Bavarian Flying Corps. Aspects of the Aerial War 1914–1918, "*Over The Front*" 2000, No. 15, p. 101) states that in 1914, the German air force numbered 232 planes, aviation c.k. Austria-Hungary - 48 planes, French aviation - 165 planes, British aviation - 63 planes, Belgian aviation - 16 planes, and Russian aviation - 263 planes. According to Die Entwicklung der deutsche Luftstreitkräfte im ersten Weltkrieg, Diplomarbeit von Berthold Josten, Karl-Marx Universität 1963, collection of the Zentrum für Militärgeschichte und Sozialwissenschaften der Bundeswehr in Potsdam (p. 24) as of August 2, 1914, the French air force on the Western Front had 156 aircraft (in January 1915 there were about 600 of them), and the British air force at the beginning of the war, it had a force of 160 aircraft (as of August 19, 1914, on the Western Front there were four Squadrons, three Squadrons (Flight) with four planes each - 48 aircraft in total). The Russian aviation at that time was to have about 400 aircraft, of which, according to staff reports, about 250 could be used in combat. A. Inattentive (Zapomniana Wojna 1914-1918. Eastern Front, Olszanica 2015, p. 78) states that in 1914 the German air force had about 250 aircraft (half of which were of the Etrich-Rumpler "Taube" type). W. Bihl (*Der Erste Weltkrieg 1914–1918*, Wien-Köln-Weimar 2010, p. 151) states that the German aviation in 1914 had 228 aircraft, Russian aviation 224 aircraft, French aviation 165 aircraft in 27 Squadrons, British aviation 48 aircraft in four Squadrons, and the ck aviation had seven Flicks. on the Eastern Front and two on the Serbian Front. In preparation by Der Erste Weltkrieg 1914-1918. (op.cit., p. 91) there is another list of air forces in 1914: The German Empire - 232 aircraft, the Republic of France - 165 aircraft, the United Kingdom of Great Britain and

Above: In the years 1914–1915, during the hot days of fighting in the Kotor area, the Imperial-Royal army used observation balloons "raised" from the decks of KuK Kriegsmarine ships to cooperate with the artillery shelling the positions of the Lovćen battery. In the photo there is an observation balloon type M 98 put in the morning over the waters of the Bay of Kotor in readiness to work for own artillery and a balloon type M 96 put from the deck of a vessel. source: "*Das interessante Blatt*" 1916

Ireland - 63 aircraft, and the Russian Empire - 263 aircraft. P. Kettel (Kampf um das Luftmeer. Geschichte der Luftfahrt von den Anfängen bis zur Gegenwart in zeitgenössichen Berichten und Dokumenten, Ebenhausen-München 1937, p. 222) states that in August 1914, the German aviation "went out into the field" with 49 aircraft units. Geschichte der Luftwaffe. Eine kurze Darfstellung der Entwicklung der fünften Waffe, mit einem Geleinttwort des Reichsluftfahrtministers Göring von major a. D. Frhr. V. Büllow (Frankfurt am Main 1934, p. 42-43) states that the Great War was started by the German air force in 1914 with the forces of 34 FFA - including five Bavarian and eight Etapen-Flugzeugparks assigned to the 26 Army Corps and eight armies - and Eight FeFA assigned to fortresses on the eastern and western borders of Germany: Königsberg (today Kaliningrad), Boyen (today Giżycko), Graudenz (today Grudziądz), Posen (today Poznań), Köln, Metz, Straßburg, Gemerstein without assigning any air force to the reserve Army Corps. Each FFA has six planes, six pilots, seven observers, 116 soldiers, five cars and 11 trucks; for each FeFA - four planes, five pilots, four observers, 87 soldiers; for each Etappen-FlugzeugPark - three planes, two pilots, one observer and 62 soldiers. In addition, the German air force after mobilization had eight airship field air units (Luftschiff Abteilung) - each field army had one airship unit - and 15 such for fortresses - Lt. Luftschiff und Luftschihhahrt in Vergangenheit, Gegenwart und Zukunft, Berlin 1926 - passim;

Above: Front "Sausages" ... Festungsballonabteilungen KuK Luftschifferabteilungen before and after 1914 used observation balloons on the imprisoned KuK Militärballon M 98 (Drachenballon) of the German construction Maj. Dr. Eng. August von Parseval and Hptm. Rudolf Hans Bartch Sigsfeld (Parseval Siegsfeld M 98.), which superseded the older army, spherical balloons (Kugelballons M 96.). In the photo Drachenballon M 98 used in the German army. source: collection of the Polish Aviation Museum in Krakow

Luftfahrt...., Op. cit., p. 206-207. The British aviation industry in the years 1917-1918 produced 30,761 aircraft, including 1,865 seaplanes (see W. Murray, J. Keegan, op. Cit., P. 72) and the number of pilots, e.g. in republican France, increased at the same time: in 1914 - 134 pilots, in 1915 - 1484, in 1916 - 2698, in 1917 - 5608 and in 1918 - 8000! - see L. Kennett, La premiere guerre aerienne 1914-1918, Paris 2005, p. 129. By 1918, the German air force had increased to 2,500 aircraft, and by the end of the war, the German aviation industry had produced 48,626 aircraft - see S. Rosenboom, op. cit., p. 6; H. Klüver, H. G. Nippe, Marineflieger. Zur Geschichte der See-und Marineflieger in Deutschland, Dusseldorf 2002, p. 7.

16 O. Groehler, op. cit., p. 58.

17 Ibidem, p. 8, 15, 120.

18 Ibid. Surprisingly, in the above-mentioned study under the term Tannenberg 1914, only the ground operation of the German 8th Army is mentioned, but ... the key role played by the crews of the German aviation in its victory was not mentioned, contrary to the famous phrase Ohne Flieger kein Tannenberg. The aviation activities in this battle are presented by G. P. Neumann (op. Cit., p. 463-464). When the Great War ended in the fall of 1918, Generalleutnant Ernst von Hoepner - Kommandierender Genaral der Luftstreitkräfte (Kogenluft), i.e. the commander-in-chief of the Kaiser air force, who left behind interesting memoirs (Deutschland Krieg in dr Luft 1921), Leipzig descended from ... infantry and in the past commanding the 75th Infantry Division, and his deputy Major / Obstlt. Hermann von der Lieth-Thomsen - Chef des Feldflugwesens and then Chef des Generalstabes der Luftsteitkräfte - had a staff service in the 2nd Railway Regiment and in 1914 as a staff officer of the "Battle of Tannenberg".

19 O. Groehler, op. cit., p. 21.

20 Ibidem, p. 53.

21 Ibidem, p. 26. In February 1915, the Huth company produced an aviation radio weighing 15 kg - the weight of subsequent copies soon increased to 30 or even 42 kg.

22 Ibidem, p. 31

23 Ibidem, p. 34

24 Ibidem, p. 36.

25 SAW-KA, Radio-Artil. Schiesen eingeführt erstmals durch Flik 1. Artilerie vorbereitung der Durchbruchschlacht bei Gorlice. Frühjahr 1915 vom Hptm. Ritter Fiala v. Fernbrug, s. 8; Ł. Chrzanowski, A. Olejko, Tarnów..., Tarnów...., wyślijmy tam nasze maleństwo, „Archeologia wojenna" 2014, t. 4, passim Observation of own artillery fire, slowed down for a long time by the difficult conditions of communication, was

Above: Another flight, another reconnaissance ... In the photo a Lohner type flying boat in the colors of KuK Seeflugwessen during a flight over the waters of the Bay of Kotor - source: "Das interessante Blatt" 1916

not as successful. The first shooting was performed with the help of airframe movements and fireworks. These methods originally caused errors and confusion. Radio telegraph settled the matter properly; it allowed the air observer to talk to the battery, which corresponds to the cloth shields- Lieutenant Major Orthlieb, Air Fleet. Translated the French , with a preface and appendix written by Adam Stebłowski, Major of the General Staff, Warsaw 1928, p. 139. Among the recognized authors of the so-called Western schools dealing with the history of aviation, e.g. W. Murray, J. Keegan (op. cit., p. 10) state that the first use of the radio to correct artillery fire took place in December 1914, but without specifying where it was how and what air forces did this experiment for the first time!

26 O. Groehler, op. cit., p. 11–12.

27 Ibid, p. 73; https://dzieje.pl/aktualnosci/100-lat-temu-pierwszy-raz-zbombardowano-anglie-z-powietrza-wejscie from 8 May 2019. GP Neumann (op. cit., p. 583) gives the following losses among the British for the period from January 20, 1915 to June 17, 1918, as a result of 51 airship raids: 1,913 people who died, including 171 women and 110 children, and 431 women and 218 children were injured.

28 O. Groehler, op. cit.; G. P. Neumann, op. cit., p. 391–395.

29 O. Groehler, op. cit., p. 68.

30 Bundesarchiv Militärarchiv (BMA) Freiburg, Feindliche

Above: Gallery of the losers ... In the photo, the military governor of Bosnia and Herzegovina in 1914, General Oskar Potiorek, responsible for the security situation in this part of Austria-Hungary and ... for the defeats of the Imperial-Royal army in the Balkan campaign of 1914, the field of the Imperial-Royal 5th and 6th. Source: Fotoarchiv SAW-KA Wien

Above: During flight training, each candidate for a pilot was a student pilot (Pilotenschüller), after its completion he became a pilot (Flugzeugführer) and "in the field" each of them became a Feldpilot (Fp.). In the photo, before the next flight in a full air gala, the Flik crew. 4 on the Italian Front in 1915, before taking off for a combat flight - source: Fotoarchiv SAW-KA Wien

Flugzeugverluste einschl. September 1915 bis einschl. September 1918 - copy in the author's collection.

31 The National Archiv (TNA) London Kew, Air/1/8, Summary of British German battles 1915-11 XI 1918

32 В. М. Ткачев, op. cit., p. 310–312.

33 The airborne ram of Captain P. Niewierow in the area of Wola Wysocka near Żółkwia in the Lviv region, the victims of which on August 26 (September 8) 1914 fell

Right: The time of the pioneers… Namely, on October 13 [1914], three planes with officers [on the route] from Wiener-Neustadt flew over Schneeberg, a mountain 2070 m above sea level. Oberleutnant Béla von Losonczy [Flik 15] took off at [hours] 7.30 and flew over the mountain at an altitude of 2,200 m. Flying over St. Poelten, Tulln and Baden reached [Wiener] Neustadt at 9.30am. 10 minutes later Oberleutnant Alfred Heine and Leutnant Rudolf Henke [both from Flik 15], flew over Schneeberg at an altitude of 2,700 m, headed to Aspang and, having passed over Ödenburg, returned after a two-hour flight… War time… In the photo, a Flik 4 Albatros B.I type during a flight over the mountains on the Serbian Front in 1915. source: Fotoarchiv SAW-KA Wien

Above: August 3, 1914 the crew of Flik 5 made the first combat flight over Belgrade, commencing regular reconnaissance and bombing flights over the territory of the Kingdom of Serbia ... In the photo, the capital of the Kingdom of Serbia seen from the third dimension through the eyes of the Flik 6 crew's camera. Source: Fotoarchiv SAW-KA Wien

Above: The power of war photography ... Obrenovac railway station, photographed from the air. source: Fotoarchiv SAW-KA Wien

Above: The first shots in the Great War were fired by the monitors of the Imperial-Royal Danube Flotilla (KuK Donauflotille) SMS "*Themes*", SMS "*Bodrog*" and SMS "*Szamos*", in the immediate vicinity of Belgrade, on the first night after the declaration of war to the Kingdom of Serbia ... The photo shows a period engraving that shows the shelling of Belgrade by K.u.K. Donauflotillas and field artillery of the Imperial-Royal army. source: B. Hamann, Der Erste Weltkrieg. Wahrheit und Lüge in Bildern und Texten, München-Zurich 2014

both the pilot and his opponents Oblt. Friedrich von Rosenthal and Zgsf. Franz Malina from Flik 11.includes in his study Lee Kennett (op.cit., p. 161, 191), based on information from the epoch (P. Kritskii, Podwigi Russkich aviatorov, Jarosławl 1915, p. 5-6) calling the opponent of the Russian aviator Baronen Rosenthal as and these data are provided by S. Rosenboom (op. cit., p. 70). W. Bihl also mentions the ram with dated September 8, 1914 (op. Cit., P. 47).

34 A. Chwalba, *The Suicide of Europe*, Krakow 2014, p. 500.

35 O. Wedigen, Deutschlands Luftkrieg und Helden Flieger 1914-1917, Regensburg 1918, p. 86-97. German aristocrat, baron, a native of Świdnica, Manfred Freiherr von Richthofen, pilot of the Great War, German fighter ace of the Western Front with 80 kills on his account - see J. Batchelor, Ch. Chant, Flight. The history of aviation, New York 1990, p. 32-33; J. Winchester, Fighter. The world's finest combat aircraft – 1914 to the prezent day, Bath 2004, p. 27, 35-41; Prvni svètova válka, Praha 2010, p. 116. E. Rosenboom (op. Cit., P. 81) emphasizes the role of M. von Richthofen in the fighting on the Eastern Front, who in his memoirs appreciated the role of the victories of the German 11th Army in Galicia in 1915, but also wrote about the ferocity of Russian airmen - see M. von Richthofen Rote Kampfflieger, Hamburg 1990 (Berlin 1917), p. 38, 62; Manfred Freiherr von Richthofen, Familie und Kindheit, von Bolko Freiher *von Richthofen,* Berlin1933 – passim; „Kriegs-Echo", nr 28, Jahr 1915. The person of the red baron has become the subject of many feature films, computer games or even comics, but also in the form of internet nicknames, it has become a new tradition among Polish "academics" dealing with the subject of the Great War in Galicia and the application of scientific critique at a level that has little to do with science! Recently, we can observe the process of "emergence" among authors dealing with the subject of aviation, authors who did not bring anything new to science (Manfred von Richthofen (1892–1918) and his Polish relatives. On the hundredth anniversary of the

Above: The first shots in the Great War were fired by the monitors of the Imperial-Royal Danube Flotilla (KuK Donauflotille) SMS "*Themes*", SMS "*Bodrog*", and SMS "*Szamos*", in the immediate vicinity of Belgrade, on the first night after the declaration of war to the Kingdom of Serbia ... The photo shows the shelling of Belgrade by K.u.K. Donauflotillas and field artillery of the Imperial-Royal army. source: B. Hamann, *Der Erste Weltkrieg.* Wahrheit und Lüge in Bildern und Texten, München-Zurich 2014

Above: The Sava and Danube bridges were a bridge to the Serbian shore ... The photo shows the destroyed bridge on the Sava near Belgrade, photographed by the Flik 6 crew. Source: Fotoarchiv SAW-KA Wien

end of World War I and the death of the Red Baron", Rzeszów 2018) in its texts, based on the findings of multilingual authors and compiling commonly known facts (K. Mroczkowski - Manfred von Richthoffen jako obiekt muzealny, [w:] Manfred von Richthofen…, op. cit., p. 107-126; P. Korzeniowski (Archival materials about Manfred von Richthoffen can be found in The National Archives in London, [w:] Manfred von Richthofen…, op. cit., sp 126-139; J. Centek („Red Baron" - https://www.tvn24.pl/wiadomosci-ze-swiata,2/czerwony-baron-zginal-100-lat-temu-ekspert-trudny-do-okreslenia-dar,831296.html https://www.dzieje.pl/aktualnosci/j-centek-richthofen-niewatpliwie-byl-asem-lotnictwa-i-wojny

36 O. Wedigen, op. cit., p. 43, 54.

37 M. von Richthofen, Wspomnienia Czerwonego Barona, Wrocław 2011, p. 48–50.

38 Ibidem, p. 52.

39 Major Orthlieb, Air Fleet. Translated from French, a foreword, and an appendix by Adam Stebłowski, major of the General Staff, Warsaw 1928, p. 139.

40 Ibidem; s. 73; A. Whitehouse, Flieger Asse 1914-1918, Stuttgart 2014, p. 405; M. Immelmann, Der Adler

Above: Habsburg wings 1914 ... The photo shows the prototype of the Albatros B.II type aircraft (20.01) in the collection of the Heeresgeschichtliches Museum Wien. source: author's collection.

von Lille, Leipzig 1934, p. 179-182; R. Italiaander, Wegbereiter deutscher luftgeltung, Berlin 1941, p. 248; F. L. Weber, Das wunder des Fliegens, München 1936, p. 148; Luftfahrt ist not! Herausgegeben von Ernst Jünger, Leipzig-Berlin 1935, s. 206-207, p. 228-241; G. P. Neumann, op. cit., p. 585; F. L. Neher, Das Wunder des Fliegens, München 1936, p. 148; M. Pöhlmann, H. Potempa, T. Vogel, Der Erste Weltkrieg 1914-1918. Die Deutsche Aufmarsch in ein kriegesreiches Jahrhundert, München 2014, p. 89–92.

41 P. Ham, 1914. Rok końca świata, Warszawa 2015, p. 508.

42 In addition to the five existing school units (Flieger-Ersatz-Abteilung - one was Bavarian), there were also 10 private schools under military control. In 1914, there was no mobilization plan for the German aviation industry and the entire air force was under the Inspektion der Fliegertruppen. Each FFA was commanded by an Abteilungsführer with the rank of Hptm., Its adjutant was an officer for special tasks, moreover, in each FFA there were six or seven observer officers and as many officer pilots and non-commissioned officers as well as 150 ground personnel. Each FFA was equipped with six aircraft and one reserve aircraft, and in the state of each FFA there were four aircraft of the following types intended for reconnaissance: two-seater Gotha-Etrich and Rumpler "Taube" monoplanes, or later Albatros B, Rumpler B and LVG B two-seater biplanes with 100 HP Mercedes engines. All frontline FFAs, as well as those in Germany, were under the command of the Chef des Feldflugwesens (March 11, 1915), and then, from October 1916, Komandierenden General der Luftstreikräfte. In individual German field armies, the air force was commanded by Stabsoffizier der Flieger (from March 11, 1915), and later by Kommandeur der Flieger - see TNA, London, Air / 1/8, Copy of a German document made on October 6, 1921.

43 E. von Hoepner, op. cit., s. 15; G. P. Neumann, op. cit., s. 462; H. Jäger, Erkundung mit der Kamera. Die Entwicklung der Photographie zur Waffe und ihr Einsatz im 1. Weltkrieg, München 2014, p. 98–99.

Above: Under the wings, next to the red and white [red-white-red] war colors of the united [ck] army are painted with an iron [black] cross and long red-and-white war banners on their tail ... In the photo the color of KuK Luftschifferabteilung on the wings and fuselage of the Albatros B. (20.01) in the collection of the Heeresgeschichtliches Museum Wien - source: author's collection.

44 Geschichte der Luftwaffe, op. cit., p. 56-57; G. P. Neumann, op. cit., p. 462; S. Rosenboom, op. cit., p. 7.

45 Война 1914-1917 г. Из личного фотоалбома генерала от кавалерии графа Ф. А. Келлера, Харков 2013 - passim.

46 J. Livesey, Wielkie bitwy I wojny światowej, Warszawa 1998, p. 20–22.

47 H. Münkler, op. cit., p. 126; W. Bihl, *Der Erste Weltkrieg 1914–1918*, Wien-Köln-Weimar 2010, p. 47.

48 H. Münkler, op. cit.

49 Aviation activities in the Battle of Tannenberg were presented by GP Neumann (op.cit., p. 463–464), and surprisingly, O. Groehler (op.cit., p. 8, 15, 120) in his study which was the history of German aviation, under the term Tannenberg 1914 he mentioned only the German 8th Army's ground operation in question, contrary to the well-known phrase Ohne Flieger kein Tannenberg. A similar state of knowledge was presented

Above: The next airports were built in Fischamend near Vienna and in Görz, located away from the capital, at the foot of the Julian Alps ... In the photo Görz - now Gorizia in the Republic of Italy - through the lens of the Flik 6 observer. Source: Fotoarchiv SAW-KA Wien

by P. Szlanta (Tannenberg 1914, Warsaw 2005 - passim) and J. Rubach (Activities of the German 8th Army in East Prussia in August and September 1914, [in:] "Scripta Historica" 2015, No. 21, p. 79-112). The first of the above-mentioned authors, in his popular science study, mentioned aviation issues in the period preceding the Battle of Tannenberg in two entries (p. 37, 52), while during the battle itself he devoted only two short entries to them (p. 82 and 92)! In the description of the battle on the Masurian lakes, he raised the aviation issues in the next two entries (p. 166 and 183), while the signature in the photo annex under the photo showing the symbol of the battle, i.e., the plane - as a Russian plane - is incorrect because the Voisin LA plane presented on it ... did not take participate in this battle! This study, in relation to the above-mentioned well-known words of General P. Hindenburg about the role of aviation in the Battle of Tannenberg, does not confirm their content at all! Similarly, the author of the second, scientific study did not notice the role played by aviation in the Battle of Tannenberg, but in the annex covering the OdeB of the German 8th Army, he mentioned air units using Polish terminology in the naming of units and in military ranks, which should be as follows: FFA 16. Hptm. Wilhelm Schmoeger and the airship Z 5 Hptm. Grüner, subordinate to the command of the army (p. 103); FFA 14. Hptm. Heinrich, subordinate to the command of the 1st Corps (p. 104); FFA 17. Hptm. Dincklage was subordinate to the command of the XVII Corps (p. 105); FFA 15. Hptm. Donata and FeFA 7. Hptm von der Goltz from the Boyen Fortress (today Giżycko) subordinate to the command of the XX Corps (p. 106), 8. Luftschiffabteilung - Airship unit subordinate to the command of the I Reserve Corps (p. 108); FFA 5 and the airship Z 4 from the Królewiec Fortress (German: Königsberg, today Kaliningrad) subordinate to the command of the Królewiec Fortress (p. 110).

50 For the first time this problem was widely presented only in 2015 by A. Olejko and M. Bakun (Ohne Flieger - kein Tannenberg! Participation of aviation in the military operations of 1914 on the northern section of the Eastern Front, including the Battle of Tannenberg, [in: World War in East Prussia, ed. N. Kasparek, D. Radziwiłowicz, Olsztyn 2015, p. 85-110).

Above: A four-legged pet ... There have always been mascots, superstitions, and animal favorites in aviation. Also, in the KuK Luftschifferabteilung. In the photo Oblt. Ferdinand Cavallar von Grabensprung from Flik 4 with his four-legged favorite on the Serbian Front. Source: Fotoarchiv SAW-KA Wien

51 The study includes the dating of events on the Serbian and Eastern Front, used on the basis of the Gregorian and Julian calendars.

Western Front - Battle of Mons, Air Reconnaissance for the First Time

On August 23rd, 1914, transported on decks of transporters under the cover of Royal Navy ships, unloaded in the ports of northern France, Le Havre and Rouen on August 12-17, 1914, British Expeditionary Forces (British Expeditionary Forces) - BEF - in 1914 160 thousand soldiers, seasoned in battles in South Africa and Egypt - 80 thousand soldiers were sent over the English Channel in the first wave), fought their first battle in Europe in the Great War. The clash was to support the German army of the Kingdom of Belgium, which was in a difficult position after the invasion, and to stop the attack of the II Field Corps of the German 1st Army of General Alexander von Kluck through Belgian territory into the territory of the Republic of France. These forces, together with the neighboring German field army of General Bernard von Bülow, constituted the right wing of the German forces implementing the Schlieffen Plan and performed the maneuver to encircle Paris. As the days went on, the half of the German 1st Army was about to strike the left wing of the French Army on Brussels, Mons and Le Cateau, towards the Somme Line at Amiens. The first "bucket of cold water" German right-wing forces received in the clash near Liege on 4–15 August 1914, where after 11 days of heavy fighting, the Belgian army was forced to retreat and close in the area of the fortress city of Antwerp and to take Namur, and on August 22nd , 1914, Brussels. However, the abovementioned resistance and maneuvering activities forced the commanders of both above-mentioned German field armies to change the direction of the attack, and thus the entire Schlieffen Plan was changed[1]. The commanding officer of the BEF, General Sir John French, in accordance with earlier arrangements with the French High Command, directed his troops to the left wing of the French 5th Army, which were concentrated on a distance of 40 km, between

Right: Brandenburg C.I 61.13. The crew is checking their map before takeoff on their mission. The Brandenburg C.I was built in many series and was adapted to more powerful engines as they were developed, so replaced earlier two-seaters of modest performance and reliability.

Left: Brandenburg C.I 61.53. An order for 10 C.I aircraft of the 61 series was placed in October 1915. The 160 hp Mercedes powered this batch. The Brandenburg C.I was sturdy and had good flying qualities, making it popular with its crews. It was so successful that it comprised 26.9% of all Austro-Hungarian aircraft and was thus the iconic Austro-Hungarian aircraft of the war.

Maubeuge and Le Cateau. Four British DPs went to Mons on August 22nd, 1914, where it turned out that French forces had been defeated in the clash at Charleroi with a larger opponent ahead of them, who continued their offensive towards the BEF position. As a result of the shifting position of the allied forces, British troops found themselves in a dangerously exposed position. Despite the overwhelming superiority of the opponent's forces (400,000 German soldiers: 80,000 British), General J. French

he promised to hold positions for 24 hours, giving the French time to regroup and then retreat. In this way, it was intended to prevent a breach in the front. On the morning of the next day, August 23rd, 1914, all the power of General Alexander von Kluck's army collapsed against the forces of the [British] *Expeditionary Corps. The British, however, withstood the blows of the overwhelming enemy forces. For a long time, they defended themselves fiercely, fending off one attack after another. Only in the afternoon they did retreat upon the news that the French army had retreated so far that they had exposed their* [right] *wing. For the cost of 1,600 troops, the [*British*] Expeditionary Force gave the Allied commanders what they needed most - time to reorganize and regroup*[2].

On August 23, 1914, BEF forces fought their first battle in the Great War on Belgian territory. It happened on the anniversary of the British victory in 1346, during the 100 Years' War at Crecy (this battle was included in the 18 battles that would decide the fate of the world). Belgian troops retreating under the pressure of overwhelming German forces were supported, stopping the attack of the German II Corps, and tactical success was achieved after the defeat of the French forces at Charleroi, preventing the outflanking of the French 5th Army[3]. After the clash at Mons BEF, he was forced to retreat on the night of August 23/24, 1914.

Above: Zemun 1914, port at the mouth of the Sava to the Danube. This is where Austria-Hungary ended or began... In a photo taken from a bird's eye view by the Flik 6 crew present suburb of Belgrade. source: Fotoarchiv SAW-KA Wien

To prevent all Ententa forces from flanking, four DP BEFs withdrew from most of the territory of the Kingdom of Belgium[4]. For *the Germans, the battle of Mons was like a cold shower. They had won this war quite easily so far, now they proved helpless against hard defense, and lost over 5,000 people in unsuccessful attacks*[5]. After the clash at Mons, the right wing of the German forces encircling Paris turned out to be "unlucky" because OHL on August 25, 1914 just weakened them, directing two Army Corps and a Cavalry Division (DK) to the Eastern Front, which were to stop the attack of Russian field armies in Prussia Eastern. As time has shown, these forces on the Eastern Front arrived too late to play their part in the decisive victory of the German 8th Army at Tannenberg. The decisive success in 1914 on the Eastern Front, achieved by its forces, unfortunately "suffered a hiccup" by OHL because the above-mentioned forces lacked in the decisive clash over the Marne on the Western Front[6].

The "silent hero" of the land battle - at Mons - was the plane! RFC crews as well as French-British mixed crews, performing individually air reconnaissance tasks with British type B.E. 2, Avro 504 and the French type Bleriot XI and Farman M.F. 7., after combat flights, they provided precise reports, which a well-functioning communication network directly delivered to the BEF headquarters. It was the "aviator", so far underestimated by British staff officers, who became the supplier of the fastest and most accurate information about the movements of the German 1st Army at Mons and it was thanks to air reconnaissance that the BEF commander, previously belonging to the "avia sceptics", was warned about enemy movements as well as the attempt made during the battle to encircle the British forces [7].

In 1914, when the Great War broke out, 63 RFC planes were ready for use under the BEF[8] (RNAS had 95 planes, including 55 seaplanes, six airships and two observation balloons), and the BEF observation plane was the basic plane BE 2[9]. Contrary to criticism of the time, *the newly established British Air Force was as war ready as time, budget constraints and the state of knowledge allowed* [10]. The BEF was supported by four – 2nd, 3rd, 4th, and 5th - RFC Squadrons, which together numbered 48 aircrafts out of a total of 63 RFCs ready for action. Interestingly, the BEF forces had the highest rate of efficient airplanes per one DP. The 2nd and 4th RFC squadrons were equipped with B.E. The 2nd, 1st and 3rd RFC squadrons were equipped with French Bleriot XI and

Above: Located on the Adriatic Sea, Castelnuovo - now Herceg Novi, hosted Flik 6. pilots during the Great War…. In the photo, Castelnuovo is photographed from the third dimension - source: Fotoarchiv SAW-KA Wien

Farman M.F. 7 and the 5th Squadron RFC was equipped with Avro 504 aircraft at that time considered the most modern in RFC and B.E 8[11]. All *four types of aircraft were selected mainly for stability in the air and ease of piloting, which was an extremely important factor because, before being sent to the front, the pilots had only 3.5 hours of flight time*[12]. The BEF was therefore adequately equipped with the air force, both in terms of the number of aircraft and their equipment. Shortly after the start of hostilities, on August 13, 1914, the 2nd, 3rd, 4th,, and 5th RFC squadrons were transferred to French territory (12 aircrafts of several of the above types, 24 pilots and observers, and 130 servicing soldiers of various military ranks were transferred from Dover to Amiens) and the RNAS Squadrons were tasked with the defense of the British Isles. For the RFC, the war started in an unfortunate way, as the first losses were incurred during the transport of units (one plane crashed shortly after takeoff from the main RFC base in Larkhill) but eventually his forces joined with BEF in Amiens, henceforth moving with the units by land towards the borders of the Kingdom of Belgium. In the first group of commanders, which included from the 2nd RFC Squadron, Maj. C.J.B. "Pregnant Percy" Burke and Maj. J. F. A. "All bum and E.Y.E. Glass" Higgins from the 5th Squadron, there were planes with crew mechanics on the observers' positions. The honor of being the first RFC pilot to land on French territory fell to Lieutenant H. D. "H'K" Harvey-Kelly, who could not be forgiven by Maj. C. J. B. "Pregnant Percy" Burke for overtaking him in this "historic event" as his commander! Lieutenant Louis Strange of the 5th RFC Squadron had a reluctant passenger in his crew who stocked up a bottle of Whiskey for the road and was arrested in Dover, but fled to town … for more alcohol, but the pilot caught him and upon arriving in Amiens, he was given 56 days in jail. The first air operation was performed by RFC on August 19th, 1914, but it had to be interrupted due to bad weather conditions and bad planning (originally it was decided not to take the observer on board, so that the lightened plane could stay in the air longer)[13]. On that day, French Capitaine Philip Joubert de la Ferté and British Lieutenant G. W. Mapplebeck carried out the first reconnaissance in the Mons area for the BEF command, detecting German forces from the air, for which the commander expressed his thanks to both airmen. On the eve of the battle, on August 22nd, 1914, the observer of the RFC Capitaine L. E. O. Charlton detected the approach of the leading German field forces of the 1st Army to the right flank of the BEF forces, because of which the BEF commander was able to regroup his forces and successfully face a much

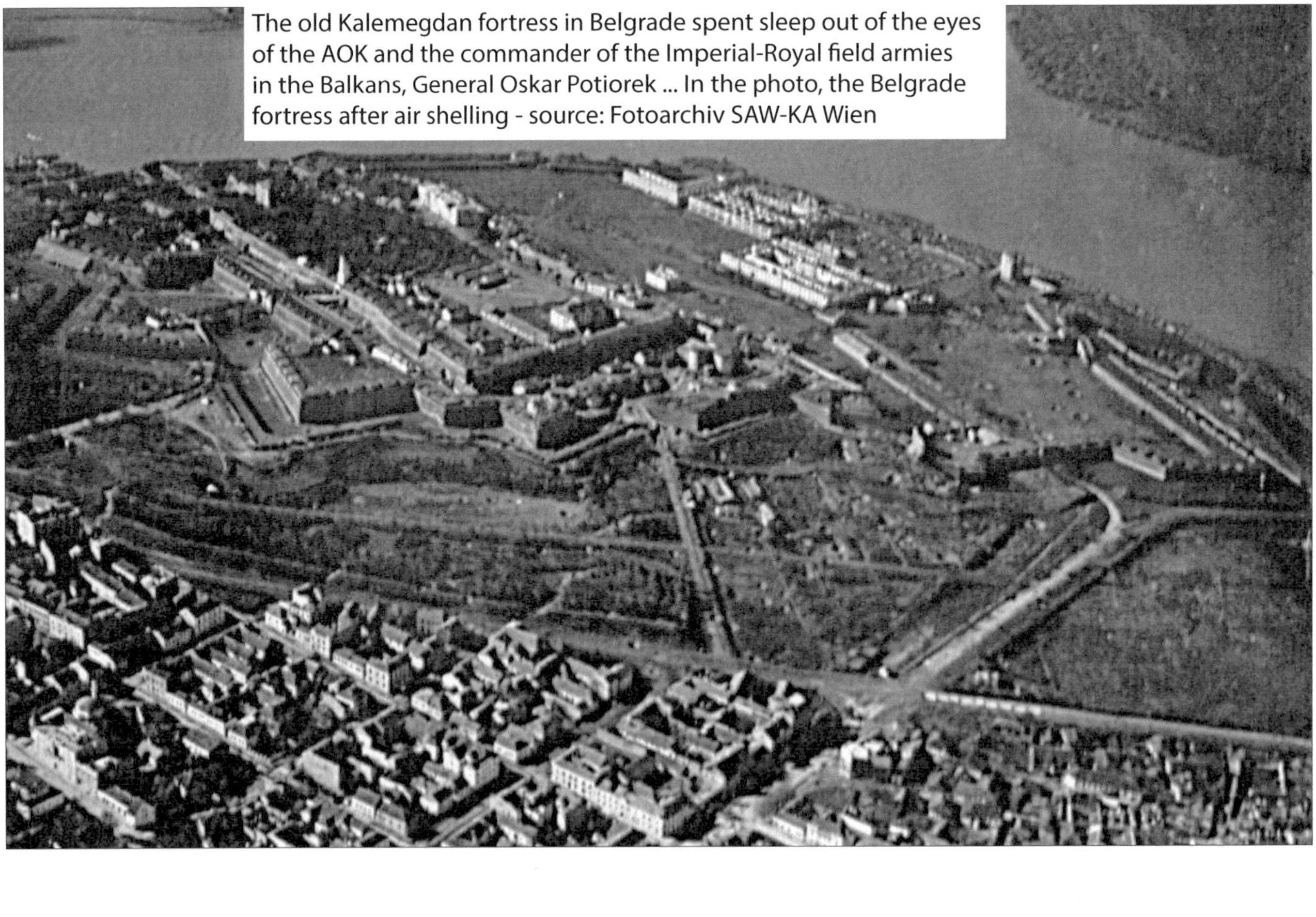

The old Kalemegdan fortress in Belgrade spent sleep out of the eyes of the AOK and the commander of the Imperial-Royal field armies in the Balkans, General Oskar Potiorek ... In the photo, the Belgrade fortress after air shelling - source: Fotoarchiv SAW-KA Wien

Above: This is how Tirana looked from the third dimension in 1915, from 1913 the capital of the Principality of Albania ... The photo shows the city photographed by the Flik 6 crew. Source: Fotoarchiv SAW-KA Wien.

stronger opponent[14]. Unfortunately, on August 22nd , 1914, the RFC suffered its first combat loss (the Avro 504 aircraft from the 5th RFC Squadron was shot down over Belgian territory by the German infantry) and on August 25th ,1914, the RFC achieved its first combat success (piloting the Avro 504) . Lieutenant Rabagliati forced the German Rumpler "Taube" to land in the forest near the advanced German field airport)[15]. Appreciating the role of "his airborne eyes," the BEF commander, in a message of September 7th ,1914, reported to the British Government on how the RFC crews proved their front-line utility:

I would especially like to draw your Lordship's attention to the excellent work that the Royal Air Corps [RFC] has done. They provided [the pilots] *with the most accurate and complete information that cannot be overestimated during the operations carried out. Constantly under fire by enemies* [ground anti-aircraft defense - opl.] *And their own, and who do not hesitate to fly in any weather, they continue their mission without complaint*[16].

In the Battle of Mons, air force played an equally important role for the commander of the German 1st Army, detecting British forces shortly after landing on the "old continent", contrary to previous opinions about the surprise by the presence of BEF forces of the German command. The crews of Rumpler "Taube" planes from five FFAs[17] operating individually, systematically recognized the movements of British forces, and yet the field commander of the 1st Army's battle at Mons did not end favorably for himself. Active air reconnaissance of the German 1st Army field at Mons (II, III, IV and IX Corps, III and IV Corps in reserve) after the success over the Belgian army in the Gette area, observed the retreat of the Belgian forces towards Antwerp, before the occupation of Brussels on August 21st , 1914 r.[18] . At the moment when the German field 1st and 2nd Armies directed their attack on Namur, from the German FFA group, FFA 12 was under the command of the field command of the 1st Army, FFA 30 was under the command of the 2nd Corps, FFA 9 was under the command of the 4th Corps, FFA 7 . was under the command of the III Corps, FFA 11 was under the command of the IX Corps, and from August 23rd , 1914, the newly formed FFA was under the command of 38[19]. As of August 21st , 1914, the corps of the German field 1st Army were located in the following regions: 2nd Corps - Ganshoren,

Above: The capital of the Kingdom of Montenegro was Cetinje, but Podgorica was important due to its strategic location in the Balkans ... In the photo, the present capital of the Republic of Montenegro, captured in 1915 by the Flik 6 observer. Source: Fotoarchiv SAW-KA Wien

4th Corps - Castre, 3rd Corps - Hal, 9th Corps - Braine-le-Chateau. Aerial reconnaissance for the above-mentioned forces was carried out by: FFA 12 under the field command of the 1st Army and conducting long-distance strategic reconnaissance; FFA 30. conducting tactical reconnaissance for the II Corps; FFA 7th conducting tactical reconnaissance for the 3rd and 4th Corps; FFA 11. conducting tactical reconnaissance for the IX Corps; FFA 9. from August 23rd ,1914, separated from the X Corps of the German 2nd Army Field and from August 24th ,1914 "working" for the 4th Field Corps of the 1st Army. As a result of aerial reconnaissance of the Rouen and Le Havr ports, British troops were unloaded immediately after their arrival in Europe. Having obtained this information, the German command assumed two possibilities of their operation:

1.) Cooperation of the BEF forces with the French forces (field 5th Army) on their left wing; 2.) concentration of British forces in the Tournai-Lille area and attack on the right wing of the German forces (1st Army field).

As a result of the reconnaissance of the 2nd Cavalry Corps (CC), the area of its right wing near Courtrai was considered dangerous for the German 1st Army, but its commander needed information from the southern region and on August 22, 1914, the FFA crews began a reconnaissance action over it (the crews of FFA 30 recognize the area of Audenarde-Courtrai-Lille for the benefit of the II Corps; 7th FFA crews recognized the Tournai-Lille-Douai area for the III and IV Corps; 12th FFA crews identified the area up to the Armentieres-Lille-Orchies-St. Ghislat-Mons line for the army command; 11th FFA crews recognized the Valencienes-Maubeuge region for the IX Corps)[20]. On the basis of air reports, it was established that the Alost-Ghent-Deynze-Cortrai-Lille-Tournai-Leuze area was clean from enemy forces; BEF forces were not noticed on the railroads; there was no concentration of enemy forces in the Lille area. On the other hand, enemy troops were observed marching and camping in the area north of the Conde-Mons line. Aerial reconnaissance also found that near Le Cateau, northeast of Mons, British cavalry squadrons had meet the German II KK turning back to the Mons area. On that

In 1914, the war once again visited the Balkans ... Located on Lake Skadar, in a strategic place in the Balkans Shkoder in 1913 hosted Serbian airmen. The photo shows the city photographed in 1915 by the Flik 6 crew. Source: Fotoarchiv SAW-KA Wien

The wartime Šabac on the Sava ... Troops of the Austro-Hungarian army on the streets of the city - collections Fotoarchiv SAW / KA Wien

Above: From April 24, 1912, the commander of the former air force (KuK Fliegertruppe) was a Croat - Oberstleutnant Milan Emil Uzelac ... In the photo Obst. ME Uzelac visits the Flik 6 field airport on the Italian Front in 1915 during a conversation with Major Junović. source: Fotoarchiv SAW-KA Wien.

day, the German infantry shot down the above-mentioned English Avro 504 from the 5th RFC Squadron recognizing the 5th DP BEF in the Maubeuge area[21]!

On August 22nd ,1914 the crew of the Oblt. Haena from FFA 12 in the report from 9.20-12.25 reported that the roads and railroads between Hal-Enghien-Tornai were clear of enemy forces; the Lille-Orchies route was unobstructed, as was the Lille-Wahagnies route; the enemy forces were not recognized at railway junctions, as was the whole area of Lille, which was free from enemy forces. Oblt crew. Decken from FFA 11. in the report from the period 13.35-16.20 reported that enemy columns had been spotted in the Bavai-Englefontaine area; south of Soigny, enemy railway transports were detected (at 2:10 pm a German VII Corps force was spotted on the wing of the German 2nd Army field). Oblt crew. Niemöller from FFA 30 in the report from 6.15pm - 6.30pm reported that camp fires had been observed north of the Ninove-Enghien-Grammont line; the Alost-Ghent-Deynze-Audenarde area was free from enemy forces. Oblt crew. Fink, Lt. von Gravert from FFA 7. in the report from 16.30-18.45 reported that the roads Enghien-Ath-Leuze, Peruwelz-Leuze, Vallencienes-Peruwelz by 17.40 were free from enemy forces; at. 4:35 PM was seen at Boussu, southwest of St. Ghislain, a long car column moving towards Mons; between 16.00 and 18.00 the railroads between Enghien-Lille-Courtrai-Leuze and Tournai were free from hostile forces; campfires of enemy cavalry units were noticed at Audenarde and Renaix and at 18.00 enemy cavalry brigade in the Soignies area. The 11th FFA crew reported reconnaissance Course Nivelles-Binches-Givry-Maubeuge-Bavai-Solesmes-Valenceiennes-Mons-Soignies-Braine-Le Comte to IX Corps headquarters; mixed columns of troops have been observed on the road in the Binche-Bavai area towards Mons; to the north and north-west of Meubauge fortification works on infantry and artillery positions were recognized; in addition, a hostile airport was identified on the Mons-Meubauge road - 2 km north on the road from Meubauge[22].

On August 23rd ,1914, German air reconnaissance found large British forces in the area of the railway stations above Mons - the aerial photographs were transferred to the command - their reserves were found in the Malplaquet area. Air reconnaissance of the Third Corps concluded that the Mons-Valenciennes road was empty. The command of the German field 1st Army, based on the above reports, decided to attack the BEF positions as well as on the next day, at 05.00 August 24th ,1914. Aerial reconnaissance was ordered to constantly observe the Cambrai-Bohain-Avesnes area. The crew of the FFA 11 (Oblt. Böhmer, pilot Lt. Kentscher) at 08.46 delivered a reconnaissance report of enemy forces to IX Corps headquarters about the situation east of Mons. The crew of the FFA 12 delivered a long reconnaissance report detecting enemy forces marching in the area north and northwest of Bavai. At 9 AM, the command of the German 1st Army in its order stated that the enemy had a force of 2-3 DP in the Curgies-Bavai area. Oblt crew. Böhmer of the FFA 11, performing a reconnaissance flight for the command of the IX Corps during 11.40-14.30, recognized the region of Braine-Le Comte-Soignies-Mons-La Longueville-Avesnes-La Capelle-Fontaine-Ouise-Landrecies-Bavai-St. Denis-Braine-Le Comte. At. 12.10, after recognizing the position of the artillery, enemy trenches were bombed near Nimy; the area of Mons was classified as quiet; road Mons-Villers-St. Ghislain and Mons-Barmignies were willpower from the enemy; at. 12.30 hostile column sighted on La Longueville-Mons road; at. 12.40 a second hostile column was observed on the same path; At 1:50 PM third enemy column was sighted by Query-Genly route; At 1:55 PM enemy cavalry march on Bougnies-Asquillies road was observed; At12.45 hostile transports were detected on the La Capelle-Avesnes road (3 km long column) and at 13.30 a second column of the same length on the Guise-Landrecies road. At 14.30 columns of trucks detected on the Vervice-La Capelle road; the roads in the Mons-Givry-Mons-Villers, Mons-Harmignies area were empty without the presence of enemy forces. Oblt crew. Finka from the FFA 7, on August 23rd ,1914, recognizing in favor of the command of the III Corps between hours 14.19 and 16.10 defined the Mons-Valenciennes road as clear of enemy forces as well as the area south of Rombies, Audregnies, Dour, Wasmes and towards Bavai. A concentration of enemy forces in the Mons area as well as east of Mons and artillery positions in the Mons area have been observed. The crew of Lt. von Grawerta FFA 7. recognizing in favor of the command of the III Corps

Above: In the summer of 1914, Flik 2 was transferred to the Süd-Ostfront, which as from August 2, 1914 ended up at the field airport in Brčko under the command of the Imperial-Royal 5th Army… In the photo the planes of Flik 2 on railway loras during transport to the Serbian Front. source: Fotoarchiv SAW-KA Wien

between hours. 18.10 and 18.40 found the roads Mons-Maubeuge and Mons-Valenciennes free from enemy forces - further observation was prevented by increasing fog[23]

On August 24, 1914, the crew of the FFA 11 recognizing between 07.30 and 08.15 for the IX Corps in the report from 08.45 reported that the main routes used for the retreat of British forces were Givry-Bavai, Mons-La Longueville, Mons-Genly-Bavai; Bawai-St. Waast-Jenlain; Audregnies-Angres; Roisin-Bry; Angre-Sebourg; enemy artillery positions were observed in the area south of Roman and south of Givra. During the morning reconnaissance, the crew of Lt. Mackenthuna FFA 7th for the Third Corps recognized enemy columns on the roads: Bellignies-Bavai, La Flamengrie-Bavai, Wargnies-St. Waast-Bavai, Comegnies-Bavai. 3 km from Le Quesnoy train station, a group of cars stopped to the northwest; all the roads leading from Le Quesnoy to the southwest and south as well as to Mormal were free from enemy forces - the same situation was observed in the Meubauge area. In the Le Quesnoy area, the plane was shelled by AA artillery on cars. On August 24th ,1914, the crew of Lt. Kittera FFA 12 in the report to the command of the German field 1st Army from 18.48 reported that a long column of enemy troops was detected at Le Cateau, at At 09.50, a hostile railway transport was detected on the Valenciennes-Solesmes section, and a French plane was observed over Le Quesnoy[24].

Orders for August 25th ,1914 ordered the attack to be continued. Air report to the command of the German 1st Army field from 02.00 from the crew of the FFA 7. Lt. Mackenthuna was imprecise and spoke of enemy columns of various types of troops in the Bavai and Le Quesnoy area[25]. The morning reconnaissance was directed to the Bavai-Solesmes-Le Cateau area - on August 25th ,1914, it was carried out during the fighting in the area of Solesmes and Landrecies, and on August 26th ,1914, during the Battle of Le Cateau[26].

After the Battle of Mons, the BEF forces on the Western Front increased to five DPs. German losses suffered in this clash and at Le Cateau, Nery and Villers-Cotterets, tiredness of the attacking troops and the increasingly stronger resistance of the French 5th Army's forces in Guise as well as the dilemmas of the commander-in-chief, General Helmut von Moltke (concerns about the role of the new, multiplied by Inaccurate reports of British forces

Above: In the fall of 1914, the Balkan Fliegerkompanien used more and more air bombs during combat flights ... The photo shows Veljačić bombs used in Flik 4. Source: Fotoarchiv SAW-KA Wien.

that landed near Ostend - the British Marines brigade - and the question of the size of the French 6th Army, which was just emerging), caused the German field 1st Army to attack not to the west, but to the east of Paris, stepping onto the line instead of encircling Paris. Miserable. The report of the British air reconnaissance on September 3, 1914 made it clear to the French High Command, who had been skeptical about the air reconnaissance so far, that the German field 1st Army was not heading towards the capital of France and its attack was directed to the south-east, at Compiegne. On the same day in the evening, General Joseph Simon Gallieni, the commander of the French 6th Army, asked the French commander-in-chief, General Joseph Joffre, for permission to carry out a counterattack on the Marne, for which he obtained consent only after a very emotional meeting with his superior with Commander of the BEF in Melun on September 4, 1914, on which, despite the hesitations of the BEF commander (the French government leaving Paris and considering the possibility of a retreat of British forces towards the English Channel), a decision was made to counterattack. A breakthrough for the struggle in 1914 on the Western Front, the Battle of the Marne began on September 5, 1914 at 14.30, when the front-line troops of the French 6th Army north of Meaux came into fire with units of the German field 1st Army striking south of Marne, whose commander was convinced that the British-French army with whom contact had been lost were in reverse[27]!

If it was not the plane, if it was not the aerial reconnaissance, the 1914 "miracle on the Marne", so widely publicized in historiography, would not have taken place[28]! On September 2, 1914, a report on a gap spotted from the air between the German 1st and 2nd field armies that started the Battle of the Marne was provided by Corporal Louis Breguet. Gen. J. S. Gallieni, commander of the defense of Paris and the French 6th Army, co-creator of the "Miracle on the Marne", trusted the information provided by the airman, which resulted in a turn in the fighting on the entire Western Front[29]. In this clash, OHL also received reports from its own aviation, but this did not detect the French maneuver. On 5-9 September 1914, the German FFA 30 crews performed tasks for the commands of the 2nd and 4th Corps on the right field wing of the German 1st Army, in the area north of the Marne, focusing their attention on the Paris area (the crews carried out reconnaissance on 30-50 km deep into enemy territory). The crews of the 12th FFA, under the direct command of the German 1st Army, carried

Above: Propaganda of the Great War ... This is how the citizens of Austro-Hungary were informed in 1914 about the triumphs of the Imperial-Royal field armies in the Kingdom of Serbia, while their defeats were not mentioned there. source: Fotoarchiv SAW-KA Wien

Above: In 1914, the KuK Luftschifferabteilung was based on structures made of wood, canvas and thin aluminum sheet lacquered with nitrocellulose, covering hot lubricants and fuel ... In the photo, a shiny, varnished, truly Austrian production, i.e., a Lohner B plane "Pfeilflieger". source: Jacek's Stachiewicz collection

out reconnaissance flights at a depth of 50-120 km and in the area of the right wing of the army. During the battles over Aisne on 12-15 September 1914, FFA 12 operated for the field command of the 1st Army, FFA 30 was under the command of the 2nd Corps, FFA 9 was under the command of the 4th Corps, FFA 7 was under the command of the 3rd Corps, FFA 11 was under the command of the IX Corps and the newly formed FFA 38 was under the authority of the III Corps from August 23, 1914[30].

Above: The Maritime Power of Imperial-Royal Austo-Hungary ... In the photo: KuK Kriegsmarine SMS "Franz Ferdinand" and the Donnet Leveque S 8 flying boat taking off against it. source: J. Butkiewicz collection.

Endnotes

1 A. Livesey, op. cit., p. 14-16.

2 http://www.paranormalne.pl/topic/36445-anioly-spod-mons/-wejście z 15 III 2017 r.

3 TNA London, Air/1/8, Activity gth Aerial Forces gth German First Army durning spedlins near Mons from 21 to 25 August 1914, compiled on July 29, 1921 by the Air Historical Branch Air Ministry on the basis of information from the Military Mission London based on data from the Reichsarchiv - a record from the report of British 3rd and 5th DP, which on August 22, 1914 was between the Canal du Center and Haine .

4 J. Livesey, op. cit.

5 http://www.paranormalne.pl/topic/36445-anioly-spod-mons/-wejscie z 15 III 2017 r.

6 J. Livesey, op. cit.

7 Ibidem, p. 17. The battle of Mons is associated with the legend of the Angels of Mons, because its result was so astonishing that "it was explained by a miracle and the intervention of supernatural forces." On September 29, 1914, the London Evening News published a short story by Arthur Machen entitled "The Bowmen", in which the author, "interested in the occult and paranormal phenomena, described his extraordinary vision of the battle at Mons (the name is not mentioned directly, but it is about this battle), presented as if one of the soldiers fighting there told him - see http://www.paranormalne.pl/topic/36445-anioly-spod-mons/ - entry from March 15, 2017

8 P. G. Cooksley, *The Royal Flying Corps 1914-1918*, Stroud 2014, p. 37

9 L. Kennettt, op. cit., p. 25, 37.

10 D. Trinder, *Beginning of British air war 1903-1915*, referat – copy P. Carradice, *First World War in the Air*, London 2012, p. 37-39

11 D. Trinder, op. cit.; P. Carradice, op. cit, p. 40-41.

12 D. Trinder, op. cit.; R. Barker, *A Brief History of the Royal Flying Corps in World War One,* London 200, p. 23-25

13 D. Hadaway, S. Hadaway, *The british airmen of the first world war*, Oxford 2016, p. 6-7.

14 J. Keegan, R. Holmes, *Soldiers. A history of men in battle*,

Above: Czechs serving in the Imperial-Royal Naval Aviation in 1914-1918 accounted for about 12% of its personnel ... Before 1914, Linien Schiffsleutnant Václav Woseček was entrusted with the purchase of flying equipment for KuK Seeflugwessen (fifth from the right in the photo). source: J. Butkiewicz collection.

London 1985, p. 185; D. Trinder, op. cit.; P. Carradice, op. cit.

15 D. Trinder, op. cit .; R. Barker, op. cit., p. 34. A. Whitehouse (op. cit., p. 355) states that on August 24, 1914 in the Quesnoy area, British troops received a downed Aviatik B.I plane, piloted by Feldfebel Kausen on a reconnaissance flight.

16 P. Carradice, op. cit

17 TNA, London, Air / 1/8, Kopia niemieckiego dokumentu wykonana 6 X 1921 r .; *Luftschiff und Luftschihhahrt in Vergangenheit, Gegenwart und Zukunft*, Berlin 1926 - passim

18 TNA, London, Air/1/8, Activity gth Aerial Forces gth German First Army durning spedlins near Mons from 21 to 25 August 1914, compiled on July 29, 1921 by the Air Historical Branch Air Ministry on the basis of information from Military Mission London based on data from Reichsarchiv, pp. 2–3.

19 TNA, London, Air / 1/8, the state of the German air force assigned to the 1st Army during the Battle of Aisne, p. 3 - a copy of the German document made on October 6, 1921

20 Ibidem.

21 Ibidem, p. 4

22 Ibidem, p. 10.

23 Ibidem, p. 7

24 Ibidem

25 Ibidem, p. 8

26 Ibidem, p. 9

27 J. Keegan, R. Holmes, op. cit., p. 185

28 J. Keegan, R. Holmes, op. cit., p. 185

29 Corporal L. Breguet was shot on September 8, 1914 during a reconnaissance flight over Triaucourt by the artillery of the German XVI Corps - see W. Murray, J. Keegan, op. cit., pp. 10, 26; W. Bihl, op. cit., p. 47; D. C. Cooke, *Sky Battle: 1914-1918. The Story of Aviation in World War 1, New York 1990*, p. 60; Major Orthlieb, op. cit., pp. 9–17; S. Rosenboom, op. cit., p. 7; H. Münkler, op. cit., p. 145; M. Pöhlmann, H. Potempa, T. Vogel, op. cit., p. 91; ⊠.M. Tkachev, op. cit., p. 319; W. Murray, J.

Above: After launching ... KuK Seeflugwessen flying boat No. S 22 in the Brioni area - source: J. Butkiewicz collection.

Above: Flik crew. 15 reported on 30 November 1914 a success in an air duel - the shooting down of a Russian plane over Miechów. The crew of the Lohner C "Gebirgspfeilflieger" aircraft composed of a pilot Oblt. B. Losonczy von Losoncz and observer Oblt. J. Wierzejski, noticed a Russian plane taking off from the local field airport between the clouds, and then the observer fired a rifle Mannlicher M 95 cal. 8 mm enemy plane, because of which he fell to the ground… In the photo Oblt. Johann (Jan) Wierzejski from Flik 15-source: collections: J. Butkiewicz.

Keegan, op. cit., p. 27; *Geschichte der Luftwaffe…*, Op. cit., pp. 43–44.

30 TNA, London (Kew), Air / 1/8, compiled on October 6, 1921 by the Air Historical Branch Air Ministry on the fighting over Aisne based on information from Military Mission London based on data from Reichsarchiv, p. 9. p. 4. The FFA car park consisted of 15 cars and trucks. The defensive armament of the aircraft crew consisted of pistols or revolvers for defense after a forced landing in enemy territory. The bomb armament was 5-10 kg bombs called "Fliegermäuse" dropped by hand, while air darts were to be used mainly against enemy cavalry. The average reconnaissance aircraft had 2–3 cameras on board, 25 cm long - Lieutenant TNA, London (Kew), Air / 1/8, the state of German air force assigned to the 1st Army during the Battle of Aisne, p. 3 - copy German document made on October 6, 1921. During the next clash on the Western Front - the first battle of Ypres - the field command of the German 4th Army was under the FFA 6th, the command of the XV Corps and DP Winckler was under the FFA 3rd, the command of the XXII Corps had no air support , the command of the XXIII Corps was under the FFA 40, the command of the XXVI Corps, the 37th Landwehr Infantry Brigade (BPL) and the backup 2nd Infantry Brigade (BP) under the FFA 41, the command of the XXVII Corps, the Reserve Corps and the 38th BP Landwehr were without air support as well as four spare DPs, DK Gwardii. The command of the Marine Corps - Zeebrugge - was subordinate to the Zeebrugge air station (on October 9–30, 1914, the 3rd Reserve Corps together with the FFA 38 was assigned to the field German 4th Army, and in Ghent (Polish: Ghent) Flieger Park No. 4 was

Above, Left, & Right: Debunking air myths ... The red and white checkerboard, in this arrangement of color fields, was painted during the Great War as a personal emblem on airplanes piloted by KuK Luftfahrttuppen Oblt fighter pilots. Mariana Gawła, Oblt. Francis (Frank) Linke-Crawford and Oblt. Stefan Stec, in that order. From 1918 until now, the dominant view in Polish historiography is that the authorship of the chessboard from the spring of 1918 should be attributed to Oblt. S. Stec. However, earlier, in the fall of 1917, it appeared on the fighter plane of Oblt. Marian Gaweł and thus this pilot, the earliest from the discussed "fighter triad", became the "father" of the later Polish aviation chessboard - source: Piotr Gaweł-Romaszkan's collection

organized . subordinate to the German 4th Army - see TNA, London (Kew), Air / 1/8, The distribution of the Aviation Units of the newly formed 4 th German Army durning the first battle of Ypres, 1914, p. 5.

Serbian Front - Imperial Battle, Air Reconnaissance Over the Balkans

The timeline of the 1914 campaign on the Balkan Serb Front in the light of the archives was as follows: August 12 (25) -16 (29), 1914 - forcing Drina and Sawa by the half of the Imperial-Royal 5th Army and preventive "stand with arms at the foot" of the field Imperial 2nd Army; 15 (28) -19 August (1 September) 1914 - field offensive of the Imperial-Royal 6th Army; August 15 (28) -24 (September 6) 1914 - Battle of Mount Cer (also known as the Battle of Jadar); 20 August (2 September) - 2 (15) September 1914 - the offensive of the Imperial-Royal field armies[1]. According to the assumptions of general O. Potiorek, c.k. the governor of Bosnia and Herzegovina and the commander of the Austro-Hungarian army units in the Balkans, the Serbian opponent was to be defeated because of the "Balkan Blitzkrieg" until August 18 (31), 1914, which was to be a "birthday gift" for Emperor Franz Joseph I Habsburg. As time has shown, it turned out to be an exceptionally "arrogant planning" of a military operation from the point of view of an opponent

Above: Toruń airport established in 1912–1913 was an important element as a fortress's field of ascents Festung Thorn, although it was located outside its territory in the so-called "Espalanades" on the road towards Fordon. During the Great War, many ... even "captured" elephants were seen on it! In 1917, they were used as draft animals - source: collection of the Polish Aviation Museum in Krakow

who is outnumbered, but not fighting in his own area. A few days after the start of hostilities by Austria-Hungary against the Kingdom of Serbia, on August 13 (26), 1914, half of the 2nd Army, despite earlier planning, still did not carry out any offensive operations, waiting for the movement of the Imperial-Royal field 5th Army from the Lešnica area. A day later, Serbian field troops of the 2nd Army attacked Austro-Hungarian forces south of Šabac (the Serbian field troops of the 3rd Army were to remain in defense)[2]. The main forces of the Serbian army on August 15 (28), 1914, took up positions for a counterattack, which was to take place the next morning, but their offensive actions began around 23.00 on the same day and led to the first major battle in the Balkan theater of hostilities, the Battle of Mount Cer. This battle, fought on August 15 (28) -24 (September 6), 1914, ended with the success of the Serbian army and the defeat of the Austro-Hungarian army. This "Serbian Victoria" was a big surprise for the AOK, proving that the Vienna-Teschen line (today Cieszyn - the AOK headquarters was located there) underestimated the attitude of the weaker Serbian army, which, however, had the experience of the two Balkan wars of 1912 -1913, as well as the patriotism of the Serbs, the nation defending its territory against the exceptionally unpopular and possessive Habsburg neighbor[3]. The reports of the airmen of Srpska Aviatika (Serb. Srpska Aviatika) determined the success of the Serbian army[4].

In August 1914, good weather conditions prevailed in the Balkan region, which made it possible to carry out systematic combat flights. On the Serbian Front, on August 3, 1914, the Flik crew 5th made the first combat flight over the capital of the hostile Kingdom of Serbia - Belgrade. The first air battle of the Great War took place over the Balkans on 14 (27) August 1914 (an observer from Flik 2nd tried to shoot down an unarmed Serbian aviation plane with his service pistol)[5]. The first loss in the 1914 campaign on the Serbian Front K.u.K. The Luftschifferabteilung suffered on 13 (26) August 1914, when the Lohner B "Pfeilflieger" "Bub" plane from Flik 2. after shelling from the ground, was forced to land near the village of Lesnica (the pilot Oblt. Artur Schlett tried to get to the border Sawa, eventually being captured). During the Tsers' battle, the crews of two Fliegerkompanien (Flik) 2 and Flik 4 c.k. aviation (KuK Luftschifferabteilung) made 27 sorties, i.e., 50% of all flights from the first period of fighting in the Serbian Süd-Ostfront on August 3 (16) -25 (September 7) 1914 (an average of two daily flights using Lohner planes) B "Pfeilflieger" and Aviatik B.I). However, their work did not change the frontline situation in the battle, unlike the role played by the much weaker aerial reconnaissance of the Serbian army. During the first phase of the Serbian campaign in 1914, the basic units, much less numerous than the forces of K.u.K. The Luftschifferabteilung of Serbian Aviation was the former Air

Above: Ohne Flieger kein Tannenberg 1914! Ohne Flieger kein Gorlice 1915! Ohne Flieger… The German 250 mm, 500 mm and 700 mm aerial cameras were the basic weapon of the flight crews of the Great War. In the photo, an observer of the German aviation at work - source: Wir Luftkämpfer, Bilder und Berichte deutscher Flieger und Luftschiffer, Berlin 1918 via M. Bakun

Squadron[6] and the Balloon Division, while the Kingdom of Montenegro, allied with its Serb neighbor, did not have its own air force[7]. In the period preceding the Imperial Battle, the Air Force Squadron and the Balloon Division on July 27 (August 9) 1914 were transported by rail to the Dabiča Pole near Valjevo airfield, where they were ready for military operations, and the first combat flight of Serbian aviation in the Great War was made from this airport to the area of the Sawa and Drina rivers on July 31 (August 13) 1914[8].

On the eve of the Serbian counteroffensive on August 1 (14), 1914, the air reconnaissance provided the Serbian Supreme Command with information about the location of the Imperial-Royal troops (the crew of the Bleriot XI-2 Artilerie aircraft composed of Captain Ż. Stankovič and Lieutenant V. Novičič recognized the Drina Valley[9]) and on August 4 (17), 1914 Lieutenant M. Tomič flew by the Bleriot XI Militaire "ОРЛИЋ" aircraft to the field airfield in Oseczin, under the command of the Serbian 3rd Army field commander and recognizing the area of Jarebice - Krupań - Żabljak, and on August 6 (19) 1914 he returned to the Dabiča Pole field airfield. On that day, the Air Squadron and the Balloon Division were moved to the vicinity of the front line and from August 8 (21), 1914, the Air Squadron after staying overnight at the field airport in Oseczin and leaving Brezjak, where the Balloon Division remained, on August 9 (22), 1914 it was ready for action from the field airfield in Stupničko Pole near Loznica[10]. In the days preceding the Cerska battle, on August 11 (24), 1914, the crew of the Bleriot XI-2 Artilerie aircraft composed of Capt. Z. Stankovič and Lt. V. Novičič recognized the Sava valley and the Jarak - Šabac - Obrenovac region in two flights, while on August 14 (27), 1914, Lt. M. Tomič carried out a reconnaissance of the Jarak region with the Bleriot XI Militaire "ОРЛИЋ" plane - Šabac, and then the area of Grabovac - Ogar - Michalevac - Peczinci - Buczanowci - Hrtkowci - Nikinci - Klenak. During the return flight to the home airport, the pilot encountered an enemy aircraft of the C.K. aviation with Flik 2, crewed by Oblt. O. Fekete and Oblt. V. Seebauer, from which the observer opened fire on him from … a service pistol[11]. The first air skirmish over the Balkans took place around 6 p.m. around Kupinovo - Mišarskie Pole, and Lieutenant M. Tomič, having neither

personal nor on-board weapons, maneuvered the enemy plane with sharp turns, pulling it over the positions of the Serbian, Šumadijski DP[12]. **It was the first air combat of airplanes with firearms in the Great War (!)[13].** In 1914, most often the crews of K.u.K. The Luftschifferabteilung took 7.63 mm Mannlicher infantry rifles weighing 1.18 kg, mounted to the sides of the planes, or used manually (the first shooting of a Russian plane on the Nord-Ostfront - Galicia - on November 30, 1914, was carried out with the use of by the observer Flik 15. Oblt. Johann Wierzejski - Pole - Mannlicher M 95 rifle, cal. 8 mm). The first air shot down in the Great War with the use of a machine gun (MG) took place on the Western Front on September 5 (October 5th?) 1914 [sergent Joseph Frantz and soldat Quenault, flying a Voisin LA V 24 Squadron, shot down in the area Remese German Aviatik BI observation plane from the FFA. 18 (17 hotchkiss missiles were enough to achieve this success)] in the wreckage of which the pilot Uffz Wilhelm Schlichting and observer Lt. Fritz von Zangen died. They became the first airmen in history to fall in air combat with machine guns.[14].

The sacrifice of Serbian airmen, who appreciated

Left: Front "Sausages" ... After the start of hostilities, in early September 1914, the following balloon units assigned to Imperial fortress artillery units were ready: 1 / 2.R., 2 / 2.R., 3/2. R., 1 / 5.B. assigned to FAR 3 in Krakow and on 1/3R., 2/3R. and 3 / 3.R stationed in Przemyśl, which used tethered observation balloons marked as KuK Militärballon M 98 (Drachenballon). The photo shows a balloon with an observer - Source: Fotoarchiv SAW-KA.

Below: For the bombing of the Przemyśl fortress, the Russian command also intended to use the airship "*Kondor*" from the 2nd Aviation Company in Brest-Litovsk.... In a photograph from 1914, the airship "*Kondor*" in Lvov, the refueling of which was mentioned in the intelligence data of the c.k. army - source: collections of Prof. A. Kozyck'ye

Above: French gem ... Krosno-Potok field airfield on December 19 (January 1, 1915). In the photo, the plane of the 12 Korpuśny Aviaotriad (KAO) type Henri-Farman H.F. XXII No. 18 - source: the collection of Walerij Romanienko from the Museum of Aviation of Ukraine in Kiev via the Polish Aviation Museum in Krako

the importance of the moment and the importance of information about the detected enemy targets and troop movements, is evidenced by the fact that on July 31 (August 13), 1914, Lieutenant M. Tomič landed during his return flight in the Bela Cerkvi area (Serb. Бела Црква), wanting to provide the obtained information as soon as possible ..., provided it by phone[15]. The information provided by the above-mentioned airmen to the Serbian Supreme Command allowed him to succeed in it. The Imperial Battle had a great impact on the morale of the Serbian army, and the battles defeated by the c.k. the field armies were forced to withdraw to their starting positions on August 12 (25), 1914[16]. Partially involved in combat, the field troops of the Imperial-Royal The 2nd Army could not be transferred from the Balkan to the Galician TW on August 18 (31) 1914 and remained longer in the Balkans, which meant that its troops arrived at the Galician Nord-Ostfront with a delay, and most of them *spent the most the critical period of fighting in railway carriages*[17].

Endnotes

1 SAW-KA Wien, Die Tätigkeit der österreichisch-ungarischen Luftstreitkräfte im Feldzug gegen Serbien und Montenegro im Jahre 1914. Bearbeitet von der Obst. Dr. Oskar Regele, pp. 1–2; A. Szelągowski, op. cit., part II, Warsaw 1938, p. 746.

2 Д.М. Бабац, Српска војска у великом рату, Београд 2014, p. 46–47.

3 Ibid, pp. 50-53. W. Bihl (op. Cit., P. 93) gives a different dating and different terminology of the battles in the Serbian campaign of 1914, recognizing that on August 14, 1914 the battle of Sava and Drina took place.

4 Б. Циглић, Крила Србије. Ваздухопловна команда и авијатика српске војске *1912–1920*, Београд 2009, s. 43-44; Српска авијатика *1912–1918*, Београд 1993, p. 37-39. It is worth recalling the words of a poem that circulated around the Imperial-Royal during the days of the Serbian army victories monarchy: *Die Save war stets Euer bester Freund, Ihr wütenden Vaterlandshasser, Drum haben wir Gut es mit Euch gemeint, Und schickten Euch einfach ins Wasser*.

5 SAW-KA, X. Mikic, Geschichte der yugoslawischen Luftwaffe. Übersetz vom Rittm. a. d. P. v. Spaits - passim; Српска авијатика *1912–1918*, Београд 1993, p. 39. A common phenomenon in the "third dimension" was that the crews took pistols, revolvers or rifles on combat flights - see S. Rosenboom, op. cit., p. 5. W. Murray, J. Keegan (op. cit.) enigmatically state that the first air battle during the Great War took place on August 26, 1914, without precisely assigning this event to any of the theaters of military operations. Most likely, the authors were interested in an event related to the first ever air ram.

6 According to the archives SAW-KA Wien (Die Luftstreitkräfte Serbiens von 1914 bis Herbst 1916 von Oskar Regele, p. 3–4) during the Serbian campaign of 1914, the commander of the Aviation Squadron was

Above: Field airfields of the Great War ... Őrmező, now Strážske in the territory of the Slovak Republic, KuK Luftschifferabteilung field airfield in the Laborec valley. source: M. Drobńak's collection

Above: Captured airplane...... During the Great War, captured airplanes were also used in the tsarist aviation. The photo shows a Lloyd Pfeilflieger from Flik 3 "taken prisoner with the crew" in the Carpathians by the Cossacks of the 3rd Zaporizhia Regiment, transferred to the 7th KAO. source: M. Hajrulin, 7 nazwanoj aviacionnyj otriad, Moscow 2008.

Above: The airship combines the advantages of a tethered balloon and an airframe ... It can also fight and bomb ... The airship is primarily a means of naval warfare... The photo shows the Russian army airship "*Albatros*". source: *Беларусъв году первой мировой войны (1914-1918)*. Сборник документов, Минск 2014

Capt. Miloš Ilič, and it was equipped with three planes: Bleriot XI with a 50 HP engine, Bleriot XI with a 70 HP engine and Deperdussim Type T. During the fighting, the unit, assigned to the Serbian 3rd Army, was stationed at the following airports: July 27 (July 9) -8 (21) August 1914 - Valjevo - Dabičevo Pole; August 9 (22) -17 (30), 1914 - Brezjak - Stupničko Pole; 18 (31) August-28 (10 September) 1914 - Jarebice; August 29 (September 11) -5 (18) November 1914 - Osečina; November 5 (18) -9 (22), 1914 - Mladenovac; 9 (22) Nov-6 (19) Dec 1914 - Mladenovac - Kragujevac - Niš;

Left: In 1914, the Flotten Flug Abteilung Kotor flying personnel consisted of: Linienschiffsleutnant Hugo Ockermüller - commander, pilots - Linienschiffsleutnant Franz Miklueczki, Fregattenleutnant Konstantin Maglič, Fregattenleutnant Glauko Prebanda, observes Fregattenleutnant Heinrich Fontaine von Felsenbrunn, Fregattenleutnant Erich Edler von Kunsti, Fliegergast (mat) Lautner, Fliegergast Nenador [Nenadič?], Fliegergast Brenner)...In the photo one of the Kotor aviators Fregattenleutnant Konstantin Maglič - source: collections: O. Trullei

Above: The first pilot of the Imperial-Royal of naval aviation was Linien Schiffsleutnant Victor Klobučar Rukavina von Bunič in the photo. source: J. Butkiewicz collection

December 7 (20) -18 (31), 1914 - Požarevac - Belgrade. On August 17 (30), 1914, a separate squadron force was created, which was assigned to the Šumadik Infantry Division of the Serbian 1st Army and stationed at the following airports: 17 (30) VIII-27 VIII (10 IX) 1914 r. – Cerovac; 29-31 VIII (11-13 IX) 1914 r. – Skela; 1 (14) IX-27 X (9 XI) 1914 r. – Cerovac; 28 X (10 XI) 1914 r. – Ub; 30 X (12 XI) 1914 r. – Lazarevac; 2-5 (15-18) XI 1914 r. – Arandjelovac. W spisie samolotów jednostki w nawiasie znajduje się dodatkowy zapis: *R.E.P. (turkische Beute) – Eindecker.* [it is a plane for the Turkish army that was taken over by the Serbian army in rail transport]; *bulg.*[garische] *Zweisitzer (Beute) 80 PS* [captured two-person plane of the Bulgarian aviation].

7 SAW-KA Wien, Die Tätigkeit der österreichisch-ungarischen Luftstreitkräfte im Feldzug gegen Serbien und Montenegro im Jahre 1914. Bearbeitet von der Obst. dr Oskar Regele, s. 6. *Der Luftkrieg…*, op. cit., p. 35. Według *Der Luftkrieg…* (op. cit.) Serbian aviation had only *few airplanes*

8 Cpt. Živojin Stankovič and Lieutenant Vojislav Novičič - the airmen recognized the situation in the area of the bridge under Vidojevica - and Lieutenant Miodrag Tomič on a Bleriot XI Militaire "ОРЛИЪ" ("Orzel") plane recognized the situation in the area of the bridge over Sava near Šabaca during the second combat flight – por. Б. Циглић, Крила Србије. Ваздухопловна команда и авијатика српске војске *1912–1920*, Београд 2009, p. 43-44; Српска авијатика *1912–1918*, Београ. д 1993, p. 37–39.

9 2 (15) August 1914 (according to another version the day before) the Bleriot XI Militaire aircraft piloted by Lieutenant V. Novičič was damaged (a lobe shot) with the fire of the Imperial-Royal infantry and after a forced landing it fell into her hands at Žabljak (cf. E. E. Kisch, *Zapisz to Kisch!*, Warsaw 1957, p. 47 . Soon, as a result of a counterattack by the Serbian unit, in front of which the incident took place, the plane was deflected and the mechanics arrived at the scene transported the disassembled plane on August 12 (25), 1914 (according to another version the day before) to the Dabiča Pole field airfield - see Б. Циглић, op. cit., pp. 43–44; Српска…, op. cit., pp. 37–39.

10 С.Ј. Микић, Историја југословенског ваздухопловства, Београд 1933, p. 87-88.

11 Б. Циглић, op. cit., s. 44; SAW-KA Wien, Die Luftstreitkräfte Serbiens von 1914 bis Herbst 1916 von Oskar Regele, p. 32; Српска…, op. cit., p. 39.

12 Српска…, op. cit., p 40; B. Dimitrijevič, P. Miladinovič, M. Micevski, *Jugoslovensko pomorsko vazduchoplovstvo*, Belgrad 2014, op. cit., p. 18.

13 SAW-KA Wien, Die Luftstreitkräfte Serbiens von 1914

Above: At the request of Technischekomitee Marinesektion, it was decided to build according to the plans of Eng. Joseph Mickl's seaplane, on which it was later planned to test the usability of the seaplanes in the service on the Adriatic ... In the photo from 1912, the first Mickl seaplane at the port road in Pula - source: P. Schupita, *Die kuk Seeflieger. Chronik Und Dokumentation österreichiszch-ungarischen Marineluftwaffe 1911–1918*, Koblenz 1983.

bis Herbst 1916 von Oskar Regele, op. cit., p. 32; ibidem, Die Luftstreitkräfte Serbiens von 1914 bis Herbst 1916 von Oskar Regele, p. 184; Српска…, op. cit., p. 39;

14 J. Rajlich, *Na křidlech světove vȧlky,* Cheb 2014, pp. 306-308; M. Spick, Aces of world fighter aviation 1914-2000, Warsaw 2006, pp. 23-24. T. Goworek (*Airplanes of the First World War*, Warsaw 1988, p. 9) reports that the km of the Hotchkiss system jammed during the fight.

15 Б. Циглић, op. cit., p. 42; Ч. Јанић, О. Петровић, Кратка историја ваздухойловсшва у Србиј*u*, Београд 2011, p. 12; Српска…, op. cit., p. 38–39. According to another version, during this flight, Lt. M. Tomič was forced to land in the Belej Crkvi area and on the next day on August 1 (14), 1914 at At 05.30 he returned to his home airport.

16 Friedrich, *14/18, Der Weg nach Versailles*, Berlin 2014, p. 227–228; Великая борьба народовъ, Выпуск *III-ій*, Москва 1915, p. 50–52; SAW-KA Wien, Die Tätigkeit der österreichisch-ungarischen Luftstreitkräfte im Feldzug gegen Serbien und Montenegro im Jahre 1914. Bearbeitet von der Obst. dr Oskar Regele, op. cit., p. 1–2; Д.М. Бабац (op. cit., p. 50–53) states that c.k. the army lost about 23 thousand in the Cersian battle soldiers, including about seven thousand killed and about four thousand soldiers in captivity. Serbian losses amounted to approximately 2,068 killed, 15,900 wounded and 8,823 captive soldiers. М. Радојевић, Љ. Димић (op. Cit., P. 120) report that the Serbian army lost 260 officers and about 16,500 non-commissioned officers and soldiers, and the troops of the Imperial-Royal of the army lost 600 officers and 22–23 thousand. non-commissioned officers and soldiers.

17 M. Zgórniak, 1914–1918. *Studies and Sketches from the History of the First World War*, Krakow 1987, p. 123. The Galician tdw presents precisely Великая борьба народовъ, Выпуск V-ый, Москва 1915, pp. 3–6.

Above:tThe time of testing foreign flying equipment showed that the KuK Seeflugwessen did not adopt floatplanes ... The photo shows the French type Voisin L floatplane No. 14 during the trials at the Versuchflugstation KuK Seeflugwessen on the island of St. Catarina - Source: P. Schupita, *Die kuk Seeflieger. Chronik Und Dokumentation österreichiszch-ungarischen Marineluftwaffe 1911–1918,* Koblenz 1983.

Eastern Front - Battle of Tannenberg, Air Reconnaissance Is The Key To Victory

The 1914 campaign in East Prussia (East Prussian campaign of the tsarist army), "the nursery of the Prussian Junkers" was an example of maneuvering operations by large forces on both sides using the "cannabis strategy" and the use of a large-scale railway network. On August 17, 1914, the field Russian 1st Army, commanded by General Paweł Rennekampf, launched an offensive towards Królewiec, being systematically recognized by the German air force until August 15, 1914 (photo reconnaissance was carried out on the line from the Baltic Sea to the Kowno- Grodno line)[1], and on August 19, 1914, the field Russian 2nd Army commanded by General Aleksander Samsonov (the so-called Narew Army) was ordered to attack Danzig (today Gdańsk) - starting an attack in the direction of Warsaw-Ostrołęka, as a result of which the German forces were to be cut off from the Vistula and pinned to the Baltic coast, which was recognized by the German air force on August 20, 1914[2]. The maneuver of both above-mentioned field armies commanded by warring generals, carried out in area of the great Masurian lakes separating them, resulted in the creation of a large gap of about 80 km between them, making it impossible for them to cooperate closely and allowing the enemy to attack each of them separately. The first clash took place on August 19–20, 1914 near Gumbinen (today Gusiew), as a result the defeated forces of the German field 8th Army, commanded by Gen. Max von Prittwitz und Gaffron constantly informed by air reconnaissance about enemy movements on the German southern flank, withdrew with the intention of establishing a defense on the line of the Vistula[3]. This decision caused an immediate reaction of gen. H. von Moltke, who as the commander-in-chief made an immediate change in the position of commander of the above-mentioned forces, replacing on August 22, 1914 the inept General M. von Prittwitz und Gaffron with a "combat duo" in the persons of

Above: March 18, 1914, another flying boat built in the Seearsenal Pula Workshop was commissioned for service in the Seearsenal Pula Workshop, marked as A 22 ... In the photo, flying boat No. A (Arsenal) 22 taking off from the Adriatic Sea - source: P. Schupita, *Die kuk Seeflieger. Chronik Und Dokumentation österreichiszch-ungarischen Marineluftwaffe 1911–1918,* Koblenz 1983

the dismissed from "retirement" gen. P. von Hindenburg (he retired in 1911) and a former quartermaster of the 2nd Army of General Erich Ludendorf on the Western Front, and on August 25, 1914, leading the aforementioned large forces from the Western Front (two Army Corps and one Army Corps). DK) to support the "threatened by the Cossacks" of East Prussia. As time has shown, this decision became the "nail in the coffin" of the OHL's success over the Marna, and the forces of the Russian field of the 2nd Army, which from August 21-22, 1914, began offensive operations, and in a short time, due to the soldiers' exhaustion, were not fit to continue operations offensive what Stawka expected from them. The mentioned gap between the two Russian field armies began to widen, however, because after the Battle of Gumbinen, the Russian 1st Field Army, operating towards Königsberg, stopped and its "tired neighbor" continued to attack westwards to encircle the German forces. As a result of the persuasion of Obst. Max Hoffmann, who was the operational chief of the German 8th Army, there was a reshuffle of the German forces, which focused the attention of both new commanders on the Russian 2nd Army's movements, as its neighbor - the 1st Field Army - still did not continue the attack. The maneuver to encircle the German forces was becoming more and more unreal. Due to the lack of communication between the two Russian field armies and the information about the enemy's position, as well as deciphering the reports sent from their commands to Stawka, the position of the Russian forces was getting worse day by day. On August 23, 1914, the center of the field forces of the Russian 2nd Army, stretched over a length of 97 km, entered combat contact with German forces prepared for it, and on August 24–25, 1914, having numerical superiority, the army rejected the center of the German forces, but its commander didn't have a thorough understanding of the position and maneuvers of the opponent's forces, he didn't know that he was "gradually getting into the pincers". It was only on August 26, 1914, when the forces were almost encircled by the commanding officer of the Russian 1st Army, who set out to help the endangered neighbor towards the Vistula River. On this and the next day, the fate of the battle started at Tannenberg was practically decided, because the attack by the forces of the 1st Field Corps of the German 8th Army, carried out on the orders of General P. von Hindenburg as a result of the air report of the 14th (!)

Above: The first 240 km group flight took place in January 1913 with three Donnet-Leveque flying boats from Pula to Trieste and back ... In the photo from 1913, the Versuchflugstation KuK Seeflugwessen base on the island of St. Catarina and a French Donnet Leveque type flying boat on a transport cart. Source: P. Schupita, *Die kuk Seeflieger. Chronik Und Dokumentation österreichiszch-ungarischen Marineluftwaffe 1911–1918*, Koblenz 1983.

Above: May 1913 - a blockade of the coast of the Kingdom of Montenegro ... In the photo there is a KuK Seeflugwessen, Donnet Leveque-type flying boat No. 12 moored on the Dalmatian beach of Banje near Dubrovnik - source: B. Ciglić, *Seaplanes of Bocche. The Story of Austro-Hungarian Naval Aviation in Southern Adriatic 1913–1918*, Beograd 2014 (via M. Raguž).

Above: On March 23, 1913, a makeshift air station was opened in Tivat in the Bay of Kotor which was operating until May 28, 1913, headed by Linien Schiffsleutnant V. Wosećek… The photo shows the pilot in the cabin of a Lohner M E 21 flying boat just before taking off from the waters of the Bay of Kotor. source: B. Ciglić, *Seaplanes of Bocche. The Story of Austro-Hungarian Naval Aviation in Southern Adriatic 1913–1918*, Beograd 2014 (via J. Zahalka)

One of the "Habsburg sea eagles"... Fregattenleutnant Konstantin Maglič (in the cabin of a Lohner ME 21 flying boat) On October 31, 1914, he was awarded a high distinction of the Imperial-Royal Army Signum Laudis (German Militärverdienstmedaille) for combat flights over the Montenegrin coast in the campaign 1914 - source: B. Ciglić, *Seaplanes of Bocche. The Story of Austro-Hungarian Naval Aviation in Southern Adriatic 1913-1918*, Beograd 2014 (via J. Zahalka)

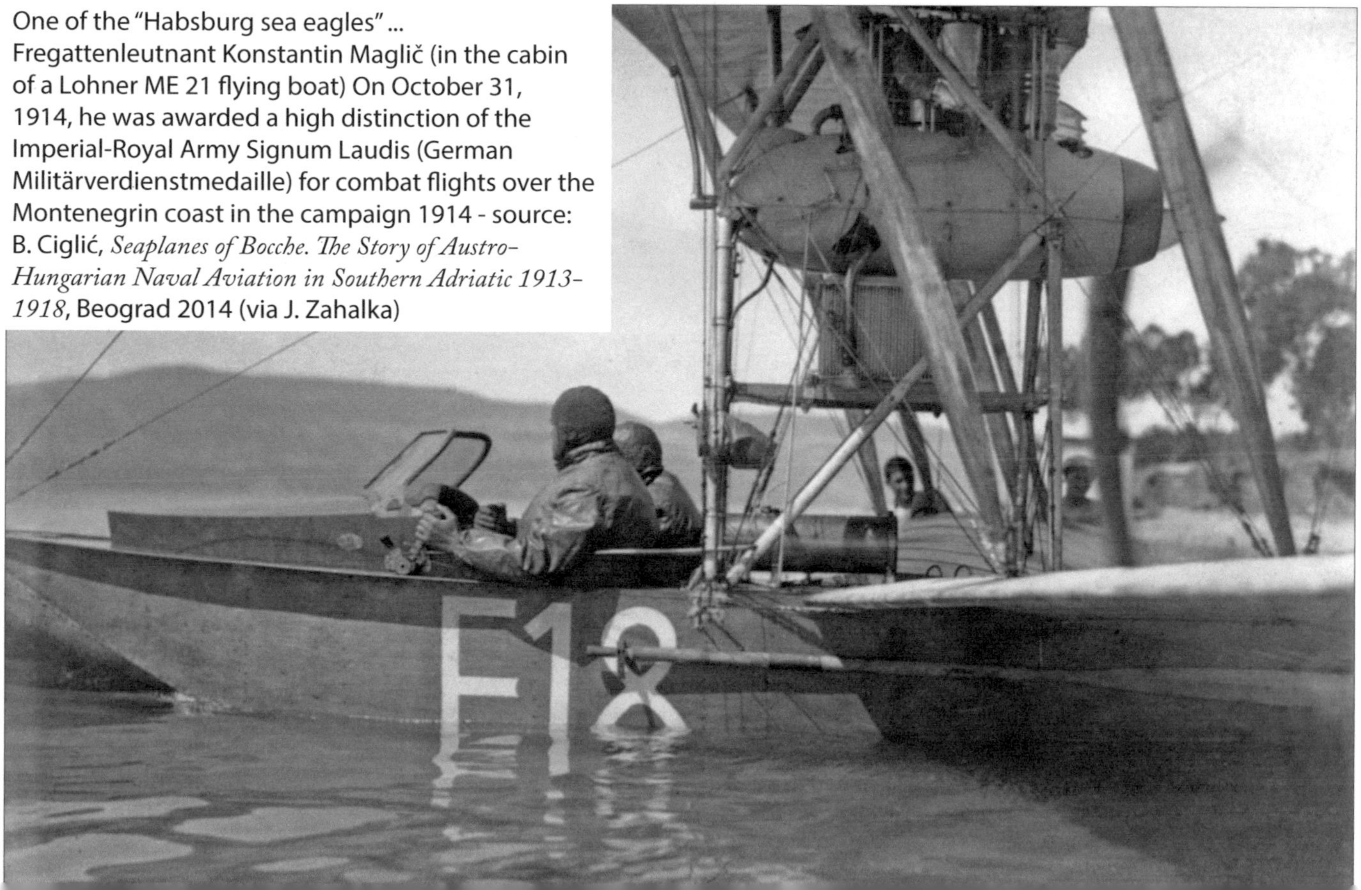

Above: At the turn of the 20th century, in the command of K.u.K. Kriegsmarine, following the example of other powers, began to consider the practical use of tethered observation balloons (Fesselballon, Kugelballon M 96.) installed in the event of military operations on board large warships (the first training use of the observation balloon took place on July 16, 1902 on the board of the pre-dreadnought SMS " *Budapest* "... In the photo SMS" *Budapest* "during the subsequent trials with the observation balloon type M 98 in Pula - Source: B. Ciglić, *Seaplanes of Bocche. The Story of Austro-Hungarian Naval Aviation in Southern Adriatic 1913–1918.* Beograd 2014 (via J. Novak)

FFA crew on 27 August 1914 on the wing of the Russian 2nd Army[4], led to its complete encirclement. On 28–31 August 1914, it led to its extermination (122,000 soldiers were taken captive by the Germans, with losses of 10–15,000 soldiers). In the next clash of this campaign - the Battle of the Masurian Lakes - on September 5-9, 1914, the Russian 1st Army (about 100,000 dead and 45,000 prisoners) was defeated because of a Cannabis maneuver, which resulted in its retreat to the East and the activities of the German aviation in the area of the Osowiec fortress and to the east as far as Berezyna[5]. The air force of the German 8th Army during the 1914 campaign consisted of four FFAs and one FeFA. Directly subordinate to the command of the German 8th Army was the FFA 16th and FeFA 4th from the Posen fortress, and another three units were located at the 1st Corps of infantry general Herman von Francois (FFA 14th), with the 17th Corps of Cavalry Gen. August von Mackensen 'a (FFA 17.) and at the XX Corps of Artillery Gen. Friedrich von Scholtz (FFA 15. additionally, from September 1, 1914, the 37th FFA was established at the 1st Reserve Corps [Reservekorps] of Gen. Fritz von Below)[6]. The words of praise of General P. von Hindenburg in favor of the German aviation for reconnaissance work, quoted several times in this study, were therefore addressed to the crews of FFA 16th Hptm. Wilhelm Schmoeger, FFA 14. Hptm. Heinrich, FFA 17. Hptm. Dincklage and FFA 15. Hptm. Donata. The "eyes of the command" of the field of the German 8th Army in the clash at Tannenberg and in the second clash at the Masurian lakes were the crews of the Etrich-Rumpler "Taube" planes, whose reconnaissance reports decided about her decisive victory[7]! Gen. P. von Hindenburg, despite his age and thus including him in the group of "traditionalists", skeptical about the possibility of a "fifth weapon", showed himself to be exceptionally confident in the reports of the "young weapons" crews from the third dimension. Thus, the term "Tannenberg Aviator" appeared in historiography[8]. Rittermeister in des Res. Elard Baron von Löwenstern recalled his flights from the Battle of Tannenberg as follows:

Near Biesellen [today Biesal] *and Deutsch-Eylau* [today Iława], *German unloading continued. German troops were coming from the north towards Pasym. The lap began to tighten. Earlier, the XX Corps under the command of Artillery General von Scholtz and his Chief of the General Staff Leutnant Hell stopped the Russians after a hard fight. Soon the hour of great victory was to come, it was to be seen whether persistence and courage would bring profits. The command wanted to find out where the Russians were. We knew about the number of enemy divisions, but suddenly we missed one. Had she already noticed our alignment and backed away? Leutnant Hell designated a bottle of champagne as found. 15 Fliegerabteilung* [FFA 15th] *found the lost division! However, as the rumors of a retreat continued, the next day we were instructed to evaluate what the Russians were doing? - what are the streets and towns in the vicinity of Gilgenburg* [today Dąbrówno], *Mühlen [today Mielno], Hochenstein* [today Olsztynck], *Muschaken* [today Muszaki], *Neidenburg [today Nidzica], Goldau* [today

Above: OFlotten Flug Abteilung Kotor ... As of August 1, 1914, there were three Lohner M flying boats E 18, E 20, and E 21 in the state of the vessel ... In the photo, flying boat E 20 in a transport trolley, taken out from a tent hangar and prepared for launch by the ground staff of Flotten Flug Abteilung Kotor - Source: B. Ciglić, *Seaplanes of Bocche. The Story of Austro-Hungarian Naval Aviation in Southern Adriatic 1913–1918*, Beograd 2014

Gałdowo]? – *will the Russians break out of their grip at the last moment? – or will they bring to the law brand new forces that the corps cannot handle? The task was entrusted to the crew of the Jeannintaube 166* [Rumpler "Taube"]. *Commander: Lt. Wittke, observer: me. The sun began to break through the morning mist. We knew it. In the vast area covered with lakes, morning mists very often turned into thick milk, so that later the clouds hung low for the whole day. It was time to hurry. We were entrusted with the task and flight as very urgent. Within a moment our steel* [Rumpler] *Taube was moving towards the front. We did not even do the standard 4-5-minute lap around the airport, which would ensure that we reached 500 m in altitude, which today's pilots would surely comment as "poor fools". Thick bits of mist flew past us. I looked around worriedly and drew a big question mark on the sign* [due to the noise of communication in the flight deck between the airmen]. *But* [Lt.] *Wittke, if you could tell by the scarf and glasses, denied his expression, and waved his hand forward. Okay, I thought if he can fly, I can do observation. We were over the battlefield, at an altitude of 600 m, "visibility is getting worse" – I wrote down in my notebook, because the stylus was blurred in the humid air.* [Lt.] *Wittke pointed to a patch of blue sky and took a course to that side. The sky was clearer over the forests. Neidenburg and a dark spot, probably Goldau, became visible. To our right, the tumult of the battle was clearly visible. 700 m – around Waplitz* [today Waplewo], *Gilgenberg lakes Groß Gardienen* [today Gardyny] *walked and rested, smaller, clearly visible, stretched units – reserves! Neidenburg and Goldau were burning in the distance. The homesteads and smaller pieces of forest were the main camps. Large chunks of forest to the rear and the streets were empty. From what could be seen, the Russians were not thinking about marching out, nor were they getting reinforcements. They were completely absorbed in the fight. Single machine gun shots tapped the cover of the plane. As they were heard only when the cloth was pierced, and yet a man only fears what he sees, they did not bother us too much. We were slowly approaching Muschaken. While focusing completely on the task, we did not pay attention to the weather. As we turned the corner, we noticed with horror that the entire sky was clouded in front*

Above: On June 28 and 29, 1914, three Lohner M flying boats E 18, E 20, and E 21 were used for flights (the crews - there were artillery officers on board as observers - they also recognized the area of the Montenegrin port in Bar) ... In the photo there is a Lohner M no. E 20 flying boat from Flotten Flug Abteilung Kotor in a transport cart during its launch into the waters of the Bay of Kotor by the ground personnel of the unit - source: B. Ciglić, *Seaplanes of Bocche. The Story of Austro-Hungarian Naval Aviation in Southern Adriatic 1913–1918*, Beograd 2014

of us. Thick clouds swirled over the Pasym lakes. It was clean only below and in front of us. I waved, shouting "higher, higher!" Wittke, however, did not want to rise and pointed to the clouds into which the morning mist had turned. This prompted us to fly to Goldau, however, to get the job done. Goldau - I wrote down. [Lt] *Wittke nodded and steered the sluggish Taube brilliantly, avoiding the pouring clouds in front of me and giving me a good view. I was just taking a sketch when* [Lt] *Wittke exclaimed "hey hey!" While waving furiously backwards. The area where it had been bright a moment ago has now disappeared into the fog.* [Lt] *Wittke changed direction abruptly. On the left side, I also recognized Usdau [eng. Uzdowo], and then we found ourselves in the middle of the rain cloud.* [Lt] *Wittke flew admirably calmly. During the sudden backwards turn, the compass needle began to rotate around its axis. The rain lashed our faces. I broke off my goggles to better understand the situation.* [Rumpler] *Taube, usually very steady in the air, was wobbling miserably now. For a moment, a piece of dirt flashed before our eyes. Too fast for our location to be recognized. But not fast enough for my pilot to check the position of the machine. The rainfall gave hope that the clouds would rise. We started falling. At an altitude of 300 m, we were moving towards the dark surface. There were crackles* [of shots] *on all sides, clearly distinguishable from the choked roar of the engine. If only they didn't puncture the radiator - it went through my head. But we are already gracefully enveloped in thick clouds. One thing was for sure, we were inside Russian-controlled territory. Following the old rules of observation, I wave very carefully, right, and left, so we don't start flying in circles.* [Lt] *Wittke tapped his fist on the chassis and pointed to the left, where there was a wide field of view for a few seconds. At the same time, he turned left. We had already passed over the piece of land that passed before my eyes a quarter of an hour ago. I patted the pilot vigorously on the right shoulder. The latter, however, shook his head furiously. Ignoring it, I waved stubbornly to the right. Finally, he turned. Talking was out of the question, so I wrote right, left, and drew a line with a snake. Suddenly, just below us, a smooth, brown water surface appeared - a lake! If we can recognize the banks, the front line can be found. But the lake died again. We shot out of the clouds and flew over the narrow tongue of land, then over the water again.* [Lt] *Wittke said it would be Wablitzer* See [today Waplewskie Lakes - Maróz Lake] *took a sharp left turn. I was speechless. I was just careful not to lose*

Above: In the waters of the Bay of Kotor ... In the photo there is a Lohner M no. E 20 flying boat during preparation for flight by Flotten Flug Abteilung Kotor ground staff - source: B. Ciglić, Seaplanes of Bocche. *The Story of Austro-Hungarian Naval Aviation in Southern Adriatic 1913–1918*, Beograd 2014 (via B. Kudlicka, P. Plattner)

Above: The strength of Bora September 14, 1914… In the photo the sailors of the Flotten Flug Abteilung Kotor next to the Lohner ME 18 flying boat damaged by a strong storm and two tent hangars destroyed by the storm. source: B. Ciglić, *Seaplanes of Bocche. The Story of Austro-Hungarian Naval Aviation in Southern Adriatic 1913–1918*, Beograd 2014

sight of the lakes which I took for the Gilgenburger Seen [today Dąbrowa Mała and Dąbrowa Wielka]. *I waved my right very carefully and acted like I knew what was going on. We made a big right turn. But suddenly* [Lt] *Wittke got mad, took the throttle, and went down 150 meters. I started pounding furiously against his helmet as the machine gun burst hit the wing sheets and the rudders on his tail. I saw tight groups of Russians, and then it got so dark around us that you could barely see the edges of the sheets. We ran deeper into the clouds. Again, very carefully,* [Lt] *Wittke went down, but still ready to retreat in case of fire. We saw the lake again, but this time there was no doubt that it was a narrow strip of land between the Gilgenburger Seen. Gilgenburger Seen, I blurted out.* [Lt] *Wittke laughed happily as if we hadn't argued at all. A huge stone fell from my heart. Our check-in could be delivered on time. Sharply fired, with 42 hits* [from opl.], *We landed. Later we heard about the courage of a German aviator who, despite the low cloud ceiling and sharp fire from the Russians, circled over them many times. If they only knew how far away our courage was then*[9]

In the light of the source materials, on August 27, 1914 it was the pilots - the FFA 14 crew - who provided the 8th Army Field Command with basic information on the location of the Russian 2nd Army Field and the possibility of attacking its wing with the forces of the 1st Corps !; information from the "third dimension" about the situation in the area of the Bojen fortress as well as from the entire battlefield near Tannenberg was systematically provided to the commanders of the 1st Reserve Corps and the 17th Corps by the crews of the above-mentioned FFA 14![10] On August 28, 1914. one of the staff officers of the XVII Corps made a combat flight as part of the orientation as to the direction of the enemy forces- FFA crew. 14. observed the direction of the maneuver of the Russian forces from the west on the left wing and from the south in the Willenberg area (today Wielbark);[11] On August 29, 1914, an observer from FFA 14 noticed a hostile column 36 km south of Nidzica, which made the German side aware of the Russian idea towards Tannenberg[12] ; from August 30, 1914, a German crew in the composition of Hptm. Martin Korner and pilot Oblt. Hans Hesse recognized that the Russian VI Corps had been cut off, which completed the extermination of the entire Russian field 2nd Army at the Battle of Tannenberg[13] ; on August 29-31, 1914 Lt. Gottfried Mertens and Lt. Ernst Canter detected a large concentration of Russian troops from the air, which they reported directly to the command of the German 8th Army field, thanks to which they managed to use the location of the Russian 2nd Army in order to encircle it and achieve a great success for OHL called the "Eastern Front Cannons"![14]. Aviation reports showed that the Russian 1st Army was approaching the area at a slow pace, thanks to which the German OHL could direct its main forces against

Above: On October 16–20, 1914, the crew of a Lohner Mn E 33 type flying boat consisting of Fregattenleutnant G. Prebanda and Fregattenleutnant H. Fontaine von Felsenbrunn performed reconnaissance the area of Ulcinj ... In the photo a Lohner Mn E 33 flying boat from Flotten Flug Abteilung Kotor moored at the shore Kotor Bay. source: B. Ciglić, *Seaplanes of Bocche. The Story of Austro-Hungarian Naval Aviation in Southern Adriatic 1913–1918*, Beograd 2014 (via C. Janić)

the Russian 2nd Army, trapping it in a snare [15].

Reports of Russian airmen from the 20th KAO flying with Farman XVI planes about the movements of German troops were completely ignored by the commander of the Russian 2nd Army, General A. Samsonov, who treated the reports of his air reconnaissance as unreliable and trusted only cavalry, and this, unlike the airmen, did not notice the movements of the German XVII Corps of General A. von Mackensen. The airmen of the 1st, 13th, 15th and 23rd KAO reported heavy traffic of German rail transports daily, making 2–3 reconnaissance flights. Lieutenant pil. Mikhail Wojewódzki from the 1st KAO, on August 13 (26), 1914, despite heavy AA fire and having his plane propeller damaged by artillery fire, made a reconnaissance in the area of Soldau-Lensk-Lautenberg [today Lidzbark] -Kelpin [today Kiełpin] -Konen-Henrichsdorf [today Płośnica] -Soldau, detecting masses of German infantry in the Lautenberg area:

*Having flown my own lines in the Tannenberg region [*Stębark*], I noticed a very large concentration of German troops, which I wrote down in the registration book and flew left down the front line. At that time, the anti-aircraft battery started firing at me, judging by the effectiveness of the fire. White clouds appeared around me and I felt a strong blow. The plane started shaking. The shrapnel broke off a piece of the propeller, which stopped working regularly and the plane began to shake heavily. My first reaction was to turn off the engine to stop the terrible shake. I guessed that the steel frame holding the Gnome engine might break from this, then the propeller blades would tear off one of the tail parts, and my Farman [*XVI*] without a tail would fly down like a stone. However, what prevailed was the fact that our Corps might be in grave danger and the request made to me by the general [*infantryman. Leonid K.] *Artamonow, I would have brought the reconnaissance results so that I decided to continue the flight. The aluminum parts of the engine and clock broke away from the shocks and fell to the bottom of the cabin. Panic took hold of me again, but after a minute I continued my course. Two more times different devices detached from the bottom of the cabin, and within a second I had to decide what to do, but I flew on. And at the very end, the city of Lautenberg* [Lidzbark] *appeared. Having covered it on all sides, recognizing its edges, I detected the* [hostile] *corps. All the backup roads and main lines were cluttered with vessels that were in a good recovery. Several columns marched towards our left wing. What I was looking for, I found in the end. My dedication was rewarded. In an instant, I entered everything into the flight book and took the shortest course*

Above: 16 October 1914, piloting a flying boat type Lohner ME 17 Linien Schiffsleutnant Hugo Ockermüller carried out bombing and shelling targets in Kuksattel and attacked the radio station of the Montenegrin army while being shelled by a Montenegrin opl…. The photo shows a Lohner ME 17 flying boat from Flotten Flug Abteilung Kotor moored in Kumbor. source: B. Ciglić, *Seaplanes of Bocche. The Story of Austro-Hungarian Naval Aviation in Southern Adriatic 1913–1918*, Beograd 2014

Below: The golden hands of mechanics ... In the photo a Lohner Mn E 33 flying boat moored at the shore of the Bay of Kotor during the inspection carried out by the mechanics of the vessel. source: B. Ciglić, *Seaplanes of Bocche. The Story of Austro-Hungarian Naval Aviation in Southern Adriatic 1913–1918*, Beograd 2014 (via C. Janić)

Above: November 2, 1914 because of heavy artillery fire of the Montenegrin opl. from Mount Ratec and from the artillery battery stationed in the Držak area, the shrapnel damaged a flying boat tyoe Lohner ME 17…The photo shows this seaplane after launching at the Flotten Flug Abteilung base in Kotor Bay. source: B. Ciglić, *Seaplanes of Bocche. The Story of Austro-Hungarian Naval Aviation in Southern Adriatic 1913–1918*, Beograd 2014 (via B. Kudlicka, P. Plattner)

Below: As a result of a strong south-west wind (Bora) and a storm up to 8 ° on the Beaufort scale, the Flotten Flug Abteilung hangar in Kumbor was destroyed, and the three Lohner M E 17 and Lohner Mn E 34 and E 35 flying boats which were in it, got seriously damaged ... The photo shows the damage done to the Flotten Flug Abteilung base - source: B. Ciglić, Seaplanes of Bocche. The story of Austro-Hungarian Naval Aviation in Southern Adriatic 1913-1918, Beograd 2014 (via B. Kudlicka, P. Plattner)

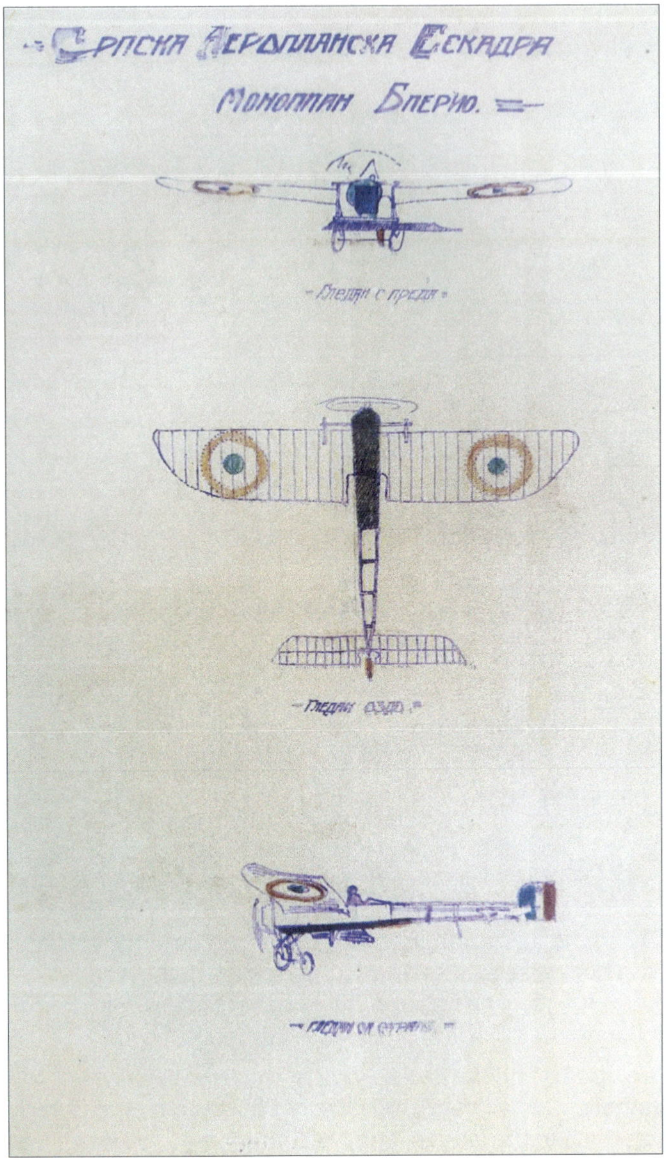

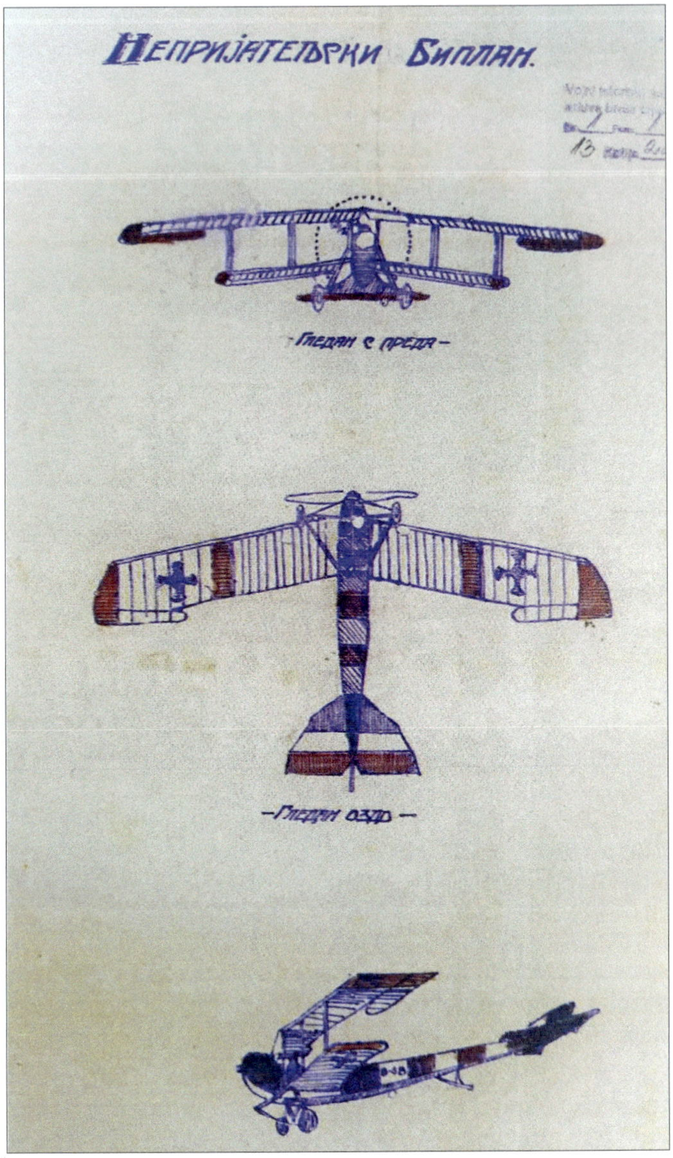

Above: Over time, the infantry troops of the fighting armies became more and more familiar with the silhouettes and markings of their own and enemy aircraft... In the photos, training materials containing the profiles and markings of Austro-Hungarian and Serbian aviation aircraft. source: *Србија у великом рату 1914–1918*, Beograd 2014

to our airport. I was at the height of 2,000. meters and I could plan any distance from this height, I calculated the distance and as soon as this moment came, feeling light and after praying, I turned off the bike [Gnome] *and started a long glider flight. The conversion turned out to be correct, I did not make a single turn, the shot plane landed calmly at its airport. The mechanics running up to me quickly started repairing the engine and very quickly found that the frame supporting the engine was broken in three places, and only a fourth part was complete. The old mechanic said that only a miracle could keep the engine in place. I went to the general* [infantryman. Leonida K.] *Artamonow to submit a report. After hearing the report, he told me:*

So, you saw everything clearly, what should we then do in your opinion? I replied that the German grouping in Lautenberg [Lidzbark] *was sure to be close to our left wing, intended to go around our flank and throw it away to the group that was located opposite our center at a sufficient distance. So why couldn't our ramps overwhelm him. According to this position, you should immediately improve the position of the Corps, rebuild its parts in a half-turn to the left, giving way, leaning on the left wing on the Soldau* [Działdowo] *line, digging in to withstand the pressure of the enemy and stop his walking around our left flank. The result of the reconnaissance should be immediately delivered to the army headquarters because the fight can start at any moment. General* [infantryman L. K.] *Artamonov summoned the chief of staff, presenting the situation to him, and ordered to write an order. This special order for the general* [infantryman. L. K.] *Artamonow and for me of August 13* [26]*, 1914 began with a description of my diagnosis and the mention of my surname*[16].

Right: In the field ... Commander Flik 4. assigned to the field Imperial-Royal 6th Army on the Serbian Front, Oblt. Matthias Bernáth - source: collections: J. Butkiewicz

In another battle - the Battle of the Masurian Lakes - in East Prussia, the command of the field 8th Army also trusted reconnaissance from the air. Ww Rittermeister in des Res. E. Baron von Löwenstern recalled his flights during this battle as follows:

We won at Tannennberg, the troops were reformed and redirected to the east against the Russians, whose positions were established on the line of Dejma, Allenburg [today Alembork], *Gerdauen* [today Gierdawa], *Nordenburg* [today Nordenbork], *and the land of lakes as far as Arys* [today Orzysz]. *Initially, the Russians set off around* [0] *7 am, then they rested around noon, to start the march again in the late afternoon. Now, however, they were almost invisible during the day. So, we assumed that they marched at night and rested in the woods during the day, invisible to the pilots. Around 7-8 September* [1914] *our command was not clear whether the Russians were withdrawing or preparing for a decisive battle. Thus, ordered the determination of the whereabouts of the enemy and reconnaissance of the terrain and roads leading from the Wirballen line* [today Wierzbołów], *Kalwarija* [today Kalwaria], *Suwalki* [today Suwałki]. *We made the flight and found that the fighting continued the Lyck line* [today Ełk], *Angerburg* [pol. Węgorzewo], *Nordenburg and to the north of them, while the barracks near Angerburg looked disassembled, and the area behind, just behind the fighting line, was surprisingly calm. So, it could be seen that the Russian*

Below: The Lovćen battery in trouble ... In the photo, KuK Kriegsmarine ships during the shelling of the so-called Lovćen battery - artistic version by Harry Heusser. source: "*Das interessante Blatt*" 1916

Above: The naval war in the Adriatic Sea was also fought at night ... In the photo, KuK Kriegsmarine during the night shelling of the enemy's on-board artillery with the use of reflectors. source: "*Das interessante Blatt*" 1916

Left: Domination over the sea… A watercolor painting showing the sinking of the French submarine "Foucault" by the crew of the Lohner L-135 type flying boat in the Oblt composition on September 15, 1916. Walther Zelezny and Lt. baron Otto von Klimburg south of Punta d'0stro. author's collection.

Above: The reality of the Adriatic air war in 1916…. On February 2, 1916, the crews of three Lohner L 81 (Linienschiffsleutnant D. Konjovič and Linienschiffsleutnant J. Ritter von Ritschl), L 83 (Linienschiffsleutnant W. Hell and Linienschiffsleutnant M. Sewer) and L 82 (Linienschiffsleutnant K.. Löw) recognized the region of Valona falling into the strong fire of the Italian Opel in the port area and over the island of Saseno and still carrying out bombing targets. As a result of the fire of the opl Lohner L 83 flying boat had to forcibly launch due to a damaged radiator. Right next to it, a Lohner L 81 flying boat was launched, the crew of which took the L 83 crew on board and, with a total of four pilots on board, happily reached the base in the Bay of Kotor. The photo shows the artist's vision of the rescue of the crew of a downed Lohner flying boat by another K.u.K. Seeflugwessen. source: "*Das interessante Blatt*" 1916

The wartime waters of the Bay of Kotor ... In the photo a U-boot of the KuK Kriegsmarine returns from a combat voyage to the Adriatic base. source: "*Das interessante Blatt*" 1916

Lohner M E21

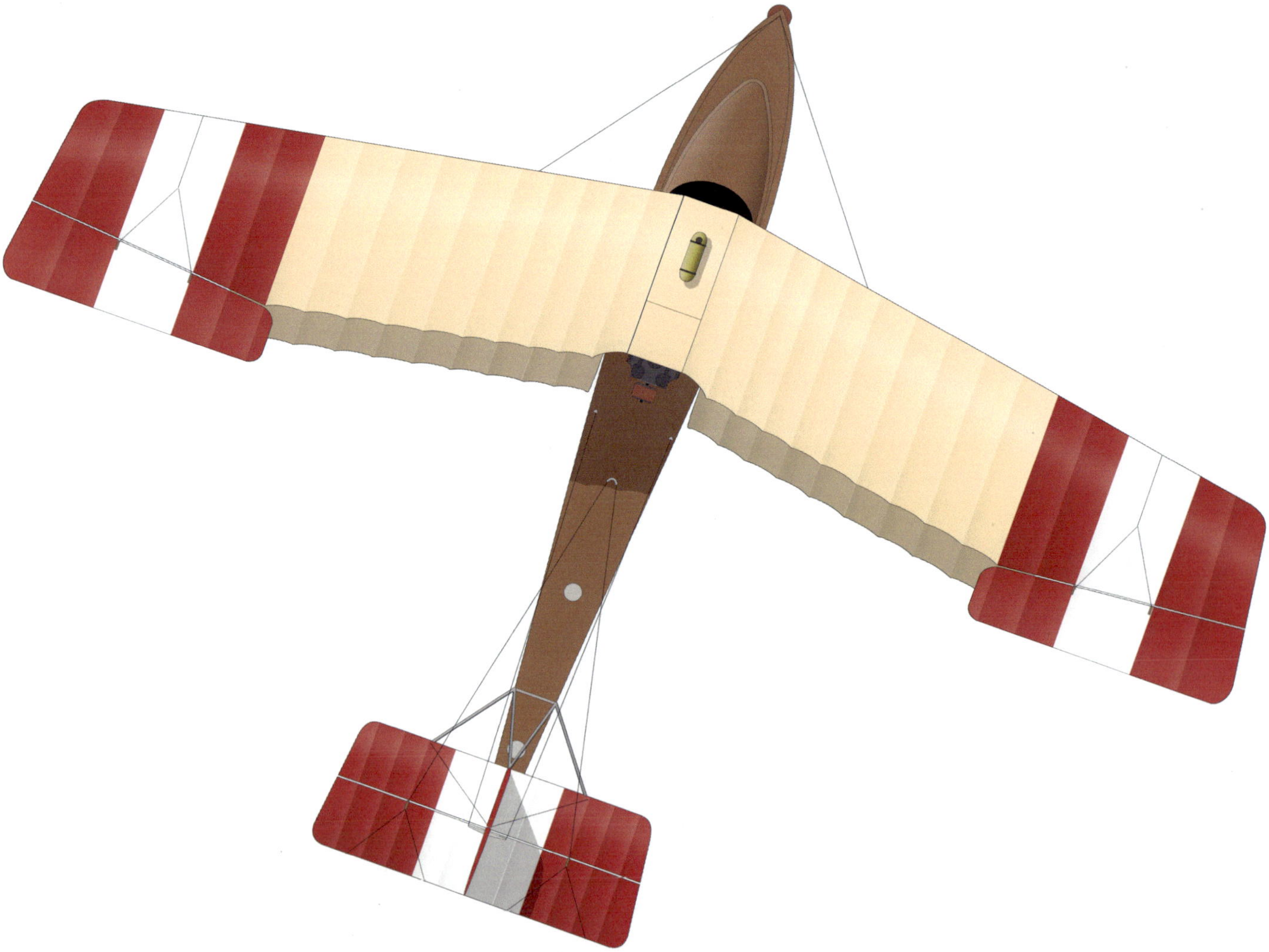

Lohner Type 'T' L47

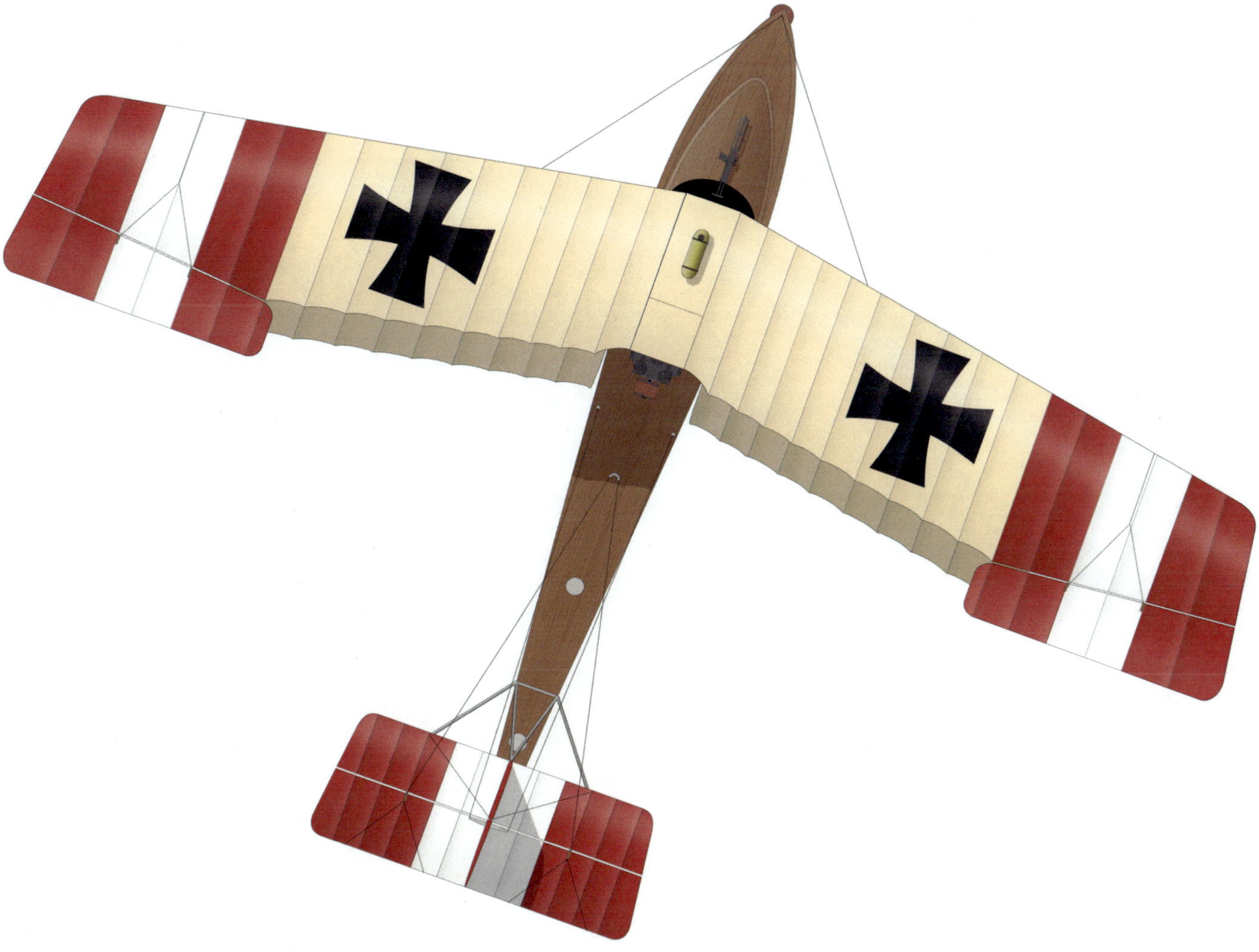

Left: Gallery of the losers ... Commander of the fieldwork of the Imperial-Royal 5th Army on the Serbian Front, gen. Liborius Ritter von Frank - source: Fotoarchiv SAW-KA Wien

rear guards were fighting fiercely while the main forces had separated. The Jeanin-Taube 166 crew ([Lt] Wittke and I) *were given a preliminary assignment. "During the fierce fighting of the rear guard, the main Russian forces broke off. It is essential to establish where they are. I'm going to Darkehmen [today Oziorsk], which should be conquered today, and I'll put up the landing arrow sign* [Ger. Landepfeil] *- we were instructed by Hptm Donat, who taught us through his own example and later through his heroic death at Lomza* [pol. Łomża] *that the unit commander, if he wants to give orders and be able to properly assess the potential of his people, should also take part in flights over enemy territory, and in the first line. We stood 900 m above Nordenburg. Before us was Darkehmen - but no, it could not be Darkehmen. There was clearly still a violent struggle here. Bright flames shot up at us. Dense clouds of smoke covered the entire town. I must have lost my way because the railroad crossed the road below me, and it looked completely different on the map*[17].

Below: The main dilemma of AOK was trying to answer the question what is on the hostile bank of the Danube? ...The task of the Flik 4 crews in the Serbian campaign of 1914 was photo recognition. The photo shows the Serbian bank of the Danube - Source: Fotoarchiv SAW-KA Wien

Above: Some plywood and cloth ... In the cabin of the Aviatik aircraft from Flik 4 crew consisting of Fähnrich von Bretfeld and Lt. Burlan -source: Fotoarchiv SAW-KA Wien

Below: The Drina was a particularly important river for the AOK and the army commanders of the Imperial-Royal army preparing to push through it... In 1914, aerial photography became indispensable for the waging of modern warfare - in the photograph of Drina's meanders through the lens of the Flik 6 observer's camera. Source: Fotoarchiv SAW-KA Wien

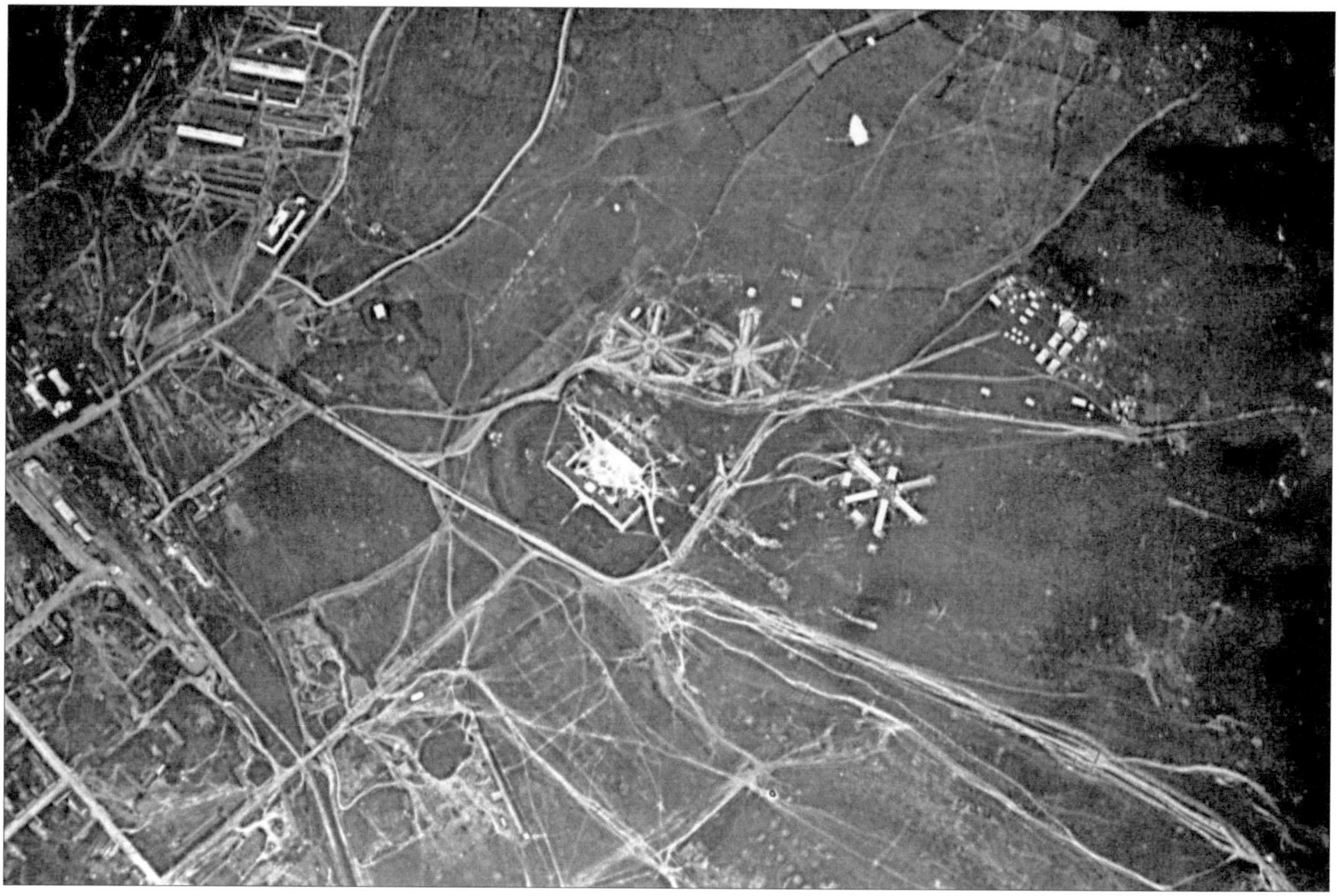

Above: Valjevo lies on the Kolubara ... As part of the photo reconnaissance, the Flik 6 crews rushed there as well. Source: Fotoarchiv SAW-KA Wien

Below: Bora - cool, dry and gusty wind that causes a drop in humidity. It was extremely dangerous for the pilots ... The photo shows a wooden hangar at the Flik 4 field airfield on the Italian Front, destroyed by this wind. source: Fotoarchiv SAW-KA Wien

Above: In 1915, the Lovćen battery, equipped with Italian heavy mortars, shelled the port of K.u.K. Kriegsmarine in Kotor and its roadstead ... In the photo, destroyed as a result of the shelling of the Imperial-Royal artillery, Italian mortar in positions taken by the Imperial-Royal branches. source: "*Das interessante Blatt*" 1916

Above: Each Fliegerkompanie KuK Luftschifferabteilung was equipped with special transport vehicles for transporting dismantled planes, and Ferrying the front Fliegerkompanie to the next field airfields in winter was not easy...In the photo, Flik 6 transport equipment at the snow-covered field airfield on the Serbian Front in 1915 - source: Fotoarchiv SAW-KA Wien

Above: Aviation fashion of the winter campaign 1914/1915… In the photo the Flik 6 crew after a combat flight on the Serbian Front in winter flight suits - source: Fotoarchiv SAW-KA Wien

Above: The reality of winter front service... In the photo the Flik 6 crew in front of a Lohner Pfeilflieger. And just before the flight on the Italian Front - source: Fotoarchiv SAW-KA Wien

AOK and the military command of the Imperial-Royal Army constantly needed up-to-date information about the enemy. Winter was not an obstacle here… The photo shows the frozen Flik crew. 6. after a combat flight on the Serbian Front in the winter of 1915 - Aviatik B II aircraft in the background - source: Fotoarchiv SAW-KA Wien

Above: The impact core of the KuK Kriegsmarine during the Great War, apart from surface ships, were also U-boats.... In the photo U-1 at the base in Kotor Bay - source: Fotoarchiv SAW-KA

Right: "Sausage" above the front ... The innocent-looking observation balloon hanging with the observer above the front line allowed for precise insight into enemy positions. The photo shows the observation balloon tethered KuK Militärballon M 98. (Drachenballon) - Source: Fotoarchiv SAW-KA.

Above: The year 1914 brought great progress in world aviation, forced by the outbreak of the Great War... In the photo the Flik 7 crew in the Albatros B.I (23.24) aircraft built at the Phoenix plant for export to the Kingdom of Greece, armed with the 6.5 mm Schwarzlose M.12 system onboard - source Fotoarchiv SAW-KA

Left: The Last Summer of Peace – July / August 1914 ... Flik 1 field airfield in Homokbálványos. Next to the Lohner D (B.III) plane, the Pfeilflieger stands from left to right: Oblt. Jillicher, Oblt. Josef Flassig, Oblt. Viktor Schünzel, seated Oblt. Benno Fiala von Fernbrug from Flik 1. source: Fotoarchiv SAW-KA.

Right: With the growing plague of being shot down by their own troops due to the ignorance of the profiles and the marking of the planes of the parties fighting in the 1914 campaign in Galicia, it was necessary to popularize them among their own troops ... The photo shows the Russian guidelines for land units regarding the marking of aircraft of the fighting parties - source: collection of the Polish Aviation Museum in Krakow

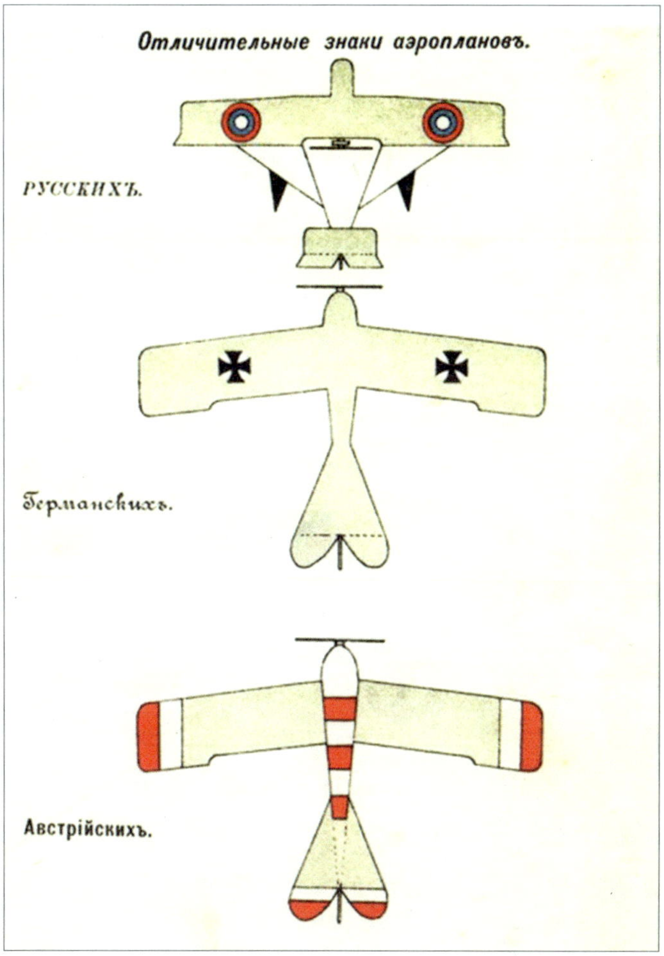

Endnotes

1 *Geschichte der Luftwaffe…*, op. cit., p. 56.
2 Ibidem
3 Ibidiem
4 *Geschichte der Luftwaffe…*, op. cit., p. 56.
5 According to other data, the Russian field of the 2nd Army was destroyed, losing 75,000. soldiers, including 30 thousand. prisoners and 150 guns, while the German losses amounted to only 9 thousand. soldiers - Lt. J. Livesey, op. cit., pp. 30-37; M. Pöhlmann, H. Potempa, T. Vogel, op. cit., pp. 18-21, 91; *Der Erste Weltkrieg im Bild. Deutschland und Osterreich an den Frontem 1914 1918,* Wien-Graz 2014, p. 76-77; G. P. Neumann, op. cit., p. 463; Исторія великой войны, Т. II, Москва 1915, p. 177–178; Исторія великой войны, Т. III, Москва 1916, p. 153-165; Великая борьба народовъ, Выпуск *IV*-ый, Москва 1915, p. 33-36; *Prvni svètova válka*, Praha 2010, p. 53–55; H. Münkler, op. cit., p. 145–149; S. Rosenboom, op. cit., p. 7. In line with the

Below: Durrës, the main port on the Albanian coast, from 1913 it was in the field of interest of the KuK Luftschifferabteiulung command... In the photo taken in 1915 by the Flik 6 crewport and city -source: Fotoarchiv SAW-KA Wien

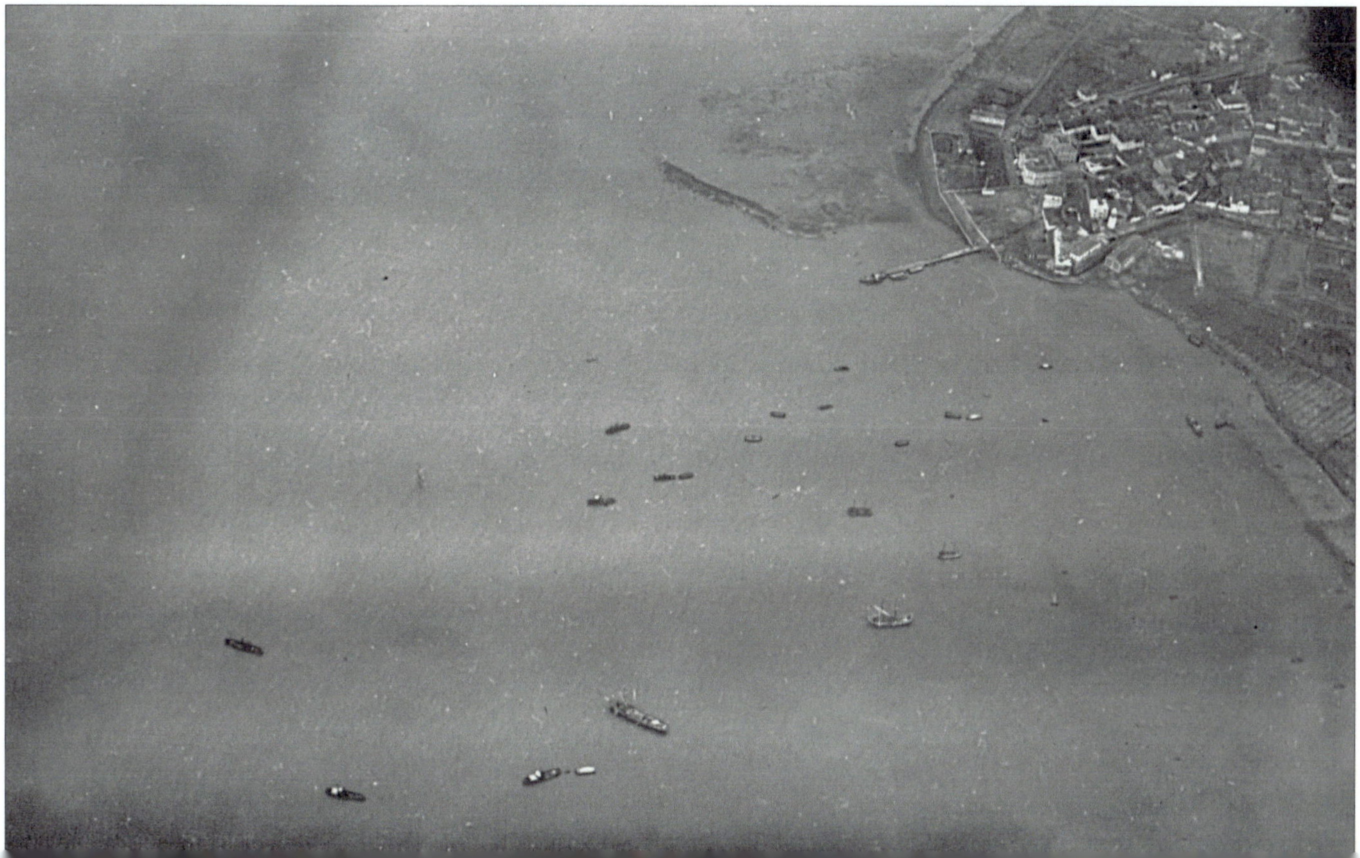

Above: A window to the world of the Austro-Hungarian Empire ... In the photo Trieste, photographed by a Flik 6 observer.- Source: Fotoarchiv SAW-KA Wien

The trained eye of an aviation observer could look into virtually any ground target….The photo shows the old fortress Peterwardein on the Serbian bank of the Danube and Újvidék on the bank of the Austro-Hungarian Empire, photographed by the Flik 5 crew- Source: Fotoarchiv SAW-KA.

Right: In the air service, every free moment was extremely precious, and a cigarette and a glass were soothing.... The photo shows the Flik 7 flying staff on the Süd-Westfront in winter 1915. From left: Oblt. Dotte, Oblt. Adolf Heyrowsky, Oblt. Arpád Gruber, Oblt. Camilo Perini, Zgsf. Josef Tóth, Oblt. Winkler, Oblt. Lodzinski, Oblt. Breiteneker, Oblt. Max Perini, Oblt. Lerch - Source: Fotoarchiv SAW-KA.

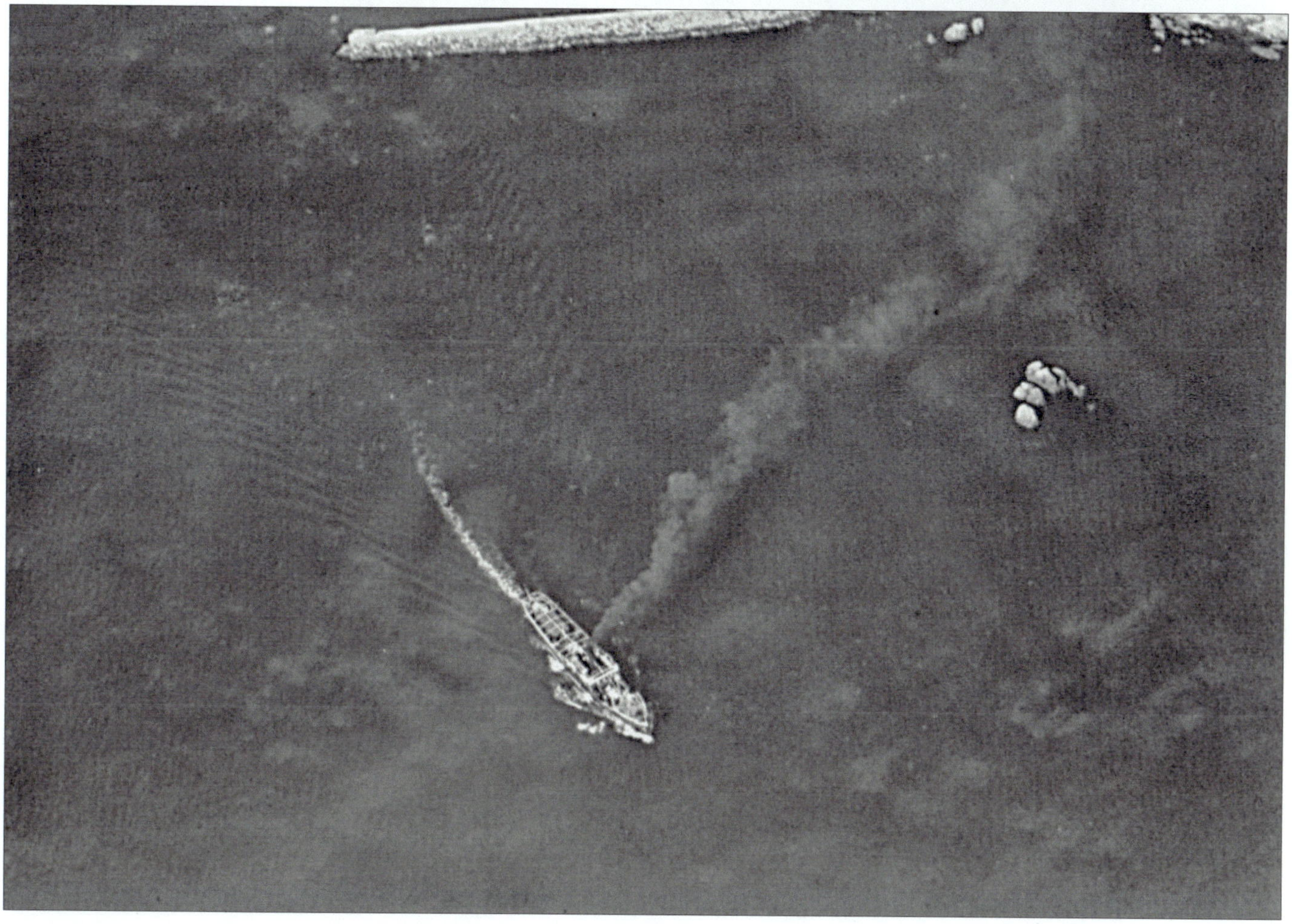

Above: Not only land objects were photographed by Flik 6 observers... In the photo a French or Italian steamer photographed by the Flik 6 crew in the region of the Montenegrin coast - source: Fotoarchiv SAW-KA Wien

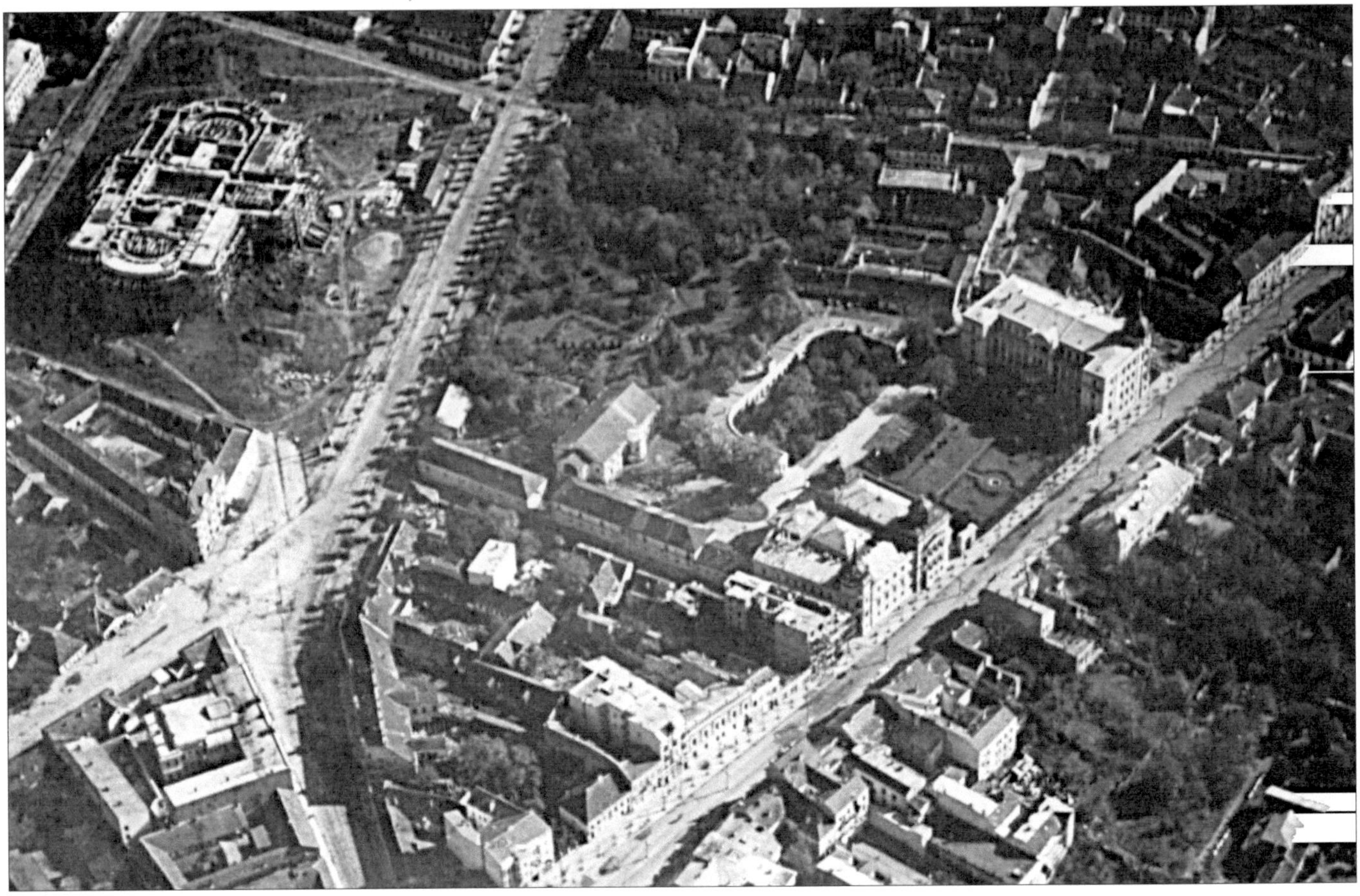

Above: Belgrad 1915 ... In the photo, the capital of the Kingdom of Serbia seen from the third dimension through the eyes of the Flik 6. crew's camera. Source: Fotoarchiv SAW-KA Wien

Above: Belgrad 1915 ... In the photo, the capital of the Kingdom of Serbia seen from the third dimension through the eyes of the Flik 6. crew's camera. Source: Fotoarchiv SAW-KA Wien

Leserbriefe

Mag. Hubert Prigl, Wien, hat in Ergänzung zu den Biographien von **Carl Banfield** und **Ignaz Stiefsohn** in den vergangenen öfh Nachrichten 4/13, Seiten 4ff und 18ff, Fotos der Grabsteine der beiden Piloten übermittelt. Banfield liegt auf dem Hietzinger Friedhof in Wien und Stiefsohn hat seine letzte Ruhestätte auf dem Wiener Zentralfriedhof gefunden (siehe rechts).

22

Unsere gefallenen Fliegerhelden:

Mitte: K. u. k. Generalstabshauptmann Oskar Rosmann. Links oben: K. u. k. Oberleutnant Josef Flassig. Rechts oben: K. u. k. Oberleutnant Albert Sanchez de la Cerda. Links unten: K. u. k. Oberleutnant Adalbert Feßl. Rechts unten: K. u. k. Oberleutnant Manfred Georgievics.

Alexander Kustan, 1030 Wien, kann zum Grabstein des **Hptm. Oskar Rosman** (manchmal fälschlich als Rosmann geführt) auf dem Wiener Zentralfriedhof, siehe öfh Nachrichten 4/13, S.38, folgende Angaben machen:
Hauptmann und Generalstabsoffizier, Fliegerreferent, Freiballonfahrer Diplom am 29. Juli 1914; im

Right: The stones say ...In the photo, the symbolic grave of Hptm. O. Rosman in the village of Czereśnie located to this day on the site of the former Flik 10 field airport (Cieszanów-Folwarki) - as of 2015 - source: author's collection

Previous Page: August 29, 1914 as an observer to Flik 10 was assigned Oskar Rosman an officer of the General Staff of the Hptm ... After take-off from the Cieszanów field airfield (Folwarki) due to pilot error Hptm O. Rosman died in an air accident. In the photo Hptm. O. Rosman - in the center - among the airmen of the KuK Luftschifferabteilung who died in the 1914 campaign - source: G. Penzl's collection.

Below: The realities of anti-aircraft defense in 1914 and historical policy… In the picture, the fire of small arms and machine guns by the Russian infantry and cavalry of the airship of the Imperial-Royal Airship. Interestingly, in the 1914 campaign, the KuK Luftschifferabteilung did not use airships, and the caption under the painting from the collections of one of the museums in Moscow proves that on 8 August 1914, near the Russian border, the Imperial-Royal airship was shot down and then delivered to Kiev on three railway platforms. The facts say that German type airship Schütte Lanz (SL II) On August 21/22, 1914 "Liegnitz" made a flight on the border of Galicia and the Russian Empire with a landing in Przemyśl at the fortress airport in Żurawica, but the airship was not shot down! It is easy to deduce that this lithograph says about this event, but the description of the event contains a serious factual error -source: M. Franz's collection

Wir suchen...

Hptm. Oskar Rosman, am 29.08.1914 in Galizien als Flieger gefallen, liegt auf dem Wiener Zentralfriedhof. Wer kennt die näheren Umstände?

Left & Above: Aviation fashion KuK Luftschifferabteilung 1914 and later ... The photos show an aviation "leather" and an aviation helmet on display at the Heeresgeschichtliches Musem in Vienna - source: author's collection.

propaganda slogan *Vorwärts mit Gott für* König und *Vaterland bis der letzte Russe besiegt am Boden Liegt (Forward with God for the King and the Fatherland until the last Russian will not be defeated)* it is worth recalling that the co-author of this great victory of the German army in 1914, which in the propaganda dimension was to suppress the role of the Slavic triumph over the Teutonic knighthood at Grunwald in 1410, was General E. Ludendorf, known from the words: *Ohne Opfer kein Sieg! Ohne Sieg kein Friede! - Without sacrifice, there is no victory! There is no peace without victory! - see B. Hamann, Der Erste Weltkrieg. Wahrheit und Lüge in Bildern und Texten,* München-Zurich 2014, s. 49; H. Münkler, op. cit., p. 153-154 . At the end of May 1915, the FFA 37 crews worked in the section of the "Nemunas" Army and in June, during the offensive on Lithuanian territory, they performed flights in the Riga section and in the south - the Rovno section - flights were made to a depth of 250 km - with use in the region Riga also has aircraft from naval units. In the field section of the 9th Army, during the fighting in the Warsaw-Modlin area in October 1914, the German air force played another significant role - on November 24, 1915, the crew of the FFA 17. made a flight for the XXV Reserve Corps in the Bzura-Rawka area, photo recognition hostile positions and minefields in the Błonie-Warsaw area. During the Masurian winter battle, the aircrews once again showed their role on the battlefield, and in April 1915, also during the field offensive of the 10th Army on Rivne - the crews of the FFA 51 coordinated artillery fire there - as well as in the Vilnius section and in the field sections From the beginning of 1915, the air crews of the 12th Army also conducted active reconnaissance activities before the spring offensive in the region of Modlin-Warszawa-Różan-Ostrołęka as far as Białystok. In the period from autumn 1914 to March 1915, the German air force actively operated on the Eastern Front in the Allenstein section (today Olsztyn) - operating from the abovementioned area within 30 days to Łomża, Ostrołęka, Przasnysz, Ciechanów, Nowogród and Białystok, 20 attacks were made bombs on Russian targets located at a depth of 280 km, dropping about 4 thousand. bombs. Until the first half of June 1915, in the field zone of the 9th Army, the crews of the FFA 28, on the section Pułtuska and FFA 45, conducted air reconnaissance in the section up to Modlin, and at the

Above: Tarnów - winter 1915 ... The position of the Russian Maxim system machine gun in the vicinity of the railway station, shooting at the Austro-Hungarian aviation plane circling over the city - author's collection

Above: Each appearance of an airplane, unknown wonders of technology, aroused the natural interest of the local community ...In the photo the plane of the Lohner B "Pfeilflieger" AD 210 type with its own name "Boy" (11.23) from Flik 8 at the Festung Przemyśl fortress airport in Żurawica in August 1914 - source: collection of Lt. col. Dr. T. Kopański

Above: The first days of the Great War ... Festung Przemyśl fortress airport in Żurawica. The photo shows the Lohner B "Pfeilflieger" AD 210 planes with its own name "*Boy*" (11.23) from Flik 8- first from left Oblt. Arthur Böhm - source: collections of Lt. Col. Dr. T. Kopański

same time General Woyrsch's forces were given the FFA 31st and 64th. started systematic reconnaissance flights on the Deblin-Warsaw line. During the fighting in June 1915 in the area of Radom-Jozefów, the FFA 31 crews from 03.00 to At 20.00 reconnaissance flights were carried out as well as at night, while the fighting near Dęblin was photographed by the crews of the FFA 64 and the attack after August 1, 1915 in the direction of Brest-Litovsk at Russian railway stations, field airfields and artillery positions - see G. P. Neumann, op. cit., pp. 464–465.

6 *Der Weltkrieg 1914 bis 1918, Bearbeitet im Reichsarchiv, Die militärischen Operationen zu Lande, Zweiter Band, Die Befreiung Ostpreussen*, Berlin 1925, pp. 358–359.

7 G. P. Neumann, op. cit. pp. 463

8 According to Die Entwicklung ... (op. Cit., P. 23) in 1914 OHL allocated 190 aircraft, 5 airships and 16 tethered observation balloons to the Western Front, and 36 aircraft, 1 airship and 6 tethered observation balloons to the Eastern Front (a total of 226 aircraft, 6 airships and 22 tethered observation balloons). During the East Prussian campaign of the tsarist army, the German field 8th Army had 58 aircraft, while the Russian field 1st and 2nd armies had 24 and 42 planes (a total of 66 - in the case of the 2nd Army, 12 of the abovementioned number of aircraft were stationed in the fortresses of Osowiec and Grodno – por. I. Рєпін і Ю. Дзюбенко, Східно-прусська операція російської армії під час першої світової війни: причини поразки (1914 р)., [w:] Перша світова війна: історія та урокі. 1914–1918 рр, Київ 2014, p. 235–236.

9 *Flieger aus Tannenberg*, collection of the BMA library in Freiburg, pp. 145–147. The composition of the FFA 15 included min. Hptm. Bahr and Fldf. Praetzmann.

10 *Geschichte der Luftwaffe...*, op. cit., p. 56

11 G. P. Neumann, op. cit., p. 462.

Above: One of the best pilots of the KuK Luftschifferabteilung in 1914 was Oblt. Rudolf Holeka from Flik 8 .... The photo shows the pilot in front of the Lohner B "Pfeilflieger" AD 205 (11.05 / 73.05) from this unit with its own name "*Bandit*" at the Festung Przemyśl fortress airport in Żurawica - the airport hangar in the background - source: Lt. Col. Dr. T. Kopański

12 Ibidem, p. 463

13 W. J. Boyne, *Wings of War. The decisive force of the 20th century*, Warsaw 2011, pp. 68–69; "Over the Front", 1996, Vol. 12, No. 2; A. Olejko, M. Bakun, op. cit., pp. 85–110; M. Bakun, *German Aviation on the Eastern Front 1914–1915 - an outline of the problem, [in:] Aviation in Przełamanie Gorlickie 1915*, ..., op. cit., p. 203; S. Rosenboom, op. cit., p. 60.

14 G. P. Neumann, op. cit., p. 463.

15 A. Olejko, M. Bakun, op. cit.

16 A. Victor v. Koerber, *Die deutschen Fliegertruppen,* [w:] *Der Soldat, Blätter für deutsche Wehr und Ehr*, z. 7, 1934, p. 19; A. Olejko, *Unknown cards of the great war or a short history of two Russian Aviaotriads of the North-West and South-West Fronts from the 1914/1915 campaign*, „Przemyski Yearbook - Military History", vol. XLVIII

Right: Camillo Perini - famous aviator with Italian roots, during the Great War an aviator of Flik 7 as well, in 1914 he made combat flights over Galicia ... In the years 1918-1919, as a Polish aviation officer, he commanded the air force gathered at the Lwów-Lewandówka airport during the Polish-Ukrainian war, and in the independent Second Republic of Poland, where he settled, he became respected aviation theorist. In the photo, freezing in an aviation uniform right after a combat flight to the Nord-Ostfront - Source: Fotoarchiv SAW-KA.

2012, z. 1. passim.

17 *From the Battle of Masurian Lakes*, the collection of the BMA library in Freiburg, pp. 147–148.

Below: In Fliegerkompanien K.u.K. In 1914, the Luftschifferabteilung used bombs weighing 18 kg, carbonite bombs weighing 10 kg, 12.5 kg, and 5.5 kg, universal "Škoda" hand grenades intended for aviation ... The photo shows the bombs suspended under the Albatros B.I aircraft from Flik 1- source: Fotoarchiv SAW-KA.

Above: Weapons of the observers - Pistollenkamera ... The German 250 mm, 500 mm and 700 mm aerial cameras were the basic weapon of the air crews of the Great War. In the photo the Pistollenkamer in the Heeresgeschichtliches Museum Wien collection - source: author's collection.

Left: A cigarette after the flight ... On August 19, 1914, the Flik 7 crew composed of pilot Oblt. Max Perini (pictured) and Observer Oblt. Georg Betka made a long reconnaissance flight on the route of 300 km - source: Fotoarchiv SAW-KA.

Above: The capricious aura on December 15, 1914 "grounded" the crews, including Flik 1 ... due to the air fog... In the photo, the airmen of this unit (Oblt. Benno Fiala von Fernbrugg on the left, Commander Oblt. Josef Smetana next to the dog, Oblt. Stefan [Pista] Horwath first on the right) during the departure by car from the Krakow field airport to the Raba river - source: Fotoarchiv SAW-KA.

Air Resume

In the twentieth century, through constant evolution[1], an aero plane, most often called an airplane, turned from a *shapeless box* into a perfect tool, obedient to man and providing him with irreplaceable services in all areas of life[2]. Until December 17, 1903, so till the first flights of the Flyer I aircraft in Kittyhawk, the brothers Orville and Wilbur Wright in the US, airspace - the third dimension, was conquered only by the aeronauts[3] who gave beginning for aviation.

Aviation uses heavier-than-air instruments: airframes; balloon troops, on the other hand, use devices lighter than air: free and tethered balloons, airships and one device heavier than air - a kite ... The slow balloon is not suitable for military use. Rising without a rudder by the gas lift, it flies towards the air currents ... The kite is an observation device that replaced the spherical tethered balloon on those days when the force of the wind prevented it from rising ... The tethered balloon remains tied with a tether rope, coiled on a car or horse jack ... The ceiling of a tethered balloon is measured by the length of the rope that it can unwind. This length, depending on the thickness of the rope and the lifting force of the balloon, is very limited ... The airframe, on the other hand, recognizes and fights ... At low altitude, on the bare ground, it catches everything down to the smallest detail; raised very high to the ends of its "ceiling", it covers the entire battlefield ... The airship combines the advantages of a tethered balloon and an airframe... It can also fight and bomb at the same time... The airship is primarily a means of naval warfare.... Each modern airframe consists of the following essential parts: supporting panels (so-called colloquially wings); feathering (or ballasts); hull; rudders (altitude and directional); landing gear for departure and landing; one or more power units (engine-propeller connections)[4].

In the "era of aerostats" (for the first time in history, tethered observation balloons and free balloons were used during the wars of the French Republic with the anti-French coalition in what is now Belgium and France in 1792-1794)[5], in the second half of the 19th century, for the first time

Above: Flik 1 airfield in Bardejov on September 30, 1914 .... In the cabins of the Albatros B.I aircraft commander of the Oblt. Josef Smetana and Oblt. Stefan Horwath - source: Fotoarchiv SAW-KA.

air reconnaissance with the use of an observation balloon determined the success during the land fights during the American Civil War 1861–1865, when in the army of the "north" - Unionists - were used to recognize the tethered aerostats. During the fighting in 1861, General George B. Mc Clellan used the aeronauts La Mountein and Ullan to inform himself via balloon observations of the location of the secessionist force camps (the reports contained information from up to 50 km from the launch site; Ullan and Prof. Lowe used an observation balloon with ... a telegraph for reconnaissance, about which Prof. Lowe informed the Union army command in Washington by telegraph). During the battle of Richmond won by the Confederate forces on August 29-30, 1862, the Union army used the observation balloon to observe again ... with a telegraph on board (during the American Civil War, he took part as a balloon observer as a volunteer Oblt. Graff von Zeppelin)[6]. During the conflicts on the "old continent", on the other hand, aerostats were used during the Franco-Prussian war, when between 23 September 1870 and 28 January 1871, the French side made 66 transport flights from Paris besieged by the Prussian army with the use of free balloons on board, 164 people, 381 pigeons, 5 dogs and 10,675 kg of various materials were evacuated from the besieged city[7]. In the war conditions of the turn of the 19th and 20th centuries, the tethered observation balloons were used for the first time in 1898 by the balloon detachment of the US Army Signal Corps to observe the fire of their own artillery (this took place during the fights on Cuba in Santiago and San Juan where aerostats were used for observation of own artillery fire in the port and the shelling of ships as well as the position of Spanish troops). In the 20th century, there is no source information about the use of observation balloons in the Russo-Japanese War of 1904–1905, while during the so-called during the Libyan War of 1910, the Italian military 1st Army used on 23 October 1910 two tethered observation balloons to observe and fire at the position of the Turkish army (Carlo Montu became the first European war balloon pilot), while in Europe, tethered observation balloons were used during the First Balkan War 1912–1913 i.e. in the Bulgarian army)[8]. During the Great War, Drachenballon or the popular "Sausages" (length 24 m, width 6.4 m, area 600

Above: German aviation technology ... At the airport in Bardejov on September 30, 1914, against the background of the Albatros B. 1st commander from Flik 1 Oblt. Josef Smetana - Source: Fotoarchiv SAW-KA.

Above: Kraków-Rakowice Airport, December 15, 1914 .... In the photo the Flik 1 crew composed of Oblt. Stefan Horwath and Oblt. Benno Fiala von Fernbrug in the cabins of the Albatros B aircraft. And during the "dry mortar" - training with an aerial bomb - source: Fotoarchiv SAW-KA.

m3), as in the front jargon, i.e. soldiers of the Imperial-Royal the Austro-Hungarian army called observation balloons on a tether type K.u.K. Militärballon M 98. (Parseval Siegsfeld M 98), which could operate at an altitude of 1000-1500 and even 2000 m, at the outbreak of the Great War, displaced the older Kugelballons M 96 spherical balloons from service. Until 1918, the French industry produced four thousand observation balloons designed by Ing. Albert Caquot (type R), and the German industry two thousand balloons in the model M 98. During the Great War, about 700 tethered observation balloons were used in the Russian Air Force (Russian: Tsarist Air Force - Императорский военно-воздушный флот), of which 138 were lost in the fighting[9].

During this period, hundreds of sky operations took place in the skies of war-torn Europe, and its inhabitants, in many cases, for the first time in their lives saw airplanes with crews flying not too high above their heads, tasked with gathering information about the enemy on the enemy side of the front line. Observers or pilots could see people on the ground with the naked eye from a height of 1,200–1500

Above: "Aviation fashion" in winter 1915.... In the photo against the Rumpler B.I Oblt. Ludwig Dumbacher and Oblt. Cizinsky of Flik 1 surrounded by German officers at the Croatian airport in Slatina, March 15, 1915 - source: Fotoarchiv SAW-KA.

m, and in the marching columns from a height of 2,500 to 3,000 m. This type of observation was ideally suited for both maneuvering and positional warfare. It became imperative to take accurate post-flight notes and quickly the use of a camera became necessary. Soon, the radio used in the third dimension made it possible to quickly transmit information, e.g., on the direction and accuracy of artillery fire, to the staffs of divisions, corps, or field armies. The first military pilots came from various types of armed forces, being ex-infantrymen, cavalrymen, and artillerymen. However, the commander of the plane was an observer, and the pilot was merely the *chauffeur*. Nevertheless, in the army and in the world of civilians, the aircraft crew was perceived as the elite and was always treated with the highest honors, and thus airmen were entitled to numerous privileges that an ordinary soldier could not count on. At the turn of the critical 1914 and 1915, atmospheric conditions were one of the greatest enemies of the *knights of the skies*, and especially in the later months of 1914, when hostilities often took place in an area dangerous not only for foot soldiers, artillerymen and cavalrymen of the Carpathians and aimed at towards winter, how often the ordered flights could often be carried out only with the use of the last forces subordinate to the commander of the KuK Luftschifferabteilung Obstlt./Obst M. E. Uzelacovi. In the long-stretched Galician tw, the weather *was not uniform, even on the same day the weather was often completely different. It also happened that there were flights also on rainy days, because they were trying to overcome the resistance all the time so that missions could be carried out*[10]. Photographs and aerial photos were taken from the air in the first days of the war[11], but not only the observation took place in *the third dimension* "dominated by Mars". A complete *novum* was the use of air force in Galicia to bomb the enemy, *and the use of bombs, built-in machine guns, darts and leaflets was used on the same terms as in the Balkans*[12].

According to the Chief of Staff of the German army, General Erich von Falkenhayn, during the 1914 campaign, the breakthrough battles of the Great War took place on the Marne on the Western Front and in Galicia on the Eastern Front[13]. The hostilities on the Eastern or Serbian Front of the

Above: We fly from Brzesko in 4 airline companies [Fliegerkompanie]. With us - a delicious bachelor, [Oblt.] Benno Fiala, the one who himself led the train with his company's belongings [Fliegerkompanie 1.], when it was suddenly surrounded by Cossacks. He ran under fire to the machine, replaced the dead mechanic, counterattacked, withdrew the train, firing at himself furiously with a service pistol. He is cheerful and often poses for photos with a cute white spitz named Muzzl. This is also how the "radio" airplane was christened [Oblt. Max] Hesse.... The photo shows a "radio" Albatros B.I No. 7 aircraft with its own name "*MUZZL*" in 1914, appearing in the Flik warehouses. 1st, 8th and 10th with the Flik crew. 1. composed of Oblt. Ludwig Dumbacher and Zgsf. Fp. Mattl before take-off from the field airport in Jasień in winter weather- source: Fotoarchiv SAW-KA.

Above: The pilot was just flying over the campsite when our team hit it. He took a picture from above just after the explosion. Mortar shells ["Škoda" cal. 30.5 cm M. 11.] - says the lieutenant - they have such a cruel power of explosion that once, when I was flying over the Russians and a thirty-five-missile exploded below me, the next moment my machine started to sway strongly, though I was eighteen hundred meters high. A strong blast hits the heavy machine even at such an enormous height, although a projectile [30.5 cm M.11 Škoda] exploded on the ground. What the consequences must have been down there, in the immediate vicinity! ... In the photo of the museum in the Kalemegdan fortress in Belgrade - next to the "Škoda" M 11 mortar, 30.5 cm caliber author - source: author's collection.

Left: 30 December 1914 in the Flik 1 workshop at the field airport in Jasień, on a plane type Albatros B.I No. 7 with the proper name "MUZZL" "radio equipment" was installed.... Despite the extremely low temperature (-27 to -30 ° C) outside, all construction and assembly works at the airport and at the airport office was supervised by prof. Dr. L. Kann (in the center). Oblt also participated in the construction and assembly work. Stefan Horwath (left) and Oblt. Ludwig Dumbacher (right) -source: Fotoarchiv SAW-KA.

Above: Along with the threat of the Krakow Fortress - the second battle for Krakow - AOK decided to evacuate the Krakow-Rakowice fortress airport via the alarm square of the main fortress reserve (now Błonia), also used as a field landing site, so far operating from the Krakow-Rakowice Fortress airport Fliegerkompanien 1st, 8th, 10th, 15th to the field airport in Andrychów ... In the photo from 1914, the field base of Flik 1 in Andrychów - the source of Fotoarchiv SAW-KA.

Great War in 1914, however, are constantly overshadowed by the "miracle on the Marne" or from Flanders on the Western Front, where the fate of Paris and access to the English Channel in the same campaign were at stake[14]. The German OHL managed to stop the Russian offensive in the Battle of Tannenberg, which was very important for the Eastern Front. It was then that the German aerial reconnaissance played such a unique role that after the battle of the words of the German 8th Army commander, General Paul von Hindenburg, a military retired from 1911 and restored to the service of the line savior of East Prussia in 1914, *there would be no Tannenberg without an airman* (ohne *Flieger kein Tannenberg or Ohne Flieger - kein Tannenberg*) were the best assessment of the use of aviation[15]. For the French command, the key to a breakthrough victory on the Western Front, the battles over the Marne, were the reports of the French aviators of L'Aviation Militaire, about the recognition of the German forces (a gap of 35–40 km was created between the German 1st and 2nd Army Field troops of the French 6th Army and some of the BEF troops, as well as the battle itself, showed that aerial reconnaissance was an invaluable aid in planning a French counterattack[16]. It is well known about these events, forgetting the role of aerial reconnaissance that influenced the success of the Serbian army in the Battle of Mount Cer, which was decisive for the fate of the Serbian Front. It ended with the defeat of the army of the Danube Habsburg monarchy and was also the first major success of the Ententa troops over the armies of the Central Powers in the Great War! The period of the Great War has undeniably become "Airplane Day". During it, aerial reconnaissance was instrumental in the success of the Serbian army in the Battle of Mount Cer, which was decisive for the fate of the Serbian Front! The next battles of the Great War, where air reconnaissance reports had a decisive impact on the tactical or strategic success achieved on earth, were the battles at Mons on the Western Front, at Tannenberg on the Eastern Front, and only in that order – at the Marne on the Western Front. As a result of both won battles at Tannenberg and at the Masurian lakes, East Prussia was saved and the success in the above-mentioned battles resulted in the reversal of the assumptions of the Schlieffen Plan. In the 1914 campaign, not only the above words of General P. von Hindenburg were an expression of appreciation for the role of the "fifth weapon" on the example of the Battle of Tannenberg, as well as the opinion of the BEF commander. On the Western Front, the OHL was not settled in its favor and was forced to go into a trench war after the Battle of the Marne, while on the Eastern Front it achieved a decisive success in the 1914

Above: Until 9 December 1914, the airfield in Andrychów became the main air base on the Nord-Ostfront.... In the photo, next to the Abatros B.I plane in Anrychów and, in the cabins, and on the lobe, Flik 1 officers. To the right is Oblt. Stefan Horwath, Oblt. Josef Smetana, in a cavalry Oblt uniform. Max Hesse; Oblt stands on the lobe. Ludwig Dumbacher - Source Fotoarchiv SAW-KA

campaign. To this day, however, the symbol of the "miracle on the Marne" is a Parisian taxi and it should be ... a plane[17]!

Endnotes

1 A. Nhum, op. cit., p. 8-14.
2 R. Kaczkowski, *Aviation in sea operations*, Warsaw 1986, p. 7.
3 J. Batchelor, Ch. Chant, op. cir., p. 9-15; R. Witkowski, op. cit., p. 54-58, 113, 118; Ch. Chant, *Pioneers of aviation*, New York 2001, p. 6-14; E. Malak, *Little human flight, but great humanity*, "Rzeczpospolita-Planes and people" 2010, No. 1, p. 20-23; *Orville Wright's Journal*, "Rzeczpospolita-Airplanes and People" 2010, No. 1, p. 28-29.
4 Major Orthlieb, op. cit., s. 1, 3-4, 154; B. Hamann, op. cit., s. 96.
5 *Der Luftkrieg 1914-1915*, op. cit., p. 11; O. Richter, op. cit., p. 5.
6 P. Kettel, op. cit., p. 126. Colonel Zygmuntowicz (*Outline of the development of air navigation and its applications, [in:] Album of Polish Aviation*, Warsaw 1930, p. 42) states that in the period from August 28, 1870 to January 22, 1871, 65 balloons were launched from Paris with 91 passengers, approx. 360 pigeons and a good deal of letters.
7 P. Kettel, op. cit., p. 129-130; *Der Luftkrieg...*, op. cit.; O. Richter, op. cit.
8 L. Kennett, op. cit., s. 3, 19; S. Pagano, *Stormi Berei e divisi oni corazzati in cooperazione*, Rome 1941, s. 10.
9 In the period June–August 1917, as many as 45 balloons were lost - see S. Rosenboom, op. cit., p. 26.
10 SAW-KA, Die Tätigkeit der österr.-ung. Luftstreitkräfte beim Auffangen der rusischen Angriffsheere im Sommer und Herbst 1914. Bearabeitet in der "kriegswissenschaftlichen Abteilung der Luftwaffe", zweisgstelle Wien, 1942/1943 von Obst. Dr Oskar Regele, p. 103.
11 Ibidem.
12 Ibidem, p. 102. The first air bomb attack in the Great War took place on August 3, 1914, when the German crew from Rumpler "Taube" dropped three small bombs on a French marching column in the Luneville area of the Western Front - see S. Rosenboom, op. cit., p. 9; O. Wedigen, op. cit., p. 29.
13 В. Литвин, Україна: доба війн і революцій *1914–1920*, Київ 2003, p. 26.
14 For example, in the study by M. Eksteins (*The Rite of Spring. The Great War and the Birth of a New World*, Poznań 2014 - passim), the issue of the Eastern Front does not exist.
15 S. Rosenboom, op. cit., p. 7; H. Münkler, op. cit., p. 145; *Der Erste Weltkrieg* 1914–1918. *Der deutsche Aufmarsch...*, op. cit. p. 91; B. M. Tkachev, op. cit., p. 319; W. Murray, J. Keegan, op. cit., p. 27; A. Olejko, M. Bakun, op. cit .; M. Pöhlmann, H. Potempa, T. Vogel, op. cit., p. 91; H. Mordawski, *Air forces in World War I*, Wrocław 2008, p. 134.
16 H. Münkler, op. cit., p. 126.
17 *Geschichte der Luftwaffe...*, op. cit., p. 45–49.

Above: Kraków Rakowice Fortress Airport in the period of December 8-16, 1914 (dates from archival records - dates from the record under the photograph on December 9-19, 1914) became the main base of Flik 1.... The photo shows tent hangars and Albatros B.I planes from Flik 1 against the background of airport buildings - source Fotoarchiv SAW-KA.

Lohner L135

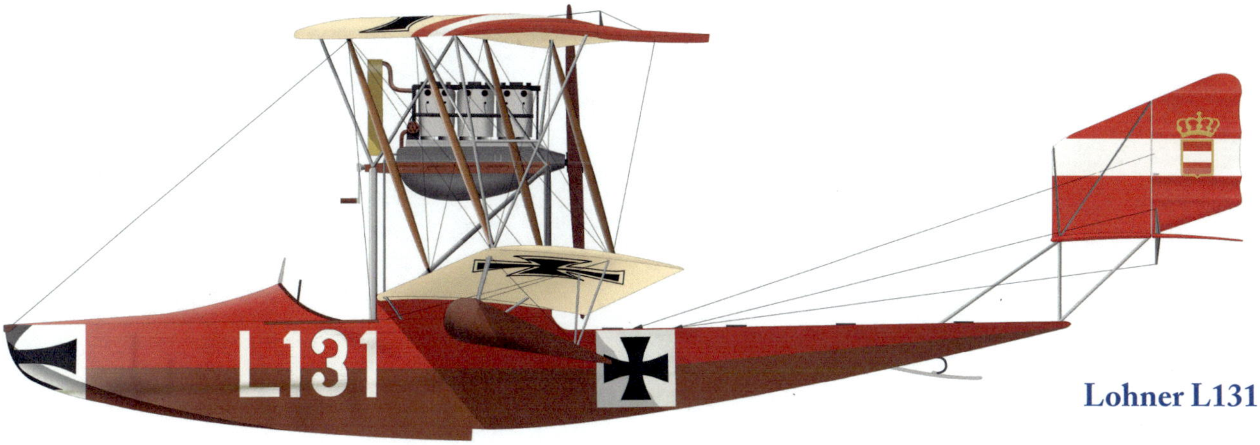

Lohner L131

II. The First Aviation Winter at the Nord-Ostfront or… Ohne Flieger kein Sieg!

On February 6 [1915*] 2 airplanes* [planes] *flew close above us, probably Prussian* [Austro-Hungarian aviation] ... *Airplanes* [planes] *run almost every day - yellow with a red circle, blue in the center of the circle, white* [Russian aviation marking] ... *The tsarist daughter in a uniform, in a soldier's cap, headed the army through Lisko* [from 1931 Lesko]. *Has 5 of them, they can drive. The second one was going by airplane* [by plane] *over Przemyśl - she wanted to see what they were there, why couldn't take it*[1].

In modern warfare, and such was the Great War, the plane became an integral part of the arsenals of the countries that led it. Helena Schramówna, a native of Olchowa, described her impressions from the flights over the Bieszczady Mountains in Galicia of both sides of the air in her diary. Air reconnaissance over Galicia in December 1914, despite daily snowfall, which sometimes even made reconnaissance flights impossible, and strong winds over the Carpathians, a systematic air reconnaissance was successfully carried out from the area of Upper Hungary (e.g., on December 4, 1914, the command of the Field Imperial-Royal 3. the army demanded air reconnaissance in the Uzhok region and in the 25 km belt from the Użocka Pass to identify the movements of the Russian cavalry)[2]. There were engine failures during the flights, which led to forced landings (this is how the reconnaissance flight of the Flik 14. crew ended too early on December 10, 1914). Particular attention was paid to the movement of Russian rail transports on the Carpathian front and in its hinterland[3]. During the December retreat to the east of the Russian 3rd Field Army from Kraków and the Field 8th Army from Beskidy, the command of the Imperial-Royal, victorious in the Łapanów-Limanowa field operation the 4th Army, needed detailed information about the enemy retreating, i.e. a constant and up-to-date aerial reconnaissance of the location of his forces. On December 12, 1914, the guidelines for air reconnaissance were as follows: *What Russian forces are in the Nowy Sącz area? Is the traffic on the Grybów-Gorlice-Jasło route considerable? What is the situation on the Dunajec River*[4]*?* In addition, the command of the fieldwork of the Imperial-Royal At the same time, the 4th Army demanded a long-distance reconnaissance around Zakliczyn-Rzeszów as far as the area of the besieged Przemyśl Fortress and in the area of Jarosław-Dębica-Krosno-Sanok-Przemyśl[5]. On December 16, 1914, four flights were performed with very good results[6].

Above: Flik 1 airfield in Secovce on November 24, 1914 .... Preparation for flight of the Albatros B.I aircraft with the crew of the Oblt. Viktor Schünzel and Oblt. Ludwig Dumbacher - Source Fotoarchiv SAW-KA.

Above: The front Fliegerkompanien was transferred mainly by rail.... Flik 1 is shown in the photo during ferrying from the field airport in Secovce to the fortress airport Kraków-Rakowice (according to archives, the unit was stationed in Secovce until November 23, 1914, and from November 27, 1914 at the Kraków-Rakowice fortress airport; according to the caption under the photograph, the transfer took place in November 26–28, 1914) - the source of Fotoarchiv SAW-KA.

Due to the worsening weather conditions, the crew of the Imperial-Royal aviation did not always return happily to their airfields[7], but efforts were made to fly in "all weather conditions" (on December 18, 1914, six reconnaissance flights were made, and on December 19, 1914 - another six, despite deteriorating weather conditions)[8]. December days abounded in fog (on December 20, 1914, the Flik 7 crew had difficulties to recognize the roads for this reason) and aircraft failures (on December 17, 1914, the command of the 1st Army's fieldwork asked Festungskommando Kraków to recognize the Stopnica-Staszów area, which it turned out to be impossible due to the lack of a functional plane and therefore the commander of the field commander of the 4th Army was asked for help). The same happened in the operational section of the military c.k. 3rd and 4th Army. The Fliegerkompanien subordinate to her command, carried out reconnaissance flights -despite increasingly "polar" weather conditions. Fliegerkompanien assigned to the field command of the Imperial-Royal 4th Army, after December 14, 1914, conducted constant reconnaissance also for the command of the Krakau Fortress (Festungskommando). They were:

- Flik 1. (December 8, 1914 - Kraków-Rakowice fortress airfield; December 17, 1914 - Brzesko [Jasień] field airfield)
- Flik 8. (from December 9, 1914 - Kraków-Rakowice fortress airport)
- Flik 10. (from December 9, 1914 - Kraków-Rakowice fortress airport)
- and Flik 15. (December 1, 1914 - assignment to the headquarters of the Krakow Fortress - Krakow-Rakowice airport; December 19, 1914 - field airfield in Brzesko [Jasień]).

Flik 3 joined this group (from December 13, 1914 - Zborov field airfield, December 28, 1914 - Prešov) and Flik 14. (December 2, 1914 - Košice field airfield; December 12-24, 1914 - Jasło - Zborov - Żmigród field airfield, December 28, 1914 - Prešov field airfield)[9].

During the flights, there were occasional air clashes with enemy planes (On December 16, 1914, Flik 3 lost a crewed plane in a fight with a Russian plane, and on December 20, 1914, a Russian Nieuport IV plane disturbed the Flik 8 crew during a reconnaissance flight)[10]. Despite the weather conditions, which more and more often resembled a "polar aura", the air war over Galicia was burning up. Until 2 May 1915, the 3rd and 8th Russian field armies had their position and activity in defense and offensive activities in the Carpathians in the direction of Budapest (the intensification of Stawka's offensive actions was brought about by the third winter battle in Bieszczady, fought on March 20-30 and the so-called The Easter Battle fought on April 1-6, 1915) to keep the AOK staff members awake at night, aware of the fact that they had beaten, but not smashed, their opponent in the December 1914 battle and that they had lost at the end of the year. These fights proved emphatically that the phrase "Ohne Flieger kein Sieg", or even "Ohne Flieger kein Limanowa-Łapanów", acquired a special meaning and sense of supporting the development of the air force in the CC. Austro-Hungary by donors who are presented with commemorative bursar medals with the inscription Für die Ősterr. [Eichische] Luftflotte in the name of the monarchical slogan Viribus Unitis. The spring of 1915 and the Gorlice Breakthrough (Durchbruchschlacht Gorlice), which disrupted the Eastern Front in its southern section, completely changing the wartime situation in the eastern theater of military operations, confirmed this thesis - "Ohne Flieger kein Gorlice"! But before that ...

The reality of winter air war over Galicia:

Singing breaks the silence. It is neither a lively breach of peasant cheerfulness, nor a melancholy "dragging" near Krakow. The high tenor leads, after a while lower voices jumped to him, there is a follow-up, a long sad note. Lots of voices, young and strong. After a while, a column of army in gray uniforms emerged from among the birches. They make their way through the fields, along the road, which bends to the left in this [spelling true to the original] *place in a great arc. They are perhaps a thousand paces away, barely visible as small figures, real lead soldiers. They are approaching; rifles and feats on their heads can be distinguished. They are Bosnian regiments. They wandered the entire monarchy, from Montenegro* [the Kingdom of Montenegro] *to Michałowice, to find themselves in the Kingdom of Poland*[11].

Probably every inhabitant of Galicia affected by military operations would be surprised by such an "oriental" view as the following, from the Miechów-Książ road from December 1914, where an "unusual" Balkan army appeared on the regular road. In November 1914, the crew of Flik 14 conducted from field airports located in Upper Hungary (today the Slovak Republic) in Zborov, Bardejov and Prešov for the benefit of the 3rd Army, flights to the Carpathian region, despite daily snowfall and strong wind[12], more and more often flying over thc Carpathians[13] and the region of the Beskid Sadecki and the Low Beskids. In December 1914, the AOK, after the victorious success in the "Marna of the Eastern Front", saw the opportunity to "roll up" the entire Carpathian Russian front and carry out the relief of the besieged Przemyśl Fortress (1 [5][14] November 1914 - 23 March 1915), ruling over the most convenient way leading from Galicia towards Hungary (in the so-called "Hungarian Gate" - towards the Dukla Pass and Użocka Pass). In addition, it could also serve as an excellent support point for any offensive undertakings towards Lviv and Lublin, and this would force the opponent to retreat beyond the Vistula and San[15]. In the opinion of Dr. Sergey Nielipowicz: *Some historians compare the significance of the Battle of Limanowa-Łapanowo with the Battle of the Marne. This is obviously an exaggeration in scale. The only similarity here is that the large-scale Russian offensive in the Carpathians and against Krakow was stopped, the Russian armies were forced to withdraw without any obvious advantage of the enemy (weakened German armies retreated behind the Marne). The withdrawal of the armies of the Southwestern Front became an acknowledgment of strategic defeat, despite valiant deeds and sometimes brilliant tactical successes. The situation in the armies of the North-Western Front, which were defeated near Łódź and suffered huge losses in subsequent battles, had a huge impact on the outcome of the battles in Galicia. Also, the loss factor, which could not be replaced by a hastily trained supplement, played a large role in the conduct of operations by the armies of the Southwestern Front. However, this operation cannot be considered completed. It started a new stage, characteristic of the Austro-Hungarian command, an attempt to achieve success, and for the army of the Russian Southwestern Front, an attempt to reverse the tide of battles by resuming the offensive in the Carpathians*[16].

This conclusion, presented by a well-known Russian researcher on the subject of the Great War, is a kind of point on the Russian side for the Łapanów-Liman operation, which brought AOK quite a success in Galicia, and the retreat of Russian forces from Krakow and the Beskids to the East clearly demonstrated it, which undoubtedly confirmed the AOK in the belief that it would be possible to "liberate" ("recover")[17] Western Galicia and come to the rescue of the "Przemyśl lions". The initial success of the counteroffensive of the Imperial-Royal the army was caused by i.e.considerable losses previously suffered by the Russian army and the bad situation of the hostile North-West Front after the Battle of Lodz. The success of AOK, however, was not entirely successful, because as the days passed, the offensive in the direction of Przemyśl met with strong resistance from the Russian troops, which caused heavy losses among the attacking Imperial-Royal army troops (in the December field offensive of the Imperial-Royal 3rd Army, its number decreased to only 47,000 soldiers). In this situation, the AOK included all the front reserves in the field CC. The 3rd Army saw the hope in its offensive operations to use the initial success achieved in the Łapanów-Limanowa operation. The new strike - the second phase of the abovementioned operation - was, however, associated with a considerable risk, as it all depended on whether it was possible to make a breach in the southern flank of the Russian troops retreating beyond the Dunajec and Biała rivers. Moreover, AOK had to reckon with encountering vigorous countermeasures on the part of Stawka to avoid the imminent "front-line catastrophe" in the south. The commander of the Przemyśl Fortress, General Herman Kusmanek von Burgneustädten, counted

Above: The first "radio" plane of the Great War, Albatros B.I No. 7 with given name "*MUZZL*", on the field airport in Jasień…. In the photo taken on January 4, 1915, Flik 1 officer in the aircraft cabins: on the left, next to the antenna retractor Oblt. Benno Fiala von Fernbrug, Oblt. Otto Jindra (Oblt. Jillicher?), Oblt is standing next to the plane. Ing. Rudolf Stanger from Flik 8 - Source: Fotoarchiv SAW-KA.

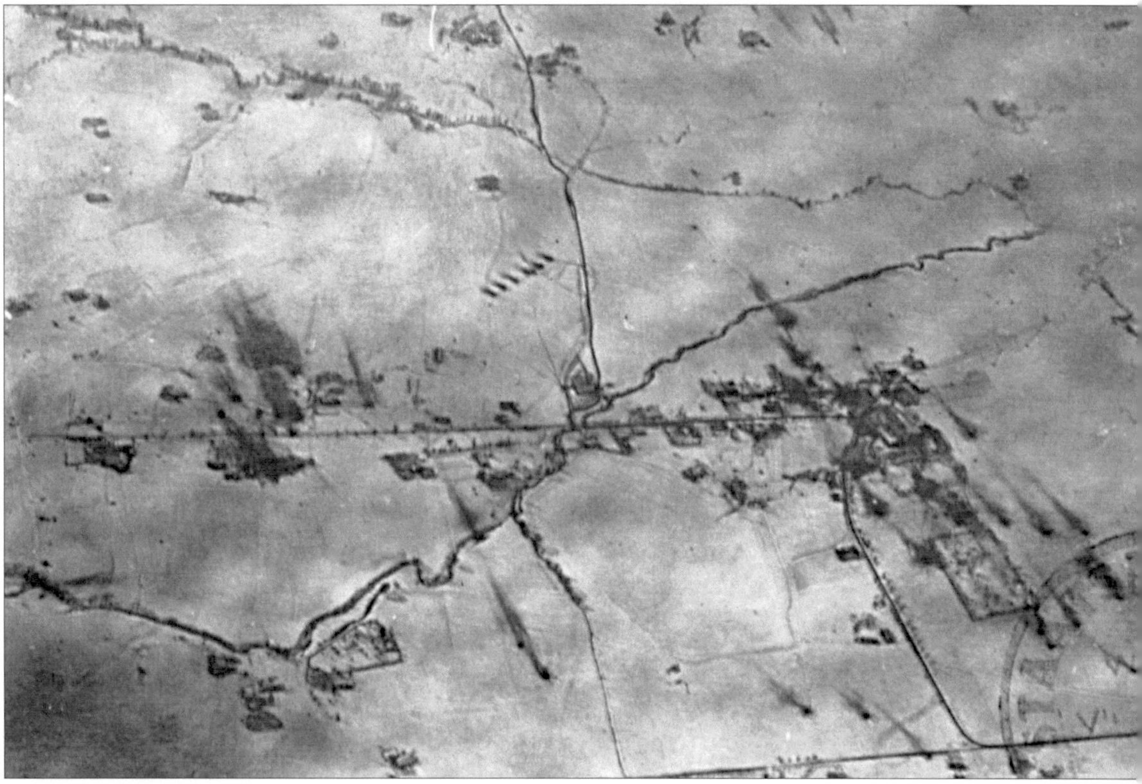

Right: War in the winter aura ... In the photo from the winter of 1915, the results of the artillery fire on the snowy Serbian territory, captured by the observer Flik 6. Source: Fotoarchiv SAW-KA Wien on another, as in the case of the first siege, a relief from the west for the fortress he commanded, which, according to the assumptions of the quartermaster, could defend itself in the "gate of Hungary" for four months[18]. The civilians confined in the fortress had similar hopes, because the news of the victory in the Łapanów-Limanowa operation was delivered to the cordon of the troops of the Russian Siege Army by air by the crews of the CC. aviation, even raised this mood. They waited with "longing" for the "people" approaching the fortress, hoping for its immediate release, however, the renewed intention to unblock the Przemyśl Fortress was unsuccessful, because on the entire front, the Russian units went on 20 December 1914 to a strong counter-offensive. The purpose of the Stawka in the so-called The Christmas Battle was the "expulsion" of the field Imperial-Royal the 4th Army in the Kraków direction behind the Dunajec, and the field Imperial-Royal 3rd Army in the direction of Budapest behind the Carpathians[19]. The success of the Russian 3rd Army's offensive actions saved the Russian 8th Field Army, facing the Carpathians, from being flanked and "thrust" into the forks of the Vistula and San rivers. On the other hand, the counteroffensive of the Russian 8th field army against the field Imperial-Royal the 3rd Army and the victory in the last days of 1914 in the Christmas battle near Jasło finally thwarted the rescue of the "Przemyśl lions". Further serious losses suffered in the December clashes caused that the commander of the field Imperial-Royal on December 25, 1914, the 3rd Army decided to move his subordinate forces to defense. In this way, the effects of the transfer of the field Imperial-Royal 2nd Army from the Carpathian line to defend the "Silesian border" and seriously weaken the Beskid section of the front, became apparent. From mid-November 1914 to the second decade of January 1915, the Russian field armies of the South-Western Front lost 348 officers and 28,421 non-commissioned officers and soldiers in the fighting in Galicia, 1,577 officers and 126,677 non-commissioned officers and soldiers, missing and in captivity 310 officers and 133,596 non-commissioned officers and soldiers. Losses in the Austro-Hungarian and German field armies at the same time on the front from Pilica to the Romanian border amounted to 698 officers killed and 25,377 soldiers, 1,769 officers wounded and 96,824 soldiers, 866 officers missing and captured and 136,155 soldiers[20]. The failure of the Stawka near Gorlice left the AOK and OHL under the control of an important starting position, the so-called "Bridgehead" for the next counter-offensive, which was to play such a large role in the planned AOK and OHL undertakings in the spring of 1915. At the end of December 1914, the stabilized Nord-Ostfront line ran along the Dunajec valley to Wojnicz, from Gromnik along the Biała river, through Gorlice and the Dukla Pass along the Carpathian line to the Użocka Pass, passing to the "Hungarian slope" of the Carpathians and running to the area of Legionów and Zielona Pass, and from there towards Radowiec in Bukowina[21]. In the direction of Budapest, the front line retreated "towards the mountains"[22], where in the Bieszczady Mountains, on New Year's Eve, troops of the Russian 8th Army captured the Użocka Pass, opening the "way to Hungary"[23] for themselves. Due to the situation created, both sides almost exhausted their strength when the "Russen ante portas" was placed, considering the end of December operations to be the beginning of a new phase of fighting - the fight for the Carpathians. The Nord-Ostfront line stabilized during the winter of 1914/1915 on the section from the Beskids, through the basins of the upper Dunajec and Biała rivers, the upper Ropa and Wisłoka rivers, and the area of the Carpathian pass in the section from Konieczna (Dujawa-Beskidek Pass) to Użok (Użocka Pass)). For several winter months, from January to May 1915, these areas became a place of fierce fighting, which took the form of a trench warfare with elements of a maneuver war in the Eastern Beskids. So, this way began the Karpathenkrieg (Winterkrieg)[24]. The miracle in the Carpathians, which the AOK hoped for in 1915, didn't take place and the second siege of the Przemyśl Fortress resulted in the failure of two attempts to break into the "Przemyśl lions" by the Imperial-Royal Fields of the 2nd and 3rd Armies - two winter battles in the Bieszczady Mountains - after a few months, on March 23, 1915, its fall[25]. The Fortress of Przemyśl fell due to hunger, just over four months after the four-month "forecast" of the possibilities of defense.

At the end of 1914, the crew of K.u.K. The Luftschifferabteilung constantly made "arctic" combat flights for the command of the Imperial-Royal field army both in the belt north of the Vistula and south of its bed, together with the Carpathian section. In the period of 15-31 December 1914, the staff of the Imperial-Royal Fliegerkompanien worked actively for the benefit of the field Imperial-Royal the 1st, 2nd, 4th and 3rd Army, providing precise information about the enemy despite the increasing number of flight days. The crew of the newly arrived at Nord-Ostfront Flik 13 from 18 December 1914 has operated from the field airport in Szczerców in the Kingdom of Poland ("Kongresówka"), and a day later they arrived at the field airport in Piotrków Trybunalski. Assigned to the field c.k. The Petröczy Group, combat flights for the benefit of the field Imperial-Royal 1st and 2nd Army together with Flik crews. 5 (6–17 December 1914 - Wieluń field airfield, 18–19 December 1914 - Szczerców field airfield; 19–31 December 1914 - Piotrków Trybunalski field airfield), flights were performed with Flik crews. 7 (from November 8, 1914 - Będzin field airfield; December 24, 1914 - Miechów field airfield) and the only German air unit on the Nord-Ostfront, which from autumn 1914 was FFA 36 (from December 16, 1914 Miechów field airfield). The Carpathian section was one of the most difficult to perform aviation tasks due to the winter weather and the typical "mountain danger" in the event of forced landing of the crews, which could lead to death or captivity. The winter days of 1914 brought "variety" to the crews of the Imperial-Royal Fliegerkompanien, whose

October 21 [1914]: the first mortar ["Škoda" 30.5 cm M. 11.] October 19 [1914] 4 shots ... after the first shot the enemy battery [artillery] ceased fire, then a change of position, confirmed by research aviation from October 21 [1914] ... The second mortar ["Škoda" 30.5 cm M. 11.] October 20 [1914] at 2.30 pm ready for fire ... very difficult observation ... fog ... 21 October [1914] rain, fog ... a mortar firing eastward ["Škoda" 30.5 cm M. 11.] not at all can shoot due to lack of observation ... The photo shows the "Škoda" 30.5 cm M 11 mortar in positions near Zemun on the Sava River during the shelling of the Serbian army in the Belgrade region -source: Fotoarchiv SAW-KA Wien

Flik 1's Tent Town- field base in Jasień…. The photo shows Flik 1 tent hangars. January–February 1915 - source: Fotoarchiv SAW-KA.

Above: I land at Błonia. Aerodrome [airport] in Rakowice, threatened by fire. The ground was cut with frost…. Krakow, November 28, 1914 the emergency square of the main reserve of the fortress (now Błonia) was used during the fights for the Fortress as a field landing site. In the photo there is a transport rolling stock Flik 1. source: Fotoarchiv SAW-KA.

members looked like "polar bears" while preparing for the flights, because they put on a few changes of underwear, sweaters and uniforms sealed with newspapers, for example. In the period of 15-31 December 1914, it is in 17 days, 61 reconnaissance flights were made, so- three and a half flight a day! It is worth noting that 2/3 days from this period were characterized by non-volatile weather, which proves the dedication of the K.u.K. Luftschifferabteilung[26]. These squads followed the retreat to the east of the Russian forces after the defeat of the Lapland-Limanowa operation, and the directions of their counterattack during the so-called christmas battle c.k. Of the 3rd and 4th Armies, south of Tarnów (Battle of Łowczówek). On December 17, 1914, the Flik 15 crew after taking off from the Kraków-Rakowice fortress airport, lost orientation in the field in the fog and having problems with the compass and encountering electrical discharges in the clouds, she happily landed in Częstochowa at a German field airport. Oblt. Fp. Wilhelm Schreiber of Flik 15. recalled this adventure flight:

Lot z Oblt. [Belą] *from Losonczy* [Losonczy from Losoncz]. *Immediately after take-off* [from the Kraków-Rakowice fortress airport] *we fell into a dense fog and electrically charged clouds, so the compass got a spin. We flew on and on, hoping to get out of this damn mist at last. We lost our direction completely, of course. At one point* [Zgsf. Joseph] *Siegel stopped the machine, turned around and said, "Listen*

Below: Popular "pop". Aviatik B.I aircraft from Flik 1wrecked on New Year's Eve 1914 at the field airfield in Jasień by Oblt. Josef Tóth during the take off to the besieged Przemyśl Fortress…. By the plane from the left: Oblt. Fp. Jillicher, ex cavalryman Oblt Max Hesse, Oblt. Benno Fiala von Fernbrug, Oblt. Ludwig Dumbacher, Oblt. Stefan Horwath and unit commander of Oblt. Josef Smetana. From the right, next to the fuselage, there are crew members of the damaged plane, including Oblt. Josef Tóth - Source: Fotoarchiv SAW-KA.

Above: On airfields with soft surface, full of uneven terrain, rollover was a frequent occurrence…. Jasień field airport December 31, 1914 - New Year's Eve turnover of the Albatros B.I aircraft from Flik 1 piloted by Fp. Max Libano. Next to the damaged plane: on a horse Oblt Stefan Horwath, standing beside Oblt. Jillicher, second from left Oblt. Josef Smetana, third from left Oblt. Benno Fiala von Fernbrug, fourth from left Oblt. Ludwig Dumbacher, fifth from left Oblt. Otto Jindra based on an unlucky pilot engine - Source: Fotoarchiv SAW-KA.

Galicia seen from the third dimension ... On October 30, 1914, the Flik crew 1 composed of Oblt. Max Hesse and Oblt. Ludwig Dumbacher made a photo diagnosis of Sanok - source: Fotoarchiv SAW-KA.

Left: By the Dunajec River ... In the photo of Zakliczyn in its winter glory. On January 20, 1915, the Flik crew. 1 composed of Oblt. Max Hesse and Oblt. Ludwig Dumbacher made a photo reconnaissance of the area of Lulkowo on the Dunajec River - source: Fotoarchiv SAW-KA.

Schreiber, we only have gas reserves left. We have to get down and think about landing. We are probably far behind the Russian front line, so be prepared to be captured. Soft landing! " - and began descending a wide spiral. I wasn't feeling well at all and wondered what to do when I came face to face with the first Russian. Let them not massacre us, as I have heard about many times; hope we will not fall into the hands of some Mongolian beasts! The fog was thinning, and the earth was slowly becoming visible. Are! - hangars! - enemy [tent] *hangars, light gray, such as I have already seen at Russian airports. Thank God we were able to land at the airport, among the combat forces related to us. After all, Russian pilots will accept us humanely, in the spirit of sportsmanship. But we were lucky!* [Oblt. B. von] *Losonczy turned and grinned at me, pointing to the hangars. I nodded with such enthusiasm that my goggles almost fell off. A few more sharp turns and our plane landed, jumped a few more times and stopped in the middle of the airport. From a distance we saw*

Below: In Festung Krakau... The photo shows the Kraków-Rakowice fortress airport, December 15, 1914 - Albatros B.I plane from the Flik 1 depot No. 13 in a tent hangar. On the right next to the plane are: Oblt Stefan Horwath and Oblt. Benno Fiala von Fernbrug - Source: Fotoarchiv SAW-KA.

Above: Just before the flight ... The photo shows the Kraków-Rakowice fortress airport on December 12, 1914 and the Albatros B.I plane from the Flik 1 depot with a crew composed of Oblt. Stefan Horwath (in the cockpit) and Oblt. Benno Fiala von Fernbrug (in the observer's cabin) - Source: Fotoarchiv SAW-KA.

people in garrison caps running towards us - Russian airmen. They were getting closer. We got out and started to throw a few grenades into the machine to destroy it. I turned to face the Russians again - but what? These are not Russian uniforms! - Losonczy, wait, look, they are - of course - Losonczy are Germans! Germany!!! - Hurray, we are at the German pilots! - but where, in God's name? Meanwhile, a few German officers and privates approached us. "Hello Austrians! - where are you coming from?". "From Krakow. We got lost in the fog. Of course, there was a great uproar. We were immediately surrounded. The commandant (it was the German 31st Fliegerkompanie [FFA 31]) also arrived, to whom we introduced ourselves. He shook our hands firmly, introduced us to the other companions, immediately giving the order to put our lumber back in the hangar. Earlier, however, it was still inspected. "Guys, it's not an airplane, it's just some cigar box!" And with this something you came here from Krakow? My appreciation! ". "And where are we actually, dear friend?" - I asked. "In the Army Group Woyrsch in Częstochowa!" - hats off, indeed. Our Lohner Pfeilflieger [B] *did a good job. However, we were coming back the next morning, ie on December 20. In Krakow* [Rakowice] *we landed smoothly, joyfully greeted by our people and richer for a new, nice experience. We arrived just in time, because the command of the 4th Army ordered the transfer of the 9th, 10th and 15th Fliegerkompanie to Brzesko* [in fact, Flik 1 was transferred to the Jasień airport from Kraków-Rakowice on 17 December 1914; 19 December 1914 Flik 15 and Flik 10; Flik 9 from November 1914 operated on the Serbian Front - in the terminology of AOK Süd-Ostfront][27].

The Austro-Hungarian field armies struggled to hold the Carpathian passes in the fall of 1914, enabling the Russian field armies of the Southwestern Front to pass to Upper Hungary. In this situation, the AOK asked for help from its German ally, however OHL, convinced that they would succeed on the Western Front, strengthened the Eastern Front with just a few divisions that created the German Southern Army[28].

Aerial reconnaissance from the third dimension was consistently the main "job" of the flight crews. It was for the AOK, after the experiences of the summer of 1914, the main source of information about the enemy, hence the crews of the Imperial-Royal During the winter of 1914/1915, Fliegerkompanien operated in Galicia regardless of changing weather conditions.

Where is Wojniczer Wald? Air contribution to the shelling of Tarnów in the winter of 1914/1915 and more

December flights of the K.u.K. The Luftschifferabteilung over Galicia was conducted not only in the fighting area, but also in the back of the front line, in constant contact with the Russian AFC[29], recognizing the situation of Russian troops[30], sometimes landing at allies' airfields[31], in weather conditions that were practically unfit for flight![32]. As part of the ferrying of the c.k. Fliegerkompanien near the front line, the Brzesko (Jasień) field airfield became the main field base of K.u.K. Luftschifferabteilung on the Nord-Ostfront[33], and the coordination of the heaviest artillery fire by radio has become the "front novum"[34]. On December 27, 1914, the commander of the Imperial-Royal aviation Obst. M. E. Uzelac delivered to the field airport in Brzesko (Jasień) together with prof. dr. Leopold Kann, the Albatros B.I no.7 aircraft from Flik 10. (Flik 8.?) Named "*MUZZL*", collected from the Krakow-Rakowice fortress airport. Next to the above-mentioned airfield, a radio station was created at that time to support the cooperation with the crews of two heavy mortars "Škoda" M 11. 30.5 cm and heavy howitzer M 14. 42 cm (Künstenhaubitze M 14 - Skodawerke Actien-Gesellschaft, Pilzno) brought from Festung Pula on the Adriatic Sea. Their hidden positions were located next to the railway line Brzesko-Wojnicz-Tuchów-Dunajec Valley, on a forested hill near Wojnicz (the so-called Wojniczerwald)[35]. Fire in the 42 cm M 14 heavy howitzer and two 30.5 cm "Škoda" M 11 heavy mortars were intended mainly for firing at the railway line running south from Tarnów to the Dunajec valley[36]. In such circumstances, the aircraft was used for the first time to coordinate their fire with a radio installed on board, which took place on the Eastern Front (shelling targets in Tarnów[37]) and near Gorlice[38]. Zgsf. Franz Tobias from Flik 8 recalled that period at the fortress airport of the Kraków-Rakowice Fortress:

One day I received an order to choose the best machine and prepare it for the assembly of the radio. It was supposed to be the first plane equipped with this type of technical innovation. At that time, I was an on-board mechanic and oversaw aircraft machines. Only Albatros [B.I No. 7. "*MUZZL*"] *with an engine of 100 horsepower was taken into consideration, machine Oblt.* [Rudolf] *H*[ole]*ka. The latter was not very happy with the upcoming change, which was, after all, an additional burden on the machine. But something else was about to happen. As soon as the radio equipment was installed, the stub pipes bridged with brass wires, the antenna mounted on the wing and the fuselage, the wheel secured with a counterweight, the dynamo pulley bolted on the propellers, and the machine itself attached to the front of the fuselage, it was finally possible to start testing . Everything was going well, also* [Obst. M. E.] *Uzelac was pleased and was the first to take to the air. The higher load and energy required to drive the dynamo contributed to the longer take-off time, but the take-off was nevertheless relatively good. And here came a surprise!* [Obst. M. E.] *Uzelac ordered that this machine is be assigned to a neighboring company for artillery observation. To this day, I saw a confused face of Oblt in fron of my eyes.* [R.] *H.* [oleki], *who was to lose his beloved machine, the best machine the company* [Flik 8.]. Oblt. R.] H [ole] *ka took me aside and ordered me to replace the propeller with another one from our fleet. As soon as I was ready with this job, the testing of the [airplane] machine began. It turned out that the start of the machine took a long time. Strenuous persuasion to leave the original propeller with the machine, however, did not work. In addition, we received a message that the machine was to be provided with all necessary operational materials, so that it would be ready for take-off in the morning of the next day. I was desperate, I tried everything possible, until I accidentally came up with the idea of reducing the angle of attack a bit and thus giving the machine a bigger "kick" at the start. It helped a little, but not much. I did not sleep all night because of worry. It was already daytime when I finished the work and started the engine. Soon the lieutenant colonel* [Obst. M. E.] *Uzelac. I reported: "Everything is ready!"; however, the feelings that tugged at me then are indescribable.* [Obst. M. E.] *Uzelac was dressed as always, got into the machine, checked the taps, ignition, etc., quickly gave full throttle and a sign to remove the wheel blocking blocks. One more wave of the hand, then a full throttle, the machine lifted, but the take-off seemed endlessly long. Finally, I was able to take a deep breath: everything went well. In the afternoon an order came from the airport in Brzesko* [Jasień] *for me to install the radio station correctly. The local commander* [Flik 1], *Captain S.* [Oblt. J. Smetana] *greeted me joyfully like a hail cloud. Lieutenant Colonel* [Obst. M. E.] *Uzelac was quite calm, he was walking back and forth as usual with his arms folded on his back, not saying a word. Commander S.* [Josef Smetana] *pointed out to me that the positioning of the wings did not match the indications on the size of the angles on the wings and that the external angle had to be corrected by a mechanic; I should take a look at it once. As soon as everything was corrected, it turned out at the start that my idea was good after all: during the take-off there was almost no collision with the factory chimney opposite.* [Obst. M. E.] *Uzelac landed after a flat round in the air and ordered me to restore my previous settings. Just one more test run and I was fired. Satisfied, I returned to Krakow*[39].

Whereas Oblt. Fp. W. Schreiber of Flik The 15th described the flight to coordinate this "little" from the Adriatic Sea from the air of fire, as the heavy M14 42 cm howitzer was called in the front dialect.

December 21 [1914]. Flight from Zgsf. [J.] Siegel on the route: Dąbrowa [Tarnowska] -Radomyśl-Mielec-Kolbuszowa and back. The purpose of the reconnaissance was to check whether the enemy's forces or their rolling stock were not moving ... We started with [J.] *Siegel at* [hours] *9 am and we took a course to Tarnów. According to the arrangements with General Horsetzky, we were to first make two laps over the Tarnów station to signal that we had arrived at our destination. They could easily observe it from the position where the 42 cm gun was located. We reached*

Above: Kraków-Rakowice fortress airport December 15, 1914.... In the photo there is an Albatros B.I plane from the Flik 1 squad with a crew composed of Oblt. Josef Smetana (in the cockpit) and Oblt. Otto Jindra (in the observer's cabin) - Source: Fotoarchiv SAW-KA

a height of about 1,800 meters above the station and began to circle. On my notebook I drew a coordinate system with the station building in its central point. We didn't have to wait long. Suddenly I felt as if it had picked up the machine and at the same moment a huge cloud of smoke spread down on the tracks in front of the station building ... the first hit. A second followed shortly after, then a third, a fourth, a fifth. Then there was a long pause, and then came the sixth and seventh. Eight shots were due, so we were still waiting. I have recorded all seven completed so far, all of them arranged along the track section; I manage to take the photo, which I will attached later. The eighth shot wouldn't come. We were under heavy fire from the bottom right and at one point the engine crashed. [J.] *Siegel immediately took a course west. Thank God we had the wind in our back, luckily, because the engine completely failed. We were gliding down towards Dunajec all the time. I looked down the road from Tarnow to the west and noticed a large car. Through my tester [binoculars] I noticed that there were Russian officers sitting inside, who were also looking at me through binoculars. So, they followed us, if we would have to land before we even reached the river. "So, Siegel, do we get back home?" I called to him. He threw up his arms. The altimeter showed 1,200 meters, the Dunajec River was still about 5.6 kilometers ahead of us. 1100 meters, 1000 meters, 900, 800, ... after all, the river was right there. I turned around, tore the rifle from the kickstand next to me, and fired the entire magazine at the car, which, meanwhile, stopped. Whether I hit it - I don't know, at least the car turned and drove back towards Tarnów. Meanwhile, safe and sound, we were already on the other side of the river and we passed a narrow strip of forest, behind which our machine dropped abruptly and sat quite firmly on the farmland. We rolled twice, then stopped the landing gear up and couldn't get out. I called* [J.] *Siegel and he replied that he was okay. Fortunately, the machine did not catch fire*[40].

In modern times, every civil and military aircraft is equipped with an on-board radio, but since when is this condition considered as standard? When was the first used of the radio on an airplane? During the Great War, it took place on January 12, 1915 at the site where Flik 1 was stationed at the airport in Jasień, but already in December

Above: Radio in the third dimension ... In the "radio" team Flik 1 at the Jasień field airport there were among others Oblt. Benno Fiala von Fernbrug (left) as Technical Officer and Oblt. Ludwig Dumbacher - as the first radio observer. The photo was taken on February 20, 1915 - Source: Fotoarchiv SAW-KA.

1914 in a workshop commanded by Oblt. Josef Smetana (from March 1915 the commander of the unit was Hptm. Otto Jindra) Flik 1, the first works involving the installation of radio equipment on the aircraft were carried out at that airport[41]. In Jasień, a radio station was created to support the cooperation with the crews of the two heavy mortars "Škoda" M. 11 30.5 cm and the heavy howitzer M. 14 42 cm, whose hidden positions were located next to the Brzesko-Wojnicz railway line and the Tuchów-Dunajec valley road[42] , on a forested hill near Wojnicz (Wojniczerwald - there are two railway lines in the Wojnicz region: Kraków-Tarnów-Rzeszów with a railway station in Bogumiłowice - approx. 3 km from Wojnicz - and the side line Tarnów-Nowy Sącz and further to the border of Galicia, distant from Wojnicz by several kilometers - during this period Tuchów was behind the Russian lines - which was within the range of the Imperial artillery from the Wojnicz region). At the above-mentioned site, a M 14 42 cm heavy howitzer and two 30.5 cm M 11 "Škoda" mortars were installed, which were to fire at the railway line running south from Tarnow to the Dunajec valley, but as time has shown not only. As the pilot of this "radio" plane volunteered to be Oblt. Max Hesse of Flik 1 and the technical officer Flik 1. Lt. B. Fiala Ritter von Fernbrug were responsible for the efficiency of the device[43]. Thus, the Albatros B.I No. 7 "*Muzzl*" aircraft, having received radio equipment, was renamed the first radio-broadcasting aircraft of the Great War. On December 30, 1914, a dynamo weighing 18 kg (max. 22 kg) with a power of 220 V was mounted on it, which was placed in the observer's cabin with the use of special rubber in a wooden box made in Brzesko, with a transformer, variometer, and capacitors from the company Siemens u. Halske (Siemiens Tönefunken System). Due to the lack of space in the cabin, outside the structure, next to the plane's radiator in the workshops of Flik 1. the antenna retractor 25–30 m long was installed (on 1 January-12 January 1915, the Mikanit and Fiber-Rohr antennas were installed). Despite the extremely low temperature (the temperature dropped to - 27–30 degrees every day) outside, the entire construction and assembly work was supervised at the airport and at the Flik 1 airport office in Jasień by prof. Dr. L. Kann, but officers and soldiers of Flik 1 also took part in the construction and assembly work.: Oblt. Pista (Stefan) Horwath, Oblt. Ludwig Dumbacher - as the first radio observer and Lt. B. Fiala Ritter von Fernbrug - as a technical officer and non-commissioned technicians Karl Wucherer, Karl Poschwitz, Karl Kositschek, Flzg. Tischler and gefr. Mazurczak. The entire radio equipment installed on the plane weighed approx. 42 kg and had a power of 4.5 Kilowatts (a room electric bell for the apparatus was commandeered from his quarters in the palace of Baron Götz in Okocim by Lt. B. Fiala Ritter von Fernbrug). Simultaneously with these works, an antenna with a length of approx. 60-75 m was installed at the airport. The first technical flights showed good operation of the device and communication skills on the plane-ground line at up to 120 km. Prof. Dr. L. Kann together with Lt. B. Fiala Ritter von Fernbrug decided on January 6, 1915, testing the radio communication on the plane-ground line, to test the equipment for cooperation with the crew of the heavy M 14 42 cm howitzer and two M 11 30.5 cm "Škoda" heavy mortars. On the initiative of Obst. M. E. Uzelac and prof. Dr. L. Kanna was installed in an exceptionally short time, on December 18, 1914 - January 10, 1915, just 16 km from the front line (!), on the Albatros B.I No. 7 "*Muzzl*" from Flik . 1 at the field airport in Jasień, the first radio station to contact the crews cooperating with the personnel of the above-mentioned heavy mortars and a ground transmitter[44].

12 January 1915 Oblt. M. Hesse as a pilot and Oblt. L. Dumbacher, as a radio operator, made the first combat flight with the use of radio-equipment for the cooperation with the service of the M 14 42 cm heavy howitzer, which was to fire at targets detected in Tarnów (railway station) and targets located south of the city. Already the second shot was on target - a junction of the tracks in the railway lines was hit - and the third and fourth shots hit the railway station. Another missile hit the bank, where the headquarters of the Russian 3rd Army of General R. Dimitriev was located, destroying it (about 200 bodies of soldiers were transported from there - including the body of a Russian officer with the rank of colonel). On January 14, 1915, the M 14.42 cm heavy howitzer was disguised in the forest, next to the Biadoliny Radłowskie railway station, about 1 km west of the railway bridge on the Dunajec River, and FML Horsetzky, the local artillery commander, was ordered to shoot at the railway station in Tarnów. At. 09.00 Oblt. Fp. W. Schreiber together with Zgsf. J. Sieglem of Flik 15. took off from the airport in Jasień to fly over Tarnów to observe the fire[45]. At 9 am, two heavy missiles fell on Tarnów, precisely hitting the building of the railway station[46]. Oblt. Fp. W. Schreiber from Flik 15.This is how he recalled the flight from January 14, 1915:

We started with [Zgsf. J.] *Sieglem at* [hours] *9 am and we took a course to Tarnów. According to the arrangements with general* [FML] *Horsetzky, we were to first make two laps over the Tarnów station to signal that we had arrived at the place. They could observe it without any problems from the position where the gun was located. We reached a height of about 1,800 meters above the station and began to circle. On my notebook I drew a coordinate system with the station building in its central point. We didn't have to wait long. Suddenly I felt as if it had picked up the machine and at the same moment a huge cloud of smoke spread down on the tracks in front of the station building ... the first hit. A second followed shortly after, followed by a third, fourth, and fifth. Then there was a long pause, and then came the sixth and seventh. Eight shots were due, so we were still waiting. I have recorded all seven completed so far, all arranged along the track section; I even managed to take a photo, which I will attach later. The eighth shot wouldn't come. We were under heavy fire from the bottom right and at one point the engine crashed.* [Zgsf. J.] *Siegel immediately headed west. Thank God we had the wind in our back, luckily, because the engine completely failed. We were gliding down towards Dunajec all the time. I looked down the road from Tarnow to the west and noticed a large car. Through my tester* [binoculars] *I noticed that there were Russian officers sitting inside, who were also looking at me through binoculars. So, they followed us, assuming that we would have to land before we even reached the river. "So, Siegel, do we get back home?" I called to him. He threw up his arms. The altimeter showed 1,200 meters, the Dunajec River was still about 5.6 kilometers ahead of us. 1100 meters, 1000 meters, 900, 800, ... after all, the river was right there. I turned around, tore the rifle from the kickstand next to me, and fired the entire magazine at the car, which, meanwhile, stopped. Whether I hit it - I don't know, at least the car turned and drove back towards Tarnów. Meanwhile, safe and sound, we were already on the other side of the river and we passed a narrow strip of forest, behind which our machine dropped abruptly and sat quite firmly on the farmland. We rolled twice, then stopped the landing gear up and couldn't get out. I called* [J.] *Siegel and he replied that he was okay. Fortunately, the machine did not catch fire*[47].

In Zbylitowska Mountain a Russian ship battery of 210 mm caliber guns, the so-called "Japanese" guns, imported from the coast of the Sea of Japan, which successfully shelled the artillery post in Biadoliny Radłowskie, damaging the loading crane of the M 14 42 cm heavy howitzer was stasioned. On January 16, 1915 Oblt. M. Hesse and Oblt. L. Dumbacher of Flik 1 prepared the next flight as part of the cooperation with the above-mentioned heavy mortars and directed the fire from the c.k artillery air very effectively. The third missile hit the position of the Russian battery, and after the fourth shot, all four 210mm guns were destroyed. Further flights with good results were made by: Oblt. L. Dumbacher, Hptm. Victor Schüntzel and Oblt. S. Horwath along with Zgsf. Max Libano and Lt. B. Fiala Ritter von Fernbrug around Zakliczyn, Gorlice (hill 410) and Bobowa for the Csurilo Artillery Group[48]. Command K.u.K. The Luftschifferabteilung quickly appreciated the results, and a decision was made to equip the Flik 10 and Flik 15 aircrafts with on-board radios (in Flik 1, in the period until February 1915, three Albatros B.I plane with 100 hp engines were to receive on-board radio stations, the fourth radio station was to be installed in the Flik 1 headquarters and in the

Above: The aviation style of the Great War ... In the photo, the Flik 1 airfield in Jasień, February 20, 1915. In the photo from the left, are standing Oblt Ludwig Dumbacher and Zgsf. Mattl from Flik 1 - source: Fotoarchiv SAW-KA.

field command of the 4th Army in Brzesko)[49]. From Flik's 1 headquarters to the field staff of the Imperial-Royal Of the 4th Army until February 19, 1915, a number of reports were received informing that: Hptm. prof. Dr. L. Kann, Lt. B. Fiala Ritter von Fernbrug and Oblt. L. Dumbacher were cooperating with heavy howitzer service M 14. 42 cm; for the Csurilo Artillery Group - Lt. B. Fiala Ritter von Fernbrug; near Zakliczyn - Lt. B. Fiala Ritter von Fernbrug and Oblt. L. Dumbacher; near Gorlice - on the section Bobowa-Nowy Sącz - Lt. B. Ritter Fiala von Fernbrug. Work for the field units of the Imperial-Royal The 4th Army of the described invention was assessed by the air force in staff talks with AOK Hptm. Father Jindra, Oblt. L. Dumbacher, Oblt. M. Hesse, Oblt. S. Horwath and Lt. B. Fiala Ritter von Fernbrug and on the artillery side of Obst. J. Janitschka (Obst. J. Janečka as a general during the Gorlice Breakthrough in 1915 was the commander of the artillery group in the 14th Corps, where there were 12 "Škoda" M 11 30.5 cm mortar battery and M 14 42 cm heavy howitzer). Flik 1 in the period until February 21, 1915, performed the described specialized flights with the above-mentioned equipment for the field command of the Imperial-Royal Of the 4th Army, and at the end of February 1915 the unit was transferred to the disposal of the field commander of the Imperial-Royal Of the 7th Army in the area of Marmaros-Sziget (Syhot Marmaroski), Nadwórna (today Надвірна), Delatyn (today Делятин) in the Eastern Carpathians, where from March 1915 tests continued with the use of on-board radio stations for artillery[50]. At the end of February 1915, heavy missiles fell on Tarnów again (mainly at the railway station, and on February 26, 1915, the targets were, among others, a Police building and a real school, turned into a grain and food store) - these were the last actions with the use of airplanes Flik 1[51].

In the period January-March 1915, the crew of Flik 8, Flik 10 and Flik 15 from the field c.k. The 4th Army specialized in radio reconnaissance, and in cooperation with artillery by radio, especially for the heavy artillery battery. Firing was directed against positions, fortifications, river crossings and important objects marked on staff maps[52]. Oblt. Fp. W. Schreiber of Flik 15. this is how he described the cooperation with the "Škoda" M 11 mortar battery, 30.5 cm:

When a distant reconnaissance finds a grateful target on the enemy and gives its approximate location, two 30.5-centimeter mortars ["Škoda" M 11] *are placed in their positions. (For a smaller caliber, planes are engaged less often). The position must be at the limit of the profitability of the shot, because at short distances mortars have a lower impact force, because then the height achieved by the projectile is not exploited, and therefore the* [kinetic] *energy accompanying its falling is not exploited. In the meantime, the pilot determines the location more precisely and reports it to the battery commander. The moment an aviator finds it easiest to observe, he lets you know: "fire!" A shot is fired. The aviator approaches the target and gives the error: "200 meters too short, 30 to the left." The second mortar* ["Škoda" M 11. 30.5 cm] *is corrected and if the airman's report is correct, the target is hit in the second shot. hus, it can be assessed to what extent aviation technology supports heavy shock artillery*[53].

Oblt. L. Dumbacher and Oblt. On February 21, 1915, B. Fiala Ritter von Fernbrug was delegated to the Luftfahrarsenal to Vienna, to transfer the experience gained during the above-mentioned flights and start the production of radio apparatus for aviation, and after an eight-day stay in Vienna, four new radio stations weighing 30 kg each were sent to Flik 1. to the Delatyn-Nadwórna airport. In December 1916, Oblt. B. Fiala Ritter von Fernbrug was sent back to Vienna to prepare two passenger cars with radio equipment, which were built at the Siemens u. Halske factory in Vienna under the direction of Eng. Schieferstein[54].

Long reconnaissance, close reconnaissance, aerial photography, radio cooperation with the ground, bombing raids, transport flights are achievements in the development of the Imperial-Royal aviation at the turn of 1914/1915. In 1915, in the rapid development of aviation, it became an exceptionally important type of c.k. armed forces in terms of support for the battlefield, propaganda, op. What was created at the turn of 1914/1915 could only be improved later[55]. Application during positional fights on the Dunajec line and in the prepared breakthrough operation near Tarnów and Gorlice - Gorlice Breakthrough as part of the Gorlice-Tarnów operation and the fight in the Bieszczady Mountains in the area of Kozialata peak - in the spring of 1915, ultra-modern as for that period of weapons and technology, i.e., long-range, heavy artillery, aviation and radio communications for fire control heralded a completely new way of conducting war. The firing of Tarnow from the heaviest "Škoda" M 11 30.5 cm mortars and the M 14 42 cm heavy howitzer was carried out with unprecedented, almost surgical precision and only destroyed carefully selected targets. It was the first in the history of wars to combat artillery fire using a radio installed on board an aircraft! The fact that no major damage was done to the urban development was due to the accuracy of the shots, mainly of the Imperial-Royal artillery, although mainly the authors of it were the airmen directing the artillery fire and spies with a special telephone communication linking Tarnów with the troops on the opposite side of the front[56]. Russian airmen urgently searched for mortars in the Tarnow area, but with no luck. For this reason, at the field command of the Imperial-Royal 4th Army, located in the palace headquarters of the Göetz-Okocimski, owners and creators of the Okocim Brewery, improvised opl[57].

The youngest, "fifth weapon" or aviation, throughout the winter of 1914/1915, was used by both sides primarily for aerial reconnaissance. The decisive winter conditions in its work, however, did not allow for the cyclical observation of the opponent's movements from the air, hence the inflow of information for AOK and Stawka was interrupted by them. Despite the wave of frost, the intensity of combat flights in the Tarnów area of Flik 15th slightly decreased, but on February 10, 1915, the crew of the Oblt. Fp. W. Schreiber and Oblt. B. von Losonczy made a long reconnaissance flight to the region of Jasło, reporting a large group of wagons at the local railway station, reconnaissance of the railway line east of Tarnów, Surochów region near Jarosław[58]. For the inhabitants of Galicia, the plane flights constantly caused a sensation - M. Kromkayowa wrote in her diary kept in Biecz:

17 II [1915 r.] ... *During last two days we were observing aeroplanes [*airplains*]. On the side and back wings there were red, quite large stripes. One flew very high, the other flew lower today. Whether they are Russian or Austrian airplanes - opinions were divided. Russian soldiers fired at what was flying low. If these are our [planes K.u.K. The Luftschifferabteilung had red-white-red stripes on the wings and on the hulls as nationality markings], we think that maybe some good news in the form of advances of our troops will follow them. And if we don't care about Russian, we don't have to draw any conclusions*[59].

Father Kazimierz Wais wrote in his chronicle in Klimkówka:

Because at the end of January and the beginning of February [1915] *enormous snow fell, therefore, on the order of the Russian authorities, people came out to dig the snowdrifts, and when the thaw came later, they were ordered to* [spelling in accordance with the original in the entire quoted text] *for weeks for free or at no cost to repair over supervision of sołdatów, highways. Jews in Rymanów were harnessed to do a similar job, and they were spared no time.... sometimes the quartermasters announced the arrival of troops, but they did not show up at all; sometimes the airplane* [the plane] *made a noise high above our heads*[60].

With the advent of early spring, the activities of the reconnaissance aviation over the mountain front, mainly from the crews of the KuK Luftschifferabteilung started - M. Kromkayowa wrote in her diary:

8 III [1915]. *Yesterday and today, we see airplanes* [planes] *above us. Ours or Russian? Harbingers of bad or good news for us? ... March 21* [1915]. *We saw and we watched the airplane* [plane] ... *March 23* [1915]. *The airplane* [the plane] *was flying, but very high ... March 24* [1915] *Muscovites were shooting at the airplane* [plane] *today*[61].

In the diary of H. Schramówna, written in Olchowa on

March 28, 1915, it was written:

Today 28 III [1915], *very close to the cannons you can hear rifles, and the forszpans* [caravans] *from Czaszyn through Tarnava Górna go to Łukowy and they said that they had retreated. The wounded from the mountains are still walking through Łukowe. In Lisk, they put up barracks on the market square for the wounded, because there is no more room. In the mountains, I do not know in which village, the mayor always let the Austrians know where the Moscow positions were. The Muscovites discovered it, hanged him, and burned his cottage. Airplanes [*planes*] are still circling. And Austrian (or maybe Prussian) with red and black stripes* [KuK Luftschifferabteilung marking was red-white-red] *on the wingtips and Russian with a red-blue circle on the white wing*[62].

The AOK needed precise information about the enemy from the back of the Eastern Front, hence the crews of the KuK Luftschifferabteilung were going deeper and deeper into the area of the Russian positions. February 17, 1915 Oblt. Fp. W. Schreiber from Flik 15. made an independent flight from the Jasień field airfield to the area of the Dunajec valley with a Lohner B "Pfeilflieger" C-47 plane, and on February 18, 1915 in the Rzepiennik-Pilzno area, reaching Jasło. Near Rzepiennik, the plane was shelled by a Russian Opel (even though it was at an altitude of 2000 m, shrapnel exploded about 50 m from the plane). Along with the inflow of information from the Russian hinterland, AOK decided to attack the concentration sites of Russian troops - 13 aircraft from Flik, 1., Flik 8. and Flik 15 bombed the Szczucin area throwing bombs at Russian units and the railway line[63]. However, not all flights ended successfully[64]. The deterioration of weather conditions resulted in the reduction of combat flights and on February 25–27, 1915, due to snowfall the crew of Flik 15 did not make any sort of combat flights, and on February 27, 1915, Oblt. Oskar Schmoczer von Kluczenzeff and Oblt. ing. Rudolf Stanger made a combat flight, but after 45 minutes they returned to the home airport because the clouds reached a height of 1500 m. Obst. ME Uzelac sent an Oblt crew on the next flight. B. von Losonczy and Oblt von Schön, whose flight lasted three and a half hours, but the reconnaissance results were poor. Gradually, newer, and newer planes were delivered to the front Fliegerkompanien, and thus on February 28, 1915, Oblt. Fp. W. Schreiber from Flik 15 received a new Albatros B.I aircraft No. 23-07, faster (110 km/h cruising speed) than the obsolete Lohner B "Pfeilflieger" airplanes (90 km/h cruising speed) and during the flight in Szczepanowice near Wojnicz, a Russian observation balloon shielded by a strong anti-aircraft gun was fired upon. (23 shrapnel exploded near the plane) and over Wróblowice near Zakliczyn, due to engine failure, the flight had to be interrupted and returned to the home airport[65]. The beginning of March 1915 in the frontline aviation activities brought not only a revival on the postal line to Przemyśl (as the Hungarian war correspondent Ferenc Molnár recalled, *During the siege of* [Przemyśl Fortress] *pilots flying from and to the fortress were distinguished*[66]) but also a temporary improvement in the meteorological conditions in the Tarnów region, where the Flik 15, crews were particularly active ,which from 20 December 1914 to 15 January 1915 was stationed at the Jadowniki field airfield; in the period from January 15 to February 12, 1915, it was stationed at the Kraków-Rakowice fortress airport, and in the period from February 12 to May 10, 1915 at the Jasień airfield[67]. On March 2, 1915, the crew of Flik 15. composed of Oblt. Johan (Jan) Wierzejski and Fp. Corp. Ferdinand Junker, despite the bad weather, made a two-hour flight, but according to the proverb "in March like in a pot", the bad weather quickly prevailed and on 3, 4 and 5 March 1915 the crew of Flik 15. have not made sorties. In addition, the poor condition of Flik 15., flying equipment caused that on 6, 7 and 8 March 1915, Oblt. Fp. W. Schreiber was sent to Krakow and Moravska Ostrava to collect flying equipment from the local Flieger Etappenpark. As the days passed, the weather conditions became nightmarish, and fogs and snowstorms prevented combat flights on 10 and 11 March 1915[68]. Due to weather conditions (low cloud position) on March 27–30, 1915 Flik 15 did not make any sort of sorties, and on April 1, 1915, Oblt. Fp. W. Schreiber made a flight to the area of Tarnów-Gorlice-Jasło-Pilzno and back, detecting a railway station in Jasło overcrowded with railway transports[69]. In the light of the Russian archives, on March 22, 1915, a high activity of the Imperial-Royal Air Force on the Dunajec and Biała sections was observed. Russian improvised anti-aircraft defense shot down one plane that landed on the enemy's side[70].

In the spring of 1915, between the airmen of the KuK Luftschifferabteilung from the airfield in Jasień and the airmen of the Russian AO (command of the Russian 3rd Army, they were subordinate to the 9th, 11th, 12th and 24th AO) from the Tarnów-Lipie field airfield an unprecedented in the realities of the Great War took place ... a chivalrous and social exchange of correspondence! 19 March 1915 due to bad weather conditions the crew of Flik 15 havn't made any sorties, not performed but the crew of the Oblt. Bratmann and Lt. Lord dropped a letter over the above-mentioned Russian airport with information about the imprisoned Imperial-Royal troops of Lieutenant Pieredków, and two days later, on March 21, 1915. Lt. Bratmann and Lt. Lord of Flik 15. they did not return from a combat flight, forcibly landing near Dębica and being captured by the Russians. 23 March 1915 Oblt. B. von Losonczy from Flik 15. dropped a report from commander Flik 15. over the Russian airport for the enemy unit commander. Here is how Oblt described the correspondence with the opposing airmen and the air battles with them. Fp. W. Schreiber:

Let me now present a very interesting correspondence between our commander, Oblt. [baron-Freiherr] *A. von Mandelsloh, and the commander of the Russian airmen in Tarnów. Today, on March 22* [April 4 - 1915], *by way of an air*

Above: Flik 1 field airfield in Jasień on January 4, 1915.... In the photo in the cabins of the Aviatik B.I "Durch", a veteran of postal flights to the besieged Przemyśl Fortress, they are sitting from left to right: Oblt. Benno Fiala von Fernbrug from Flik 1 and Oblt. Ing. Rudolf Stanger from Flik 8 and an ex cavalryman Oblt Max Hesse is standing next to the plane with his pet - a white Spitz named Muzzl - Source: Fotoarchiv SAW-KA.

drop to our airport [in Jasień], *we received the following letter: "To 15 Fliegerkompanie, to Mr. Oblt.* [A.] *von Mandelsloh: Our first report on the captured* [shot down and intercepted] *Albatros* [BI] *23-03* [March 1915] *was dropped in Brzesko on February 15/28* [1915]. *After three days, both pilots were unharmed, dressed in civilian clothes and transported to Russia as prisoners of war. Your report, received on March 8–21* [1915] *in Lisia Góra, makes us believe that the first parachute with the report did not reach its destination. Therefore, we kindly ask you to send a reply to each report. We also have a friendly and hostile request to deliver the news about our airman, Lieutenant Pieredków: Lieutenant Pieredkov was taken prisoner on December 20, 1914/2 January 1915, and he was wounded. We would like to know the details of this incident and whether he is slightly or seriously injured. March 9/21* [according to the rules of the Julian calendar it should be March 9/22] *1915 Russian Airmen". Today again the weather is perfect for flying, but not for me. Zgsf.* [J.] *Siegel flew out with Oblt. Students for artillery reconnaissance with the use of radio. However, they returned quickly, because the dynamo was not working. At 10.00 am the Oblt took off.* [B.] *von Losonczy from Zgsf.* [F.] *Benda. Two hours later they returned. In the evening, we made a reply to the Russian airmen, which was as follows: "To the Air Department in Tarnów: March 19/23* [the date should be different - March 23 / April 5] *1915. Thank you very much for your letter, which I received today. Unfortunately, I forgot to include you in the photo of the monoplane* [type Nieuport IV or Morane Parasol] *to Lieutenant Pieredkov. On March 2 and 17* [1915 in the Gregorian calendar] *we received notifications about captured airmen. So, I repeat lieutenant Jakowianko; Lieutenant Kobanko, Lieutenant Pieredkov; Lieutenant Dziedzino was not injured and, according to our information, they were transported to our beautiful crown country, to Salzburg. Both lieutenants Dziedzino and Pieredkov were shot in the ignition magneto. I spoke to Lieutenant Pieredkov in person. I haven't seen him get any wound. We will respond to each of your letters in the future. Yesterday I was very busy during the flight and I was not able to throw it off precisely. Best regards, Your forever loyal enemy, Freiher August von Mandelsloh, Oberleutnant. Commander of the Imperial-Royal 15 Fliegerkompagnie. March 23* [April 5, 1915] (Tuesday*). Oblt.* [B.] *von Losonczy left today with the above-mentioned letter and dropped it over the Russian airport near Tarnów. He watched if the Russians had run up and picked him up. March 24* [6 April 1915] *a Russian plane reappeared over our airport and dropped the following letter: "To 15 Fliegerkompagnie in Brzesko: to Mr. Oblt. A. v.* [On]

Above: Despite the bitter frost ... Oblt. Jozef Smetana On January 22, 1915, taking medicaments and other things on board his plane flew alone to the besieged Przemyśl without reaching his destination or returning to the base ... On the photo the preparation on January 21, 1915 of the Albatros B.I aircraft for the flight to the besieged Przemyśl fortress Oblt Josef Smetana at the field airport in Jasień. The pilot's winter outfit is noteworthy - Source: Fotoarchiv SAW-KA.

Mandelsloh: Thank you very much for yesterday's message, which has just dropped at our airport. Unfortunately, we are not able to state which of our countries your airmen were transported to, nevertheless, we believe that the address will be provided to you by the prisoners themselves by ground mail. Albatros [B.I] *VI heavily perforated: about 30 balls in the wings and body and one ball in the propeller, with only one hole smooth, without splitting. Both pilots, Lt. Schmatzer and* [Zgsf.] *Hermann Hold without injuries* [On March 19, 1915, the Flik crew. 11. Fp. Zgsf. tit. Feldvebel, H. Hold and Lt. Schmatzer, having lost orientation after taking off from the fortress airfield in Żurawica near Przemyśl, and having lost her way, headed east and landed near Stanisławów, being captured by the Russians [71]]. *Along with this news, we give you two letters from them. Please drop the next parcel to the following address: God bless you! Russian airmen. March 24* [April 6] *1915 (...) today* [Oblt. B.] *von Losonczy flew out and dropped the package over Tarnów. We didn't have time to make a parachute. Therefore* [Oblt. B.] *von Losonczy, as he said himself, almost went down to the ground and simply threw the package. He saw how the excited and lively gesticulating Russians took her away undamaged. An hour later a Russian reappeared* [at Jasień] *and threw a letter in which* [the Russian airmen] *thanked us and wrote that they were already waiting impatiently for our parcel.*

April 7 (Wednesday) [20 April 1915]. *The sky was overcast, a strong wind from the south-west, then from the south. I took off at 1.55pm in the afternoon with the task of dropping bombs on a Russian food warehouse near Wola Rzędzińska ... right after landing I found a great commotion and nervous atmosphere at the airport. There was an enemy airman here earlier (certainly none of those with whom we corresponded) and he dropped a bomb that hit very close to hangar II and killed our brave platoon commander (Zufsführer) Hruschka. There was no material loss.* [Zgsf.] *Hruschka will be buried tomorrow at the cemetery in Brzesko.*

Correspondence with Russian pilots from Tarnów lasted

Above: The last moments before take-off ... In the photo there is an Albatros B.I plane from Flik 1 during the preparation for the take-off for the flight with post from Kraków to Przemyśl Fortes from the airfield in Kraków-Rakowice 15 XII 1914 r. source: Fotoarchiv SAW-KA.

- on April 11, 1915, a Russian plane dropped another letter over the field airport in Jasień:
April 11 (Sunday) [24 April 1915]. *Today (maybe because it's Sunday?) A Russian aviator flew in again and threw us the following letter: "Tarnów, March 28 (April 11) 1915 Austrian Aviators! "Grüs Gott* [God bless] *Mr. Lieutenant! Thank you very much for your letter and delicious cigars. We were not at the airport when you dropped the letter. He fell next to a building where some boys took him. These scoundrels returned the letter, but the cigars hid it. We recovered the box with great difficulty. And now a request be so kind and please give two letters and a key to the German 54th Aviation Division (Fliegerabteilung)* [in the light of German sources, FFA 57 was stationed there] *at Biadolina Noble, because I do not know any airmen from this division. Unfortunately, I did not see Messrs. Gerold and* [Hptm.] *Neumann, but I was told that*

In another letter, the Russian airmen from the Tarnów airport informed their adversaries:

Lord, Mr. Oblt.[Hptm. A. von Mandelsloh] *and your airmen* [from Flik 15.] *you are extremely polite. Thank you very much for inviting us to visit you at the airport, but unfortunately, we are busy all day and we don't have time to come even for a few hours. I hope you gentlemen are not angry about this. For news gentlemen, let me inform you that we, the Russians, are very hospitable - for example, our airmen received very well the quite strangers, Mr. Gerold and* [Hptm.] *Neumann. In the merry company, everyone ate dinner, drank, smoked and chatted in a nice way. We will accept good friends even better. Answer Gentlemen, please, whether both letters were skillfully dropped and whether they fell far away from you. Cordial greetings! Russian Airmen, Farmanflugzeugkompagnie* [AO stationed near Tarnów used French Farman type planes, but no data are available whether version XVI, XX, or XXII], *Tarnów. PS The address as before - it was very well spelled; in Latin script it would be: "Letschikam Farmanistam".*

The Easter event of 1915 described below, in its pronunciation, was unheard of in the scale of the entire global conflict - from the field airport in Jasień, they sent

Above: More and more often, the crews of Fliegerkompanie 1 took bombs on combat flights ... In the photo, the start of the combat flight of the Albatros B aircraft with the crew of the Oblt. Otto Jindra and Kpl Max Libano from the snow-covered field airport in Jasień. source: Fotoarchiv SAW-KA.

their enemies - Russian airmen at the field airport near Tarnów, an Easter gift in the form of ... a piglet dropped from the plane:

The piglet, which, by the way, squealed frighteningly and had cut the whole plane before, fell slowly, then landed in the middle of the Russian airport. We were all still standing together [at the airport in Jasień] *laughing and commenting on the surprise we made to the Russians when a Russian aviator appeared on the horizon, as it soon turned out. We expected, of course, with great tension what would happen. The plane, having successfully approached, descended on a dive so that we already thought it would land. When he was about 100 meters above the center of the airport, he picked up the plane and at that moment something flew out of it - it was a letter. We all rushed in that direction.* [Hptm. A. von] *Mandelsloh picked up the letter and read it aloud: To the Austrian Friends of Aviators. Hooray, hooray, hooray! We have a pig! Many, many thanks - also for Easter wishes. Christ is risen, is truly risen! Let the gentlemen feel embraced and kissed by us as if we did it for real. And, gentlemen, live if we do. When the package was flying to the ground, we didn't know where this terrible squeal was coming from. And when we picked it up, we laughed terribly at such a friendly, "pig" gift. We have already given the pig to the chef. Tonight, when we eat it, we will remember the health and prosperity of Austrian* [Austro-Hungarian] *airmen many times. Thank you once again, gentlemen! With friendly greetings, Your always faithful enemies, Russian Airmen from Tarnów. 5/17* [should be 18] *April 1915. we will remember many times the health and prosperity of Austrian* [Austro-Hungarian] *airmen. Thank you once again, gentlemen! With friendly greetings, your always faithful enemies, Russian pilots from Tarnow. 5/17* [should be 18] *April 1915. we will remember many times the health and prosperity of Austrian* [Austro-Hungarian] *airmen. Thank you once again, gentlemen! With friendly greetings, Your always faithful enemies, Russian Airmen from Tarnów. 5/17* [should be 18] April 1915.

April 19 (Monday) [3 May 1915]. *The weather is still out of sight.* [Hptm. A. von] *Mandelsloh dropped a letter with a group photo of 15 Fliegerkompanie over the Russian airport in*

Above: Jasień field airfield ... Start for the winter combat flight from 15 XII 1914 r. of the crew from Flik 15 composed of Oblt. Ludwig Dumbacher and Zgsf. Mattl. Fotoarchiv SAW / KA Wien collection

Tarnów yesterday. Today a Russian plane flew in and dropped us a letter with the following content: April 5/18, Airport near Brzesko. "Until 15 Fliegerkomapanie: Thank you very much for the last letter (29/11) and photos. We'd love to know which Lord is Oblt. baron [A. von] *Mandelsloh. After this letter, gentlemen, have you dropped a few other flags? Nothing was handed over to us, although the peasants said that something flew out of the plane over Dąbrowa* [Tarnowska]. *We allow ourselves to send you our address, printed in Russian, to make it easier for the letters to reach their destination. Please send a message to which aviation department we should address the letters -* [Fliegerkompanie] *15 or 11, because one of the companies from Brzesko left about a week ago. Best regards, Russian Aviators. "*

April 24 (Saturday) [May 7, 1915]. [Hptm. A. von] *Mandelsloh recently dropped a letter at the airport in Jasło, which reads as follows: "Dear Russian Airmen: Where are you? We haven't seen you for over a week! What the hell??? There is war, dear gentlemen !!!". But if there were no Russians, they are not there. They know more than we do !!!*[72].

Aviation Events Over the Przemyśl Fortress in 1915.

He had to land in Bartoszów near Rymanów on March 24 [1915] *due to gasoline ignition, Austrian airplane. Well,*

Above: Jasień winter 1915 r. ... In the photo, Albatros B.I with crew Oblt. Otto Jindra and pilot Zgsf. Max Libano aircraft during the landing approach. source: Fotoarchiv SAW-KA.

Above: This was in fact the ferrying of the field Fliegerkompanien to the next airfields of Nord-Ostfront ... Along with the threat of the Krakow Fortress - the second battle for Krakow - AOK decided to evacuate the Krakow-Rakowice fortress airport via the emergency square of the main reserve of the fortress (now Błonia) also used as a field landing field, so far operating from the Krakow-Rakowice Flik 1, 8, 10, 15 fortress airport to the field airport in Andrychów, which until 9 December 1914 became the main air base on the Nord-Ostfront. Flik camp in the photo. 1 on the route Kraków-Andrychów on December 1, 1914 - source: Fotoarchiv SAW-KA.

the officers driving there and reported to Wojnicz, who, as a lieutenant colonel, was the oldest among Russian officers there, said that Przemyśl surrendered at 7 a.m. that day (there were two officers, one Czech and the other German). Austrian officers [spelling consistent with the original throughout the quoted text] there were two: one German, the other Czech. The Czech apparently enjoyed the fall of the Przemyśl fortress; while the German was talking about the fact, he was crying with regret and was so desperate that he wanted to cut his throat with the razor he took from the suitcase. Dr. Wojnicz prevented his intention, but the lieutenant presented him with a ring made of Russian missiles as a memento. Soon, after the plane landed, this characteristic detail is worth noting here, several Ruthenians from Tarnawka reported to Dr. Wojnicz that Austrian spies had fallen in the plane and, asking for their capture, clearly indicated that they only wanted Moscow's rule, because under the Austrian [spelling consistent with the original throughout the quoted text] have suffered enough[73]

The Fr. K. Wais in his memoirs described the event which, however, took place on March 23, 1915 and its heroes were the CzechZgsf. Robert Meltsch and Austrian Lt. F. Junker. Their Albatros B.I (Aviatik B.I?) Aircraft according to earlier research, after taking off from the fortress airport in Żurawica, fell under the fire of the Russian troops and was to be shot down, landing just behind the Russian lines! Until now, the place of this landing was unknown in the light of the cited source, the damaged plane flew from the Przemyśl fortress to the Beskid Niski region, beyond Rymanów - over 110 km in a straight line - where it was forced to land. From the area of Rymanow-Bartoszów, the "POW adventure" of both Imperial-Royal pilots began, from which the Czech pilot, known for his pro-Russian attitude after 1918, joined the emerging Czechoslovak aviation[74]. In the evening of March 22, 1915, sections of the defense of the Przemyśl Fortress in the northern and southern defense zones were seized by the troops of the Russian 11th Army besieging the fortress, while its command was waiting for the return of Col. M. Pniewski's general staff[75]. A day later, information about the surrender of "Przemyśl lions" flashed into the world. The besieged Przemyśl Fortress was systematically attacked from the air by the Russian air force in February (during the week between February 21 and 28, 1915, the Russian air force dropped 27 bombs on it)[76] but these were typical harassment actions. During the second siege of the Przemyśl Fortress, 1,452 people died there, and 3,538 soldiers were lost, and the total losses of its garrison in the period January-March 1915 exceeded 11,000 soldiers (3,440 killed and dead, 7,618 missing and captured). At the same time, the Russian 11th Army besieging the fortress lost (except for the liquidation of the last foray of the garrison of the fortress towards Medyka) 20 officers and 398 soldiers were killed, 119 officers and 3,455 soldiers were wounded, and three officers and 301 soldiers were missing[77]. Just before the capitulation, over 10% of the garrison (the fortress housed, in addition to over a thousand Russian prisoners of war, 30,000 civilians) were sick or weakened to such an extent that they could not fight (as of March 13, 1915, it was 12,140 wounded and sick, 7,873 exhausted and 17,000 unfit for combat), problems with

Above: Flik 1 airfield in Jasień on January 4, 1915.... In the photo in the cabins of the Aviatik B.I *"Durch"*, a veteran of postal flights to the besieged Przemyśl Fortress, they are sitting from left to right: Oblt. Benno Fiala von Fernbrug from Flik 1 and Oblt. Ing. Rudolf Stanger from Flik 8 and an ex cavalryman Oblt Max Hesse is standing next to the plane with his pet - a white Spitz named Muzzl - Source: Fotoarchiv SAW-KA.

the ammunition appeared (250 rounds for the infantry rifle, 30,000 rounds for a machine gun, 10,000 rounds for light artillery and 600 rounds for heavy mortars) and drastically reduced provisions (daily rations were reduced to 60 g of vegetables, 300 g of bread and 50 g of breadcrumbs per person, despite the mass slaughter of horses, horse meat was not available)[78], however, before the capitulation took place ...

During the second siege of the Przemyśl Fortress, an additional landing area was created, already considered on 30 December 1914, located between Potoczek, Buszkowice and Buszkowiczki, located in the arc of the San (on the so-called Shepherdess), where Flik 11 was temporarily stationed together with the balloon Festungballonabteilung 1 / 3. R[79]. On March 18, 1915, Russian artillery shells fell on the airfield south of the San: [The airmen] Flik 11 ... they had no more than an hour of peace, because the enemy kept under constant artillery fire the air devices of one airport south of the San [most likely it was an airfield in the Lviv district, but there is also an alternative airport located between the Wiar riverbed and the railway line] and the second one near Żurawica[80]. Even though the fortress airport in Żurawica, was destroyed by the fire of Russian artillery, was constantly in use. The besieged Przemyśl Fortress was attacked from the air by the Russian airport during its second siege. These raids were not large formations, but rather harassing in nature. For example, under the date of December 5, 1914, the following was written:

Bomb [aviation] *went to the officers' casino on* [ul.] *Siedliska, there were two slightly injured. Near Krówník I saw a very nice funnel for (bomb)*[81].

"Przemyśl Land ", in the issue of January 14, 1915, stated:

13 January [1915]. *Wednesday ... Yesterday a Russian airman was throwing bombs, two of them fell on Garbarzach, but they did not cause any damage.*

On January 18, 1915, another raid on Przemyśl took place - the "Przemyśl Land" on January 21, 1915: she stated:

Above: The last moments before take-off ... In the photo there is an Albatros B.I plane from Flik 1 during the preparation for the take-off for the combat flight from the airfield in Jasień - source: Fotoarchiv SAW-KA.

Russian bombs. On 18 May. [January 1915], *in the afternoon a Russian kite appeared over the city. A fortress* [plane from Flik 11] *and began to pursue him among the shots of cannons and rifles. The Russian predator escaped for some time before the chase, but soon circling in a semicircle it swam over the city from the direction of Prałkowiec and dropped two bombs. One of them fell on the corner of Ptasia and Zielona streets (opposite Mr. Fink's villa) breaking the fence and breaking several dozen windows around. The maid standing next to the house was wounded by a shrapnel in her chest and in her hand. The second bomb fell on Jagiellońska street (near the Gans warehouse). The victims were two people killed and a dozen injured, and in the whole street many broken windows and several windows torn out of the frames. Grenade shards even*

Left: The pilot was a "knight of the skies" only in propaganda "readings", which were heavily "fed" over time, the more and more war-weary society of the Austro-Hungarian Empire.... In the photo one of them - Oblt. Max Hesse of Flik 1 during flight. The aviation helmet and the "air leather" on the body of an ex-cavalryman are noteworthy - source: Fotoarchiv SAW-KA.

Above: Radio crew ... The photo shows a "radio" Albatros B.I No. 7 aircraft with given name "*MUZZL*" in 1914, appearing in the Flik 1, 8, and 10 with the Flik 1 crew composed of Oblt. Ludwig Dumbacher (on the hull) and Oblt. Max Hesse before take-off from the Krakow-Rakowice fortress airport -source: Fotoarchiv SAW-KA.

pierced many iron shutters[82].

Planes of the fortress Flik 11 and other Fliegerkompanies arriving at the airport in Żurawica, taking off from there, tried to oppose the actions of the Russian air force over the fortress. On January 19, 1915, an air fight was fought, resulting in the shooting down of a Russian plane and the death of its crew. In the diary of Oblt. S. Gayczak, under this date, there is an entry:

At 9am, a Russian airplane threw a bomb on the city that killed 5 people. He was shot down for that. As he fell, he crashed into splinters and "flugers"[the aviators] *only guts were left. Currently, our airplanes* [planes] *have a specially constructed machine gun that can shoot quite well. We saw our airplane* [plane] *chasing the Russian, trying to swim under it to shoot at it from below*[83].

The harassing raids by Russian planes on the fortress continued. About the results of the bombing of January 20, 1915, "Ziemia Przemyska" wrote:

Russian bombs. On Wednesday around 2 p.m., a Russian kite was hovering over the city again [plane] *flying from the direction of the Tatar Mound. He threw two bombs. The first one fell on Cebulanka without causing any damage, the second one at 3rd May street (in Zasanie) in the garden, right next to the sidewalk in front of tenement house 139. After the impact, a high column of smoke and fire exploded, followed by a terrible roar. The falling bomb tore a large funnel-shaped hole and at the same time a tree with roots and splashed into thousands of deadly missiles. Passers-by walking down the street fell to the ground from the air pressure. All the shafts within a few hundred paces flew out, opposite the trees; windows and shop windows with frames thrown away, iron blinds twisted like a piece of paper ... Fortunately, there are no victims in the people; there is only one soldier wounded*[84].

Another bombing took place on January 21, 1915 -Helena Jabłońska wrote in her diary kept in the besieged fortress on that date:

Today we were bombed again by a Moscow aviator. He threw a bomb on the Greenand on May 3, the maid was wounded, and the shaft fell out again. In general, panes in a huge space fall victim to. The bombs are now much larger and stronger than the previous ones, almost the size of our bucket. The power of the explosion is huge. We almost mocked them [spelling consistent with the original throughout the quoted text] *we must be serious*[85].

The end of February 1915 abounded in more raids by Russian aviation. On February 27, 1915, the Przemyśl

Above: Mountains - no matter if the Carpathians or the Alps, they always teach respect, especially in winter.... The photo shows an idyllic and at the same time menacing image of the Alps during the Great War, seen from the Flik 1 crew cabin - source: Fotoarchiv SAW-KA.

Fortress was bombed by a Russian plane, the same happened on February 28, 1915, when a bombardment ended with an air fight over the city. The above-mentioned Przemyśl diarist wrote in her diary:

February 27 [1915]. *Saturday ... I was returning amid the roar of cannons and the whistling shots of ours, thrown into the Russian kite* [of the plane]. *Many bombs were thrown today like it would disregard the arrows sent to him. When our clouds surrounded it too much, it rose very high and flew on and continued to throw bombs. February 28 [1915]. Sunday ... Again, in the morning, bombs are throwing by a kite* [plane], *our fortress* [plane] *chases after it ... March 1* [1915]. *Monday ... I am lying and listening to this ominous hurricane of an airplane* [plane], *throwing bombs and cannon shots sent to it ... 11 bombs on Sunday, a lot on Saturday, and today also a lot. One fell in the control room, severed both legs and an arm of the soldier; he died on Sunday ... They say towards Żurawica an armored train was supposed to go, and a kite would meet it* [plane] *but I doubt it* [there was no armored train in the besieged fortress]. *Most likely the target is Festungscommando* [Festungskommando - fortress headquarters*] and a railway track*[86].

Just before the capitulation of the Przemyśl Fortress, on March 21, 1915, the diarist noted in her diary:

Moscow's kites fly [planes] *and they throw bombs. They fly low because nobody is firing at them, there are no rifles, everything goes unarmed. Only the police have it yet, but they haven't fired. Made of varieties* [the forts] *also the kites* [the planes] *were not fired*[87].

From January 1915., the air communication line developed from the field airport in Jasień with the besieged Nadsań fortress (thanks to the proximity of the airport to the railway stop, the post office could be immediately dispatched by rail to Krakow), was, apart from radio communication, the only source of communication between the defenders of the fortress and the outside world[88]. Postal flights (and not only) were carried out under the supervision of the commander of the KuK Luftschifferabteilung Obst. ME Uzelac (on parcels dispatched by air mail from Przemyśl, the Flik 11 seal was imprinted). According to archival sources, the following flights to the Przemyśl Fortress took place in the winter of 1915:

- Jan 12, 1915 Lt. Rudolf Stanger from Flik 8 performed a postal flight on the Brzesko (Jasień) -Przemyśl route, the next flight on that day was made by Flik 10 composed of Oblt. A. Schindler and Oblt. Martinek;

Above: Flik 3 during the 1914 campaign, on October 15, 1914, it found itself at the airfield in Sanok, assigned to the 2nd Army field ... In the photo, Aviatik B.I aircraft during the landing approach - source: Fotoarchiv SAW-KA.

Above: Flik 3 in August 1914, it was on the Nord-Ostfront ... In the photo, a Lohner B "Pfeilflieger" plane in flight - source: Fotoarchiv SAW-KA.

64.19
Fliegerkomp. 3.

3405

Above: With five locomotives across the mountain pass next to the village of Orlov ... Flik 5 was transported by the railroad to the Nord-Ostfront and on October 15, 1914, it found itself at the airfield in Sanok with assignment to the field Imperial-Royal 2nd Army. The photo shows the hulls of the planes of the unit - source: Fotoarchiv SAW-KA.

Facing Page, Above: War tableau ... Flik 3 flying personnel- in the center the commander of Hptm. Karl Stohanzl - Brandenburg C.I aircraft No. 64.19 in the background (photo taken in 1916) - source: Fotoarchiv SAW-KA.

Facing Page, Below: Galicia 1914 .... Oblt. Guido Georgievic (first from left) and Oblt. Andreas Alexay from Flik. 5. at the Baranów Sandomierski field airport, next to the Lloyd C.I aircraft. source: Fotoarchiv SAW-KA.

Below: On 12–13 October 1914, Baranów Sandomierski became the Flik 5 airfield. . . . In the photo, there are transport crates with Aviatik B.I fuselages arranged for transport - source: Fotoarchiv SAW-KA.

Above: Vermist... December 10, 1914 Oblt. (Hptm.) Josef Bendik from Flik 5 together with Zgsf. Johann Papp disappeared during a combat flight. In the photo Oblt. (Hptm.) Josef Bendik in the observer's cabin of the Albatros B.I from Flik 5 - Source: Fotoarchiv SAW-KA.

- On January 21, 1915, the unsuccessful postal flight was made by Oblt. Josef Smetana from Flik 1.
- On February 1, 1915, he made another unsuccessful postal flight on the route Kraków / Brzesko? -Przemyśl made Oblt. Rudolf Holeka from Flik 8;
- On February 3, 1915, a reconnaissance flight without mail on the Krakow-Sanok route ended with a forced landing in the fortress area by the Flik 10 composed of Oblt. A. Schindler and Obl. Jindřich Kostrba;
- On February 9, 1915, a flight with a load of medicaments and mail was made by the Flik 1 on the Brzesko (Jasień)
- Kraków route with a crew of Zgsf. Max Libano and Lt. B. Fiala Ritter von Fernbrug;
- On February 26, 1915, he made a flight with two postal packages on the route from Brzesko (Jasień) to Przemyśl Oblt. A. Schindler from Flik 10.
- On March 17, 1915, a postal flight from Brzesko (Jasień) to Przemyśl was made by the Flik 10 composed of Lt. Ferdinand Junker and Oblt. Ferdinand Reischer;
- On March 21, 1915, two postal flights were made by Lt. R. Stanger of Flik 8 carrying five postal packages and Zgst. Fp. R. Meltsch also from Flik 8-also carrying five postal packages[89].

However, according to the research of A. Clement, a historian dealing with the history of postal flights, the winter flights of 1915 looked as follows:

- 12 January 1915 - flight to Przemyśl
- 18 January 1915 - three flights from Przemyśl to Brzesko (Jasień)
- February 1, 1915 - flight to Przemyśl
- February 7, 1915 - flight from Przemyśl to Kraków
- February 9, 1915 - flight to Przemyśl
- 10 February 1915 - flight from Przemyśl to Brzesko (Jasień)
- February 12, 1915 - flight from Przemyśl to Brzesko (Jasień)
- February 18, 1915 - flight from Przemyśl to Brzesko and from Brzesko (Jasień) to Przemyśl but without the post office
- March 17, 1915 - flight to Przemyśl
- March 21, 1915 - two flights to Przemyśl
- 22 March 1915 a flight to Brzesko (Jasień) and a second

Above: Galicia 1914…. Oblt. Guido Georgievic (first from left) and Oblt. Andreas Alexay from Flik 5 at the Baranów Sandomierski field airport, next to the Aviatik B.I aircraft - source: Fotoarchiv SAW-KA.

flight from Przemyśl ended with a forced landing and loss of mail[90].

"Ziemia Przemyska" of January 24, 1915 stated that *five soldiers can send one card, since airmen can only weigh up to 130 kg. To carry excess weight ... Besides, the military prepares balloons which, together with open letters* [spelled in accordance with the original report], *will be released during a favorable air current. The military ... will be obliged to deliver the captured letters to the nearest authority*[91].

Oblt. Fp. Years later Nikolaus Wagner von Florheim assessed the aviation epic of Przemyś at the turn of 1914/1915:

Flik 11 arranged in Przemyśl normal air mail, which was then carried out with some regularity and brought a substantial income to the relief fund for the families / relatives of fallen pilots. These funds, like all savings, fell victim to devaluation / depreciation and liquidation after the war. Of all the aviation achievements related to the Przemyśl fortress, none of them had such an echo as the first flight to and the last one from the fortress. The first one, made from Nowy Sączby Oblt. Fp. [A.] Taussiga [Tausziga] and Hptm. Of the General Staff of von Raabl [Hptm. Raabl and Oblt. A. Tauszig flying Albatros BI aircraft named "*Steffel*" from Flik 14], provided the fortress with the news of the upcoming relief *and served an agreement with the fortress's command to define the fortress's role in the relief fights. The Russians recognized the significance of this flight and flooded the plane with a real hail of missiles, it was hit many times, but it fulfilled its task perfectly*[92].

The increasingly severe lack of food in the Przemyśl fortress meant that not only mail, but also food products and spare parts for the radio were transported by planes. Hptm. Fp. R. Holeka from Flik 8 On February 1, 1915, received an order to fly from the field airfield in Jasień to the besieged Przemyśl Fortress:

On February 1, 1915, at noon time, I was given a special assignment from the higher command. The idea was to bring to Przemyśl parts for the radio telegraph, one spare propeller for the plane, sugar and sweets for the garrison and other small items. A lot of things took the observer's cabin and that's why I flew alone because I knew the Nowy Sącz-Jasło-Sanok road well. This flight

Above: Field realities ... At the Flik 5 field airport, next to the Albatros B.I Oblt. Ferdinand (Karl?) Banfield - Source: Fotoarchiv SAW-KA.

meant a lot to me. I was able to see for myself what was true and how many things were stolen from us then, as the pilots told me. The weather was bad, it was cloudy, and the wind was strong. I was flying against the wind to Przemyśl and on the way I had to lower my flight by 400 meters in order not to lose my bearings due to the clouds besieging the city. The city was in a fog. I gave less throttle and decreased altitude by diving downhill. I broke through the clouds and found nothing. The wind blew me away and turned me around. It took a long time to get my bearings on the ground and get to know the river that is familiar. I wanted to get close to Przemyśl about four times. The wind was blowing at full strength, and on the last attempt to get between the besiegers and the city, I got lost in the clouds and flew west. The fuel tank froze, and I could not give full gas. Only when I was at Jasło I did know where I am. There was nothing left for me to do but keep the altitude at which I was already - around 400 meters. I grabbed the front line and headed home between the trenches. I wanted to land at the nearest airport, fix the defect and repeat the flight the next morning. The engine coughed constantly. A failed flight, winter weather and exhaustion made me indifferent. I did not notice at all that I was flying lower than the Russian trenches, which were on the crests and slopes of the mountains on one side of the valley and opposite the trenches on the other side. They were shooting at me from the second floor. It didn't matter to me that I hear the plane fire more and more often. I was already about 100 meters above the ground, when the engine stopped, the gasoline ran out. It was already dark, so I had to end up somewhere anyway, even if I didn't find the airport. After five hours of this miserable flight, I landed on the nearest, level, snowy field and thoroughly plowed the ground. The nose of the plane was completely smashed After five hours of this miserable flight, I landed on the nearest, level, snowy field and thoroughly plowed the ground. The nose of the plane was completely smashed After five hours of this miserable flight, I landed on the nearest, level, snowy field and thoroughly plowed the ground. The nose of the plane was completely smashed[93].

12 March 1915 Oblt. J. Smetana did not return from the flight to Przemyśl[94]. Having taken medicines (including cocaine) and other things "ordered" from the Przemyśl fortress on board his plane, he took off from the field airport in Jasień, but the area of the Przemyśl fortress was in fog and a strong wind was blowing (at 1.30 p.m. a plane engine was

Above: Flik 5 airfield in Baranów Sandomierski, October 12–13, 1914 .... In the photo in an Aviatik B.I plane: in the observation cabin of Oblt. Ferdinand (Karl?) Banfield in the cockpit Oblt. (Hptm.) Josef Bendik - Source: Fotoarchiv SAW-KA.

heard over the fortress) to which the Russian troops opened fire). Oblt. Fp. W. Schreiber from Flik 15 described the circumstances of that event as follows:

Oberleutnant [J.] *Smetana from Fliegerkompanie 1, the commander of the same, got lost in the fog during his flight to Przemyśl and has been missing for 5 days. Probably he was captured. According to the radio telegram from Przemyśl, the sound of the engine was heard from the clouds twice, one after the other, but the plane itself was not noticed. If it was* [Oblt. J.] *Smetana, it would be bad luck - to be on your destination and miss it. Let's hope he's at least landed somewhere and he's okay*[95].

The pilot was considered missing[96]- by Hptm. E. Kahlen, as part of the exchange gentleman's correspondence with Russian airmen, information was soon received from Tarnow that it was not a combat loss but a desertion[97]. On March 16, 1915, the AOK sent the commander of the Przemyśl fortress information about the general location on the frontline and they influenced the final decision to make another attempt to break through the Przemyśl Fortress, which, however, ended in failure and with exceptionally high losses of the attacker towards Medyka 23 DP Honved. On March 17, 1915, despite the terrible weather conditions, the crew of Lt. F. Junker and Oblt. F. Reischer of Flik 10 delivered by air urgent orders to the command of the Przemysl fortress[98]. On March 17, 1915, the AOK ordered the Przemyśl fortress headquarters in the order Op. No. 8104, to send the most important documents by air (on March 18, 1915, all logs of higher commands and observation maps of the fortress command were handed over to the airmen in secure packages), besides, all the so far unsettled award applications for officers and the last mailings of the crew. The remaining files with restricted content (service books, field elaborates, maps that are no longer needed, etc., all manuscripts and prints), which should not fall into the hands of the enemy, on the same day, on the eve of the attempt to break through the fortress, were burned at the behest of its command. and the remaining files were handed over to the square's headquarters, which deposited them in chests. 19 March 1915 at 07.00 despite a raging snowstorm, visibility below 100 m and the shelling of Russian artillery, six planes from Flik 11 were to take off from the fortress airport in Żurawica. Due to a snowstorm, five planes returned to the

Above: On 12–13 October 1914, Baranów Sandomierski became the Flik 5 field airfield.... In the photo in the cabins of the Aviatik B.I - in the observation cabin of Oblt. Ferdinand Cavallar von Grapensburg, in the cockpit Oblt. Eduard Fritsch - Source: Fotoarchiv SAW-KA.

home airfield and the pilot of the sixth, Zgsf. Fp. Feldvebel Hermann Hold, lost his orientation and having lost his way, flew east, and had to land near Stanisławów on the Russian side, getting into Russian captivity. Five planes that turned back were burned at the Przemyśl airport and their crews made a second attempt to depart from the fortress in baskets of four free balloons, but unfortunately, the winds were not favorable to the aeronauts and no one managed to reach the position of the Austro-Hungarian troops (the balloons were pushed towards Russian, and 12 airmen were captured by the Russians)[99]. The Commander of the Festungballonabteilung 1/3. R, Oblt. Julius Nimmerrichter described the "balloon epic" as follows:

Late afternoon on March 16 [1915], *captain* [Hptm. P.] *von Blaschke* [von Zwornikirchen - commander of Flik 11], *as the oldest officer of the Przemyśl air force, handed over the order to the fortress command, according to which an excursion was to be made on the night of March 18-19* [1915]. *One plane was to be used on his behalf for reconnaissance and communication, the other five planes and all the balloons could leave the fortress, but not to panic, take off before the start of the raid. If, however, bad weather conditions, snowfall, low cloud base and heavy fog at the ground were to persist, the air force officers could be deported from the fortress by my officers (balloon unit) in slowly released balloons, using the then moderate eastern wind. The last balloon or the last plane, was supposed to leave the fortress before 11 a.m., because there was a fear that in the event of a failure of the trip, the Russians could take the fortress by storm. After a detailed discussion of the situation with my officer and the gentlemen of the 11th Aviation Company [Flik 11.], the action program was compiled. If the plane failed, the fortress was to be lowered by balloons, together with the officers flying it, a total of 12 people. We had four balloons, including the service balloon of my department (Odkolegballon, 1000 m3) and balloons: "Austria" (1000 m3), a layered balloon (Schichtballon, 600 m3) and "Stryja" (Steiermark, 600 m3). The first three of those mentioned had to be picked up from the municipal fortress artillery store in Przemyśl and transport to Buszkowice. Then all the balloons had to be inspected, equipped, and filled. At the same time, blasting and destruction were to be prepared. It was also necessary to take*

Above: From the photo album of Flik 5… Oblt. Johan Mandl (first from left) and Oblt. Max Macher of Flik 5 at the Baranów Sandomierski field airport, next to the Aviatik B.I aircraft- source: Fotoarchiv SAW-KA.

care of the soldiers' uncertain future as far as possible. Despite all the shortages, I managed to get uniforms, shoes, underwear, as well as food for 2 to 3 days for my people, so that each of them went into captivity well-dressed and protected against the most severe hunger in the first days. I had less than 40 hours to complete all this work and preparation; it was a great achievement for my squad of only 60 people. Without my gas apparatus [Silliziumgasapparat] *all this would be useless anyway. It was used to inflate three balloons and I had a 600 m3 layered balloon filled from gas canisters (my emergency supply). Even in the evening of March 18* [1915], *business parcels were distributed among the balloons. 9 similar packages were assigned to my balloon. Everything went according to the plan. In the morning of March 19, 1915, 6 planes took off. Black, opaque, and very low-standing clouds (cumulus clouds)* [the cumulus] *made orientation impossible despite the use of compasses. In opaque clouds, the aviator easily loses his sense of the horizon and becomes nervous, which causes the camera to jerk up nervously. Together with the electric charge of the clouds, this causes the uninterrupted spinning of the magnetic needle. However, even if the magnetic needle was pointing north, the clouded pilot would not have a fixed landmark on the horizon with which he could correct the drift caused by air currents. A similar fixed point could then not be obtained either by flying over or under the clouds. Used in Przemyśl planes (Albatros* [BI] *1914) could reach a ceiling of just over 2,000 m, while, as I explain in my account, the clouds reached a height of over 3,000 m. The flight below the base of the clouds, due to its extremely low location and significant wear and tear of planes, especially the engines, he had no hope of success. From here all pilots immediately turned back. Only one of them did not manage to find his way back and was captured by the Russians in Stanisławów. As can be seen from this, it was already flying in the opposite direction to that intended. As a result of balloon take-offs, the balloons were fiercely fired by Russian artillery. So, I took my leave of my faithful companions briefly and gave the order to launch the balloon. Amid the terrible thunder of the* [Russian] *cannons, my balloon slowly rose westward. As the Russians noticed him, he immediately became the target of a furious cannonade. The remaining balloons took off in line with the plan under increasingly intense artillery fire. The second to climb was Odkolegballon led by captain von Blaschke* [Oblt. P. von Zwornikirchen], *whose passengers were lieutenants Schescherko* [Sesergo (?)] *and Reicher and Artilleriemeister Gruttschnig. The third was a layered balloon led by Lieutenant Ficker* [Fischer?], *whose passengers were Captain* [Fp. G. von] *Lehmann and lieutenant* [Oblt. A.] *Schindler. The 140 gas canisters that were used to inflate the layered balloon were still from the times of peace and had lost some of the contents, so that the balloon did not stretch properly and could not bear the accounting weight (3 passengers). Hence, Captain Lehmann stayed on the spot. Fortunately for him: three days later, when the weather improved and the Russians did not attempt to take the fortress by storm, despite the failure of the trip, a lieutenant* [Oblt. Eng. R.] *Steiger and one rank-and-file pilot* [Zgsf. R. Meltsch], *each with one plane. While the latter, returning, fell*

Above: Military Galicia… In the photo, the Flik 5 airfield in Baranów Sandomierski. Next to the Aviatik B.I, there are members of the flying personnel standing and sitting: the first on the left Oblt Ferdinand (Karl?) Banfield, second left Oblt. Andreas Alexay, first from right Oblt. Guido Georgievic - Source: Fotoarchiv SAW-KA.

into Russian captivity, Lieutenant Steiger, together with Captain Lehmann, as the last member of the fortress's crew, managed to escape just before the Russians entered and happily reach his homeland. The fourth balloon - "Styria" - led by Lieutenant [Lt.] *Dietzius could not be used because a piece of the Russian grenade damaged its carrier net while it was still anchored, causing it to jump off and fly away. Of the remaining two balloon units, Captain* [Oblt.] *Grebenz* [Grabenz?] *decided to fly freely and took off at night. My balloon staff then gained general admiration for their almost heroic behavior during uninterrupted 48 hours of work at the airport heavily under fire by artillery. But let's return to my free flight: to save at least some of our private equipment, instead of the equivalent sand ballast, we took clothes, underwear, my saddle with the rest of the equestrian props and one camera. In addition to all these and 9 service packages, 11 more 20 kg sandbags, which gave a total of 350 kg ballast. Within seconds of its release, my balloon was among black clouds, and I lost all eye contact with the ground. Only the terrible rumble and the thunder of the cannons let us know that we were still in the fortress area. In the meantime, we had reached 1,800 m and we were still amid opaque snow clouds. Half past two in the afternoon, as we just broke through the second layer of clouds and the first rays of sunlight reached us, the distant fire of cannons and machine guns could be heard. We assumed that we were approaching the front line of the field forces, but as we still did not see the land, we were not sure if we had reached it near Tarnówor maybe in the south, in the Carpathians. Repeated whistling and greasy smoke from the locomotives made us assume that we were on the Galician side of the Carpathians. The cannon fire was heard no more clearly and did not last long. As we approached an altitude of 3,000 m, we reached the third cloud layer where we sailed for an hour in temperature -25 degrees* [C], *which cost us half the ballast. It was after 3 p.m. and apart from a few clothes, a camera, basket equipment and service packages, we only had 4 sandbags left, which I had reserved in case of landing. At 4:30 PM we finally broke through this layer of clouds and the sun warmed our stiffened limbs as well as the balloon, which allowed us to keep the ballast, but also melted the snow covering it, which, as the balloon was not equipped with a rain collar, we were soaked to the skin. After less than half an hour, the balloon, which had so far lost almost half of its gas, began to descend. To stop the rapid descent, it was necessary to sacrifice everything that was not necessary, the equipment of the basket, Planchen, Drittel, almost all clothes and underwear. Since it was already late afternoon, which made us think about the imminent landing, the descent of the balloon was not entirely in our hand. A few minutes after 5 p.m. we descended to a height of 900 m for the first time since leaving Przemyśl we saw the ground. The sight of her, however, was not encouraging. As far as the eye could see, there were mud covered with a thin layer of ice. We considered this area as the Gródek marshes and assumed that the wind had turned us in the opposite direction and drove us to the east. Landing in similar terrain was impossible, so I tried to fly over the swamps. We were extremely desperate to know that we had fallen captive to the Russians. After about half an hour, after 5 PM, at this time of the year, at dusk, apart from the service packages that we were*

Above: Nord-Ostfront 1914…. Flik 5 in the "full gala" at the field airport in Baranów Sandomierski (?). In the photo there are six Rumpler B.I planes with their crews and ground personnel standing next to them - source: Fotoarchiv SAW-KA.

carrying on ourselves, we only had 1.5 bags of sand ballast left. We fell to a height of 800 m and we saw the ground once more. To our profession, we saw muds like those previously observed, so that we were forced to assume that rising above the clouds, in the prevailing windless weather, we didn't move from the place. Then I decided to land at all costs. I noticed a small, but sufficient to land on it, piece of brown ground, on which I wanted to sit, taking advantage of the strong wind blowing near the ground. I gave the necessary order, undertook all preparations, and managed to land smoothly on that islet. Was [h] 5.30 in the evening. Our deadly debris left the basket, and we immediately began destroying the balloon and service packages. Despite the strong ground wind and snowfall, we finally managed to set the balloon on fire. In the meantime, peasants came running from the nearest town, whom we called together with Lieutenant [Oblt.] Schneider in Polish, Czech and Ukrainian to cut and take ropes, cords, and other items useful on the farm. In this way, civilians helped us complete the balloon's destruction by cutting it into pieces. As the basket burned down completely, with nothing left but smoldering pieces of paper smaller than a walnut, the

Below: In the fall of 1914, Albatros B.I planes were delivered to the field Fliegerkompanien by the German ally…. The photo shows the Albatros B.I No. 28 aircraft from Flik 5 in front of the hangar with ground service - source: Fotoarchiv SAW-KA.

Above: On the combat trail Flik 5 appeared on August 28–September 9, 1914. Kraśnik…. In the photo Hptm. Miller surrounded by German soldiers on the city street - source: Fotoarchiv SAW-KA.

growing mass of viewers realized that there was something wrong with the destruction. The mayor, who was one of the last to arrive, successfully prevented us from continuing. Soon each of us was bound and summoned to explain our actions. They took our weapons, ammunition, money, valuables, and luggage from us. When we made it clear that we were Austro-Hungarian officers, we were taken from the mud to the village, more specifically to an inn. The primitive guest room was soon crowded with peasants, pulling on all sides to watch over us as well as out of sheer curiosity. Several curious women peeking into the interior could be seen through the windows. Now is the time for the first hearing, conducted by the mayor and some of the crowding peasants without any plan or purpose. When we asked where we are, the local pop responded only after four hours of torment. We found out that we landed not on the Gródeckie but the Rokitnieńskie marshes, near Łuków, 32 km southeast of Brzesc into Russian captivity. Our landing was reported by a police messenger and the next day, March 20, at 6 am, the first Russian military patrol arrived at the scene. We had to endure a second interrogation, which began with asking who my deceased father was and ended with another personal search. Before noon, the Russian captain of the general staff, the captain of the military police and a police officer arrived at the scene by car. Until we were transported, we were guarded by almost half a squadron of Cossacks. The officers interrogated us and sold us as thoroughly as possible. Initially, we were considered spies because of the nearby Brest Fortress.

After all, around [h] 4 in the afternoon the officers took us by car to the headquarters of the fortress in Brest. There were plenty of curious Russian officers here, many of whom, mostly senior general staff officers, rejoiced at our downcast faces. On this occasion, we were treated with all possible lies: Bulgaria, Romania, Greece, Italy, and God know what other countries were supposed to declare war on us and start an offensive any day now. At last, after two hours of this torment, we were summoned by the Chief of the General Staff, to the command of the fortress. This distinguished gentleman gave the impression of a chivalrous and jovial man. After a few questions, to which I refused to answer, the commandant of the fortress noted that he did not expect anything else from the Austro-Hungarian officer. "Assuming that the enemy's praise is at least equal to the praise of his own superior," he said, "I would like to say to you, that you managed to carry out the destruction of the balloon in a way that must be highly appreciated by my officers". He let me go, wishing that the imminent end of the war would allow me to return to

Above: "Eyes" of the field command of the 1st Army .... The photo shows a field airfield in Nisko where Flik 5 was stationed on 18–28 August 1914 and 10-12 September 1914. Visible tent hangars and Aviatik B.I planes in the colors of KuK Luftschifferabteilung - source: Fotoarchiv SAW-KA.

my homeland. Then, at 8 PM, we got the first one since leaving Przemyśl- meal. About [at] *9 PM we went with the same general staff officer to the Russian stage headquarters in Chełm. The next day, passengers of other Przemyśl balloons gradually joined us. In Brest only Captain Blaschke remained* [Oblt. P. von Zwornikkirchen - his balloon landed south of Brest], *who during the landing of the remaining balloons: Lieutenant* [Oblt.] *Dr. Henryk Ficker* [Fischer?] *landed near Sokal, captain* [Oblt.] *Grabenz near the city of Nowogród Wołyński (in Russian Volhynia), and Captain Blaschke* [Oblt. P. von Zwornikirchen] *a few kilometers south of me. From the above-mentioned directions in which the balloons sailed, one can assume that when we reached the cloud layer, the wind changed its direction and instead of pushing us to the west (towards our motherland) as at the beginning, it turned us to the south-eastern course, so that the sounds cannon fire and machine guns that we heard could have come from the nearest section of the front in this place, more or less in the vicinity of Stanisławów then the wind changed once more, pushing us northward. My journey lasted 7 hours, during this time the balloon covered 255 km in the air, which gives an average pace of 36.4 km per hour*[100].

Obst. Fp. N. Wagner Edler von Florheim years later described the balloon flight from the Przemyśl Fortress:

The rest of the pilots and balloon "drivers" tried to escape [from Przemyśl] by *free balloons. At first it seemed to be successful. The wind pushed the balloons towards the Carpathians, but then turned the other way, deep towards Russia. Commander of Flik 11 Hptm. Fp. von Blaschke* [P. Ritter von Zwornikkirchen*]* *landed near Równe, was attacked there by the Cossacks and was taken prisoner of war with severe wounds. Together with him also Hptm. Dr. Grebenz* [Grabenz?] *and some others. Oblt. Ikawetz, thanks to skillful hiding managed to avoid captivity, he was waiting for saviors in Przemyśl*[101].

The gold reserves were also evacuated from the besieged Przemyśl Fortress by air. Oblt. Fp. Nikolaus Wagner Edler von Florheim wrote in his notes:

The last flights had a sad purpose. Machine [the plane] *was to transport the gold supplies from the fortress to a safe place, it was controlled by Stabsfeldfebel Meltsch* [Zgsf. R. Meltsch], *was overloaded, was shot down and burned down. Second machine piloted by Oblt. Ing. Fp.* [R.] *Stanger, along with Oblt. Von Lehmann* [Hptm. G. Edler von] *as an observer, was sent* [from Jasień] *to Przemyśl only after the planes of the fortress were destroyed by shelling, which took place in the last days of the siege. He hid the log of the fortress and other important service orders, as well as the commander's last report to His Imperial Majesty, Emperor Franz Joseph I*[102].

March 18, 1915 in the Oblt. Fp. W. Schreiber from Flik 15. Having received the order to fly to the besieged fortress, he flew from the field airport in Jasień to Przemyśl by Albatros B.I (Aviatik B.I?) and after two and a half hours of flight along the corridor over Jasło and over San, he landed at the airport in the besieged fortress, noting in his account: *I was supposed to deliver* [to Przemyśl] *Chinina and take away the gold in bars.* On March 19, 1915, the same pilot made a flight on the route to Przemyśl for the second time, recalling: *I was in Przemyśl for the second time. Again, I was to deliver the medication and take the gold away.* March 21, 1915 Oblt. Fp. W. Schreiber reached Przemyśl for the third time by air with

Above: Aviation "pop" at the field airport in Kraśnik.... During landing, Oblt. Max Perini of Fli 7 (the unit was stationed at this field airport on August 14–August 21 and September 11–19, 1914) put a Lohner B "Pfeilflieger" plane on its nose - source: Fotoarchiv SAW-KA.

the above-mentioned plane with the above-mentioned Oblt. Eng. R. Stanger[103]. Oblt. Fp. N. Wagner von Florheim, years later described the last days of the air epic of the Przemyś Fortress, as follow:

Flik 11 arranged in Przemyśl normal air mail, which was then carried out with some regularity and brought a substantial income to the relief fund for the families / relatives of fallen pilots. These funds, like all savings, fell victim to devaluation / depreciation and liquidation after the war. Of all the aviation achievements related to the Przemyśl fortress, none of them had such an echo as the first flight to and the last one from the fortress. The first one, made from Nowy Sącz by Oblt. Fp. [A.] *Taussiga* [Tausziga] *and Hptm. of the General Staff von Raabl* [Hptm. Raabl and Oblt. A. Tauszig flying Albatros BI aircraft named "Steffel" from Flik 14.], *provided the fortress with news of the upcoming relief and served an agreement with the fortress's command to define the fortress's role in the relief fights. The Russians recognized the significance of this flight and flooded the plane with a real hail of missiles, it was hit many times, but it fulfilled its task perfectly ... The launch took place exactly when the explosions began. The debris of blown-up bridges, forts and other objects flew into the air, sometimes even exceeding the flight altitude of the machine. But the flight was successful. Oblt.* [Fp. G.] *von Lehmann was received by His Majesty* [Franz Joseph I of Habsburg] *at an almost hour-long audience, in which our old emperor, deeply moved, took the image of heroic struggles and an honorable fall. The rest of the pilots and balloon "drivers" tried to escape with free balloons. At first it seemed to be successful. The wind pushed the balloons into the Carpathians, but then turned the other way, deep towards Russia. Flik 11 Commander Hptm. Fpvon Blaschke* [P. von Zwornikkirchen] *landed near Równe, and was attacked there by the Cossacks and was taken prisoner of war with severe wounds. Together with him also Hptm. Dr. Grebenz* [Grabenz?] *and some others. Obtl. Ikawetz thanks to skillful hiding, managed to avoid captivity, in Przemyśl he expected saviors*[104].

The F. Molnár, drawing from the accounts of the defenders of the Przemyśl fortress, described its fall as follows:

Captain Lehmann [Rittermeister G. Edler von] *described in detail the sad act when, on the orders of the fortress commander,*

Above: "Bird's eye view of Újvidék in all its glory ... During the 1914 campaign Flik 5 assigned to the field Imperial-Royal 2nd Army was stationed on August 6–7, 1914 at the home airport in Újvidék, from where on August 12–17, 1914 it was transferred by rail to the Przemyśl-Żurawica fortress airport in Galicia - source: Fotoarchiv SAW-KA.

the pride of our artillery was destroyed, along with other modern guns - the 30.5 cm Skoda mortars ["Škoda" 30.5 cm M 11. mortars]. *There were two batteries in the fortress, four in total. Thirty-five* ["Škoda» 30.5 cm M 11 mortars] *for transport are placed in two parts on two automobiles* [cars]. *And now, for execution, they were assembled as if they were to be prepared for* fire. *If the mortar barrel is not directed at the right angle to the sky, it must not be fired, because then the bullet will explode in the barrel, exploding it. To destroy mortars, it would be enough to leave the barrel in the loading position, i.e., horizontally, and simply fire it. But for complete destruction, a more effective method was used. The mortar barrel was filled with sand, ecrasite and grenades were put into it, a bullet was inserted from the rear, a very small charge of gunpowder was given, the barrel was placed horizontally and then fired. Thirty-five* [the "Škoda" 30.5 cm M 11 mortars] *broke into pieces so small that a fist-sized piece was rare. At the same time, the rest of the ammunition, mountains of carts and automobiles were blown up near the city. The crew, who were marching from the trenches towards the city, broke their handguns While all this was happening in the city, Russian cannons thundered behind the forts, spewing furiously with incessant fire. The Russians were still pretended they had something to attack in Przemyśl*[105].

In response to the shelling from the fortress on March 21, 1915, the aim of which was to dispose of artillery shells, the Russian field artillery of the 11th Army began firing 11-line mortars (11-inch - 280 mm Schneider Mk. 16 copies from the French ally were ordered by the Russian side in 1913 - a total of 26 were delivered) and from 120 mm guns - fort 15 "Borek" received successful hits, where the ammunition warehouse exploded, and the radio station was destroyed. At that time, the Russian air force dropped 28 bombs of various weights, including 21 boxes (344 kg). During the day, numerous fires were observed in the fortress and its forts, and the captives, taken prisoners, confirmed the preparations for blowing up the fortress works and bridges. On the night of March 21/22, 1915, in the northern and eastern sections of the siege, fortress units fired on the positions of the besieging troops very strongly (37 thousand shots - on March 21 and on the night of March 21/22, 1915, artillery of the fortress fired approx. 50 thousand rounds)[106]. The pilots of the Imperial-Royal Air Force were ordered to

Above: Fokker B.I 03.22 was a type M10. Iron cross national insignia have been applied over white backgrounds on the wings and rudder. It is at *Flik* 8, Aisovizza airfield. (Peter M. Grosz collection/STDB)

Below: Another image of Fokker B.I 03.22. A downed Caproni bomber is in the background. Fokker 03.22 of *Flik* 8 with crew consisting of pilot *Oblt*. Wedige von Froreich and observer *Kadett i.d.Res.* Richard Maurig Ritter von Sarfeld was one of eight aircraft sharing the victory over the Caproni, which was shot down during a raid on Laibach. (Peter M. Grosz collection/STDB)

Above: Fokker B.I 03.10 was a type M7. Here it is at *Flik* 6, Igalo airfield. (Peter M. Grosz collection/STDB)

Above: Fokker B.I 03.11 was a type M7. An iron cross has been applied to the rudder. (Peter M. Grosz collection/STDB)

Above: About Tennenbaum ... Christmas 1914. In the photo, officers from Fliks 5 and 13 during the Christmas meeting - source: Fotoarchiv SAW-KA.

fly from the airfield in Jasień to the dying fortress at 02.00 and the next day, March 22, 1915, at 05.00 am they found themselves over Przemyśl, landing successfully at the airport in Żurawica, which was heavily damaged by Russian artillery, with their Albatros B.I (Aviatik B.I?) planes, reporting to Hptm. Fp. P. Blaschke Ritter von Zwornikkirchen with orders from the field command of the 4th Army. Returning to Jasień, each of them was to take one of the officers, but Hptm. Fp. P. Blaschke Ritter von Zwornikkirchen, decided to fly in a balloon, Hptm. G. Edler von Lehmann decided to fly with Oblt. Eng. R. Stanger, no one came forward to fly with Oblt. Fp. W. Schreiber. Early in the morning of March 22, 1915, among the monstrous detonations of the forts of the fortress, three KuK Luftschifferabteilung planes took off from the fortress airport in Żurawica, which was being destroyed before surrender. The plane of the Zgsf crew. R. Meltsch and Lt. F. Junker after the take-off, was shelled and damaged by the fire of the Russian Air Force, but the pilot managed to lead the plane as far as the Rymanów area, where he forcibly landed on the so-called Bartoszowy, and the crew were captured by the Russians[107]. Oblt. Eng. R. Stanger and Oblt. Fp. W. Schreiber took off eight minutes before 6 am, they rose to a height of about 500 m above the city center, where they were hit by the Russian opl. and infantry squads. After about 30 minutes, happily gaining altitude, they took a course to the airfield in Jasień, where planes with the crew of Oblt. Eng. R. Stanger and Hptm. G. Edler von Lehmann and the Oblt. W. Schreiber landed around 09.00[108]. Last flight to Przemyśl was recalled by Oblt Fp. Eng. R. Stanger, in his account of the Sunday edition of March 30, 1925, "Grazer Tagespost":

In the evening of March 20, 1915, three telegrams from Przemyśl were sent to our air camp. In the first it was stated that the air force officers who belonged to the garrison, among them also a centurion [Capt.] *Blaschke [*Hptm. Fp. P. von Zwornikkirchen] *and Professor Ficker, on March 18* [1915] *in four balloons* [five of them took off, but one was destroyed during take-off] *they left the fortress at the same time, but there is no information about their landing. The second telegram told us if we knew anything about the landing of Fire Chief Holda* [Zgsf. Fp. H. Hold], *who left with his plane on March 18* [1915]. *The third telegram said that two planes were sent to the fortress. That same evening, I got with Meltsch* [Zgsf. R. Meltsch] *an order that I should be ready to fly to Przemyśl*

Above: "Eagle of Trieste"... The photo shows the best naval aviator of Austro-Hungary Fregattenleutnant Gottfried Freiherr von Banfield in the cabin of a flying boat during its transport through the city streets - source: Fotoarchiv SAW-KA.

tomorrow morning [the author does not mention the preparations for the flight of the third aircraft piloted by Oblt. Fp. W. Schreiber from Flik 15, which in its account mentions preparations for the flight from Oblt. Fp. Eng. R. Stanger]. *On March 21* [1915] *- the day was nice, bright, right after 11.00 a telegram came from the fortress, instructing us to take off, because we were to take a friend from the fortress with this flight.* I also took some Przemyśl post office with me - *approximately 140 kg - it was the last post I took there. With mixed feelings, I got into the cabin of the plane because the content of the telegram showed that the situation of the fortress was not good. To avoid anti-aircraft fire over the Dunajec, I headed north by plane, and then flew through Rzeszów straight to Przemyśl. He took off after me* [Zgsf. R.] *Meltsch, but after a short flight he had to land because his engine was not working well. He repaired the failure and made it to Przemyśl without any problems where he landed two hours after me with three working cylinders in the engine. My flight on the 200 km route took 1 1/4 hours because I had a strong tailwind, and I could reach a speed of 180 km / h. When I saw Przemyśl, at an airport that I know well* [in Żurawica] *I saw the foggy ruins of the air barracks in the smoke* [hangars in Żurawica], *and when I landed in short spirals, I was greeted by the well-known sounds of exploding projectiles.* In this violent artillery fire, I landed in one field and with the help of few people, I immediately took care of my plane [Albatros BI - Aviatik B.I?]. *After arriving at the fortress, I immediately went to its headquarters, where I was welcomed with great joy, because I was carrying mail, the last mail from my homeland for the besieged heroes*[109].

The flight of Oblt. Eng. R. Stanger was the last flight from Przemyśl, which the pilot recalled as follows:

I came to the airline building [in Żurawica] *and I went rest, because I was ordered to leave tomorrow morning before five in the morning or after six in the morning. Before departure, between five and six o'clock, I was closely guarded, because during these hours the fortress was to be blown up. I couldn't even think about sleeping. The situation got even worse when suddenly I heard two terrible explosions so that the windows of our apartment shattered into a thousand pieces. Two 180 mm* [120 mm] [artillery] *grenades fell into the facility and the windows shattered under the pressure of the blast. I was preparing the plane* [Albatros B.I - Aviatik B.I?] *for departure by noon, although the Russians, who could watch our entire airport from a tied balloon* [in Żurawica], *fired 200-300 shrapnel at the airport,*

Above: On February 14, 1916, ten [12] Lohner biplanes [Lohner B VII] from three airline companies [Fliegerkompanien] bombed the Milan power plant that supplied power to the entire war industry of northern Italy. The squadron's flight to Milan, 190 km away, took five hours. The squadron flew over the alpine peaks at an altitude of 2,300 m ... In the photo of the crew of Flik 7 in leather suits after a happily return from Milan - Source: Fotoarchiv SAW-KA.

and then 138 grenades [cal.] *180 mm* [120 mm] *were fired – my plane* [type Albatros B.I-Aviatik B.I?] *was, however, flawless. In the evening, the spark telegraph station, which was located close to the airport* [in Żurawica], *was damaged. The consequence of its destruction was the fact that from 4 am on March 22* [1915] *the contact with the world was broken. At 3 am I was awakened from a light sleep by a friend of Captain Lehmann* [G. Edler von] *who came to me for a while. I was supposed to make the last flight from the fortress with Capt. Lehmann* [G. Edler von]. *Before I got dressed, I sent the car from the airbase with Melč* [Zgsf. R. Meltsch] *and field pilot Junker* [Lt. Fp. F. Junker] *to their plane* [Albatros B.I / Aviatik B.I?], *which was located at the second airport* [in Buszkowiczki]. *Melè* [Zgsf. R. Meltsch] *saw a terrible fire from the airfield where I landed* [in Żurawica], *and therefore landed in a different place* [in Buszkowiczki]. *I also ordered him to get ready to take off, but not to wait longer than till five o'clock. The car must be sent back to me because I need it to get on my plane.* Me and Capt. Lehmann [G. Edler von] *we got dressed and waited for the car to return. It was already four o'clock, half past five, and after the car there was no sign. Finally, the driver came to our apartment and reported that he had an accident – he drove into the trench and cannot get the car out without help. I was somewhat outraged by this matter, but this adventure, which was played out by fate, meant that later I had the opportunity to witness scenes that, with their amazingness, were especially remembered by me. After this driver's report, we hurried to the fortress's headquarters so that we could get a new car. On the road, we saw many soldiers dropping their weapons to the San or breaking and destroying them. All of them felt an unsettling tension. However, order and discipline prevailed among the soldiers in that hour, when each of them stood before the unknown, when none of them saw what their nearest fate would bring. It all must have aroused admiration. And on top of that, this terrible night. You could see the glow of the explosions in the sky, here again the luminous dust caught the spotlight while driving through the city and all around you could hear the rumble of cannons, whistles and hiss of bullets and shrapnel explosions. In the headquarters* [of the fortress] *we got a new car and "with the wind in the competition" we went to the airport* [in Żurawica]. *While driving, we heard the roar of the plane's engine* [Zgsf. R.] *Melč* [Meltsch], *who, despite the darkness, started at four in the morning* [from Buszkowiczki], *it was a masterful feat, and for me – the senior pilot, aroused respect. Behind the brave pilot, we sent wishes of a happy result. At 5:15 we arrived at the airport* [in Żurawica]. *I gave my machine* [type Albatros B.I / Aviatik B.I?] *to prepare for*

During the flights over the Alps the crews felt bitter cold ... In the photo Hptm. Gruber from Flik 7 in a leather suit and "polar" footwear after a happily return from a combat flight - Source: Fotoarchiv SAW-KA.

I met proud air force officers, the combat pilots appeared (to me) as Olympic Gods ... In the photo one of the Flik 7 pilots in a flight suit after a combat flight - Source: Fotoarchiv SAW-KA.

Above: For the AOK during the fights near Krakow and the Łapanów-Limanów operation in 1914, the constant reconnaissance of Bochnia was of particular importance.... In the photo taken by the Flik 8 crew in autumn 1914, the railway station in Bochnia - source: Fotoarchiv SAW-KA

take-off - I filled the cooler with water and was ready. Now when with Capt. Lehmann [G. Edler von] *we talked about the take-off, wondering if we should wait until six o'clock, or if we should fly out now, shrapnel started hissing over the airport* [in Żurawica]. *We didn't want ourselves and our plane* [the Albatros B.I / Aviatik B.I?] *to be put in danger, there was no time for stupid discussions. We decided to leave now, because the near explosions were dangerous for us and the destruction of the fortress had not yet happened, which made us suspect that it was somewhat postponed. I said goodbye to the commander and the few officers who escorted us with a polite handshake, and we took off. Here we had a terrible vision - uniquely beautiful though sad, (...) reminiscent of the destruction of Pompeii and the fall of Herculaneum. As if the fortress crew were waiting for our departure, to show us with what visible duty they were fulfilling the last orders, they started shooting from the city when we were at a height of one hundred meters, the flames were very high, almost under our feet and then it began to thunder, roar, explode on all sides. The stench of fire and smoke filled the air. Chunks of clay, parts of buildings and fortifications flew into the air and produced clouds of dust there that had yet to dissipate. All the time, the explosions were throwing more and more debris and dirt into the air. I changed the flight course* [of the Albatros B.I / Aviatik B.I? Aircraft] *and was flying through the city, 300 meters above the bridge, when suddenly the bridge blew up with a menacing thud below us.* It will be true if I say that the parts of the bridge were very close to our aircraft [Albatros B.I / Aviatik B.I?]. *A few seconds later, on our right hand, a powder magazine exploded beneath us, the explosion of which had a terrible effect. I have no words to collect the images that appeared to us from all sides into a complete impression. I will never forget it in my life. After a short flight through the city, the wind in our face forced us to rise higher and change the direction of flight towards the outer belt of fortifications. After a few minutes we were behind a vanishing city and those minutes were full of events we had never experienced before. We were already flying over the outer rampart. I looked back again. I saw the city* [Przemyśl] *immersed in a huge fire. The city and the fortress that had resisted its enemies heroically for so long was dying in these flames, which stained the morning cloud with blood. At an altitude of 600 meters, we passed the Russian line, but no missile followed us, although we had expected it. I had the feeling that the Russians forgot to shoot*

Above: The quality of the reconnaissance photo ... This is what the area of the Vistula and Dunajec forks looked like in the fall of 1914, seen from the board of the Flik 8 plane- Source: Fotoarchiv SAW-KA

at us. So, we left the dangerous streak of conviction without any obstacles, though about its last task it fulfilled in a heroic way. Me and my companion landed after a three-and-a-half-hour flight to the delight of all colleagues in our air base [in Jasień].

In another report, the same pilot added:

After landing, Captain Lehmann hugged me [Rittermeister G. Edler von] *thanking for taking him from the place doomed to collapse. In these last hours of heroism of the* [Przemyśl] *fortress, he avoided captivity and death, for him they were the most beautiful proof of friendly service that I could show him. We conveyed friendly greetings from the fortress. After a while, these greetings, brought from the fortress, caused a profound silence. We worshiped those who had fulfilled their duty to the end. We quickly heard the news that after* [Zgsf. R.] *Melču* [Meltschu] *who started ahead of us has no trace. I do not know what fate befell him to this day. But whatever fate it was, one cannot be denied: the glory and for the courage and heroism with which they carried out their orders, their heavy duty as soldiers*[110].

The landing of the last planes from the Przemyśl fortress at the airport in Jasień ended the history of the world's first postal airline[111]. On the next day, the airmen - the heroes of the last flight - stood before Emperor Franz Joseph I of Habsburg to report on the situation at the front and the last moments of the most famous Galician fortress, to which the crew surrendered[112] and this event had a strong echo in the Galician region. Obst. Fp. N. Wagner Edler von Florheim noted it this way:

The take-off was exactly when the outbursts began. The debris of blown-up bridges, forts and other objects flew into the air, sometimes even exceeding the flight altitude of the machine. But the flight was successful. Oblt. von Lehmann [G. Edler von] *was received by His Majesty* [Franz Josef I Habsburg] *an almost hour-long audience, during which our old emperor, deeply moved, took the image of heroic struggles and an honorable fall ... Oblt. Ikawetz thanks to his skilful hiding, managed to avoid captivity, and he was waiting for saviors in Przemyśl. Flik 11 was blown to dust, but its members could be proud to have fulfilled their duties, as befits brave pilots, to death and life. A glorious page in the history of the old brave army is created not only by the defense of the only fortress which, besieged during the World War, remained unbeaten by the enemy, and defeated only by hunger, but also by the achievements of the pilots in this young, beautiful aircraft weapon*[113].

On March 23, 1915, a plane from the Imperial-Royal Air Force took off from the airfield in Jasień, the crew of which during a ceremony organized by the Russian side on the occasion of the capture of the Przemyśl Fortress, carried out a propaganda campaign, dropping instead of bombs over

Above: In the fields near Lipie in Tarnów, in the fall of 1914, a field airfield was established, used by the Fliegerkompanien KuK Luftschifferabteilung, and at the turn of 1914/1915 by the Russian aviation ... In the photo taken by the Flik 8 crew view of Tarnów from the third dimension - source: Fotoarchiv SAW-KA

the market square in Tarnów on the heads of the gathered Russian officials,
in panic leaving the market area for fear of aerial bombardment ... advertising leaflets with the words - *Drink Okocim beer*[114]. *During the entire world war, on all its fronts, it was the only case that the fortress, on its own, played such a large role and completely changed the strategic situation*[115]. This opinion of the historian F. Forstner is extremely accurate when it comes to assessing the fall of the Przemyśl Fortress. When the fortress capitulated on 23 March 1915, over 100,000 soldiers of its crew were captured and transported to POW camps in Siberia, while over 100,000 "free soldires" were transferred to the Carpathian front from Przemyśl. Together with the Gorlice Breakthrough and with the retreat of the Russian Southwest Front to the east, the "Habsburg wings" returned to Przemyśl. An artilleryman from the 10th Corps, returning to Przemyśl, wrote in his notes:

There was to be a big one on Sunday Angriff [assault] - *our future* [Škoda mortars M. 11.] *30.5* [cm] *can start to think seriously about conquering the fortress ... the forts were smoking - the batteries were now given T-Grenades with poisonous gases - cheers to Geneva Convention! ... in one word: hell*[116].

Endnotes

1 Archives of the Folk Architecture Museum (AMBL) in Sanok, *Helena Schramówna, notes and diary from the war campaign 1914/1915 in Olchów*, op. cit., pp. 42, 45, 49.
2 S. Rosenboom, op. cit., p. 26.
3 On December 10, 1914, in the report of the field command of the Imperial-Royal The 3rd Army recorded that *the airman at noon also found one infantry regiment near Łużna, no strength on the Gorlice-Biec-Jasło-Żmigród route, slight rail traffic in both directions on the Biecz-Jasło route* - lieutenant SAW-KA, Gruppe 157 1 g, Russland II. Sommer-Herbst Feldakten der oe.-u. Wehrmacht im Heeresarchiv Wien - Przemyśl-Russen file, p. 41.
4 Ibidem.
5 Ibidem, p. 43, 46-47, 80.
6 Ibidem, p. 70. On December 16, 1914, the command of the fieldwork of the Imperial-Royal The 3rd Army reported the AOK detection by air reconnaissance of 12 enemy infantry battalions and 10 artillery batteries near Tarnów; a large camp of enemy troops was detected

Above: During the Great War, the KuK Luftschifferabteilung field airfield was also located in Sanok…. In the photo from autumn 1914, Sanok photographed through the eye of the camera of the Flik 8 crew - Source: Fotoarchiv SAW-KA.

near Zglobice; on the Zakliczyn - Tarnów route, eight infantry battalions and three artillery batteries were detected in movement, and 10 infantry battalions and a large artillery group were detected on the march to the east; Russian forces on the march were also detected in the direction of Tarnów and Dąbrowa Tarnowska, Radomyśl and Lisia Góra.

7 On December 17, 1914, the Flik crew. 15. She lost her orientation in the fog and having problems with the compass and encountering electric discharges in the clouds, she happily landed in Częstochowa - cf. Ibidem, p. 59-60.

8 The Festungskommando Krakau on December 19, 1914 reported to the AOK that Flik 15. performs flights for the field c.k. Of the 4th Army, for which thanks were given to its crews from the commander of the fortress, General K. Kuk - Lieutenant Ibidem.

9 Ibidem, p. 72.

10 Ibidem, p. 69. December 22, 1914 in favor of the field command of the Imperial-Royal The 4th Army made five flights in the runway from Staszów to Kolbuszowa-Rzeszów-Jawornik-Przemyśl and in the direction of Dunajec.

11 *War in the Kingdom of Poland. A car trip through the battlefield*, Kraków 1915, p. 41–42.

12 SAW-KA/Luftfahrttruppen, Gruppe 157 1 g, Russland II. Sommer-Herbst Feldakten der oe.-u. Wehrmacht im Heeresarchiv Wien – teczka Przemyśl-Russen, p. 27.

13 Ibidem, p. 41.

14 In Russian historiography, the second siege of the Przemyśl Fortress began on 1 (14) [the double dating in the text refers to the use of both Julian and Gregorian calendars in Russian sources, i.e. "old" and "new" styles] November 1914 (lasted 133 days) and in the light of Western historiography it began on November 5, 1914 - see Из истории Карпатской операции 1915 г. Сборник документов, Москва-Санкт-Петербург 2016, p. 7; F. Forstner, Fortress Przemyśl, Warsaw 2000, p. 147.

15 Z. Palski, op. cit., p. 240–241.

16 С. Г. Нелипович, Операции армий Юго-Западного фронта в ноябре-декабре 1914 года, Наступление 3-й и 8-й русских армий к Карпатскому и Бескидскому хребтам 30 октября – 19 ноября (12 ноября – 2 декабря) 1914 ⊠.- copy in the author's private archive.

17 This concept was presented by Dr. Wiesław Bator on September 12, 2015 in his speech during the popular

Above: Systematic photo reconnaissance in the fall of 1914 was carried out mainly on Galician railroads ... In the photo taken by the Flik 8 crew railway station in Krosno - source: Fotoarchiv SAW-KA.

science conference "Revival of Galicia, or the First Sanok Meeting with History" organized by the Galician Historical Society together with the Folk Architecture Museum and the Historical Museum in Sanok.

18 Z. Palski, op. cit., p. 240–241.

19 *The Great War ...,* op. cit., pp. 273,276-277; *The course of military operations in Western Galicia at the turn of 1914/1915, [in:] Local roads to independence*, ed. E. Juśko, Tarnów 2006, p. 15-16.

20 С. Г. Нелипович, op. cit.

21 M. Rauchensteiner, op. cit., p. 182; T. A. Olszański, J. Szponer, *II Brigade of Legions in the Carpathians*, [in:] *The First War ...,* op. cit., p. 111; *The Great War ...,* op. cit., pp. 277-278; A. Szelągowski, op. cit., p. 758, 766-768.

22 *The Great War ...,* op. cit., 273, 276-277; Великая.. Выпуск *V*-ый, op. cit., 11-12.

23 J. Bator, *The Galician War*, wyd. II, Kraków 2008, s. 171.

24 H. Magenschab, *Der Grosse Krieg. Österreich im Erste Weltkrieg 1914-1918,* Wien 2013, p. 102-104; М.Д. Бонч-Бруевич, Потеря наши Галици в *1915* году, Т. I, Через Карпаты в Венгрю зимою *1915* года, Москва 1920, passim.

25 G.A. Tunstall, op. cit., p. 49.

26 SAW-KA, Gruppe 157 1 g, Russland II. Sommer-Herbst Feldakten der oe.-u. Wehrmacht im Heeresarchiv Wien – teczka Przemyśl-Russen, p. 72.

27 SAW-KA/Luftfahrttruppen, Gruppe 157 1 g, Russland II. Sommer-Herbst Feldakten der oe.-u. Wehrmacht im Heeresarchiv Wien – teczka Przemyśl-Russen, p. 6–7.

28 Ibidem.

29 SAW-KA, Auszug aus dem Kreigstagebuch des Oberleutnant Feldpilot Wilhelm Schreiber der k.u.k. Fliegerkompanie No. 15 aus dem Weltkriege 1914-1918, p. 34. December 16, 1914 Oblt. Georg Edler von von Lehman from Flik 11. recognized the area of Radymno from the height of 2300 m, being heavily shelled by the Russian AFC.

30 SAW-KA, Die Tätigkeit der österr.-ung. Luftstreitkräfte beim Auffangen der rusischen Angriffsheere im Sommer und Herbst 1914. Bearabeitet in der „kriegswissenschaftlichen Abteilung der Luftwaffe", zweisgstelle Wien, 1942/1943 von Obst. Dr Oskar Regele, p. 6-7. The Flik 15.crew. Oblt. Fp. W. Schreiber and Zgsf. On December 16, 1914 J. Siegel recognized the Działoszyce-Słomniki region, detecting the concentration of Russian troops. On December 14, 1914,

Above: Former foot soldiers, cavalrymen, artillerymen, sappers in 1914 as KuK Luftschifferabteilung airmen in the eyes of journalists and newspaper readers became idols of the society of the Danube Habsburg monarchy…. In the photo, "Olympic gods", Flik 8 officers - Source: Fotoarchiv SAW-KA.

the Flik crew. The 7th recognized the Skała - Biskupice area and the railway lines south of it, detecting, together with the German FFA crew, 36 Russian columns in the Miechów and Wolbrom area.

31 Ibidem. December 17, 1914 Oblt. Fp. W. Schreiber with Oblt. Due to fog and running out of fuel, B. von Losonczy landed at the German FFA 31 field airfields.

32 Ibidem, p. 59. In the reports of the field command of the Imperial-Royal It was recorded in the 2nd Army *that no flights were performed between 12 and 17 December* [1914] *due to fog, precipitation, and cloud cover.* Report from the Flik crew. 10. of 24 December 1914 reported*: Pilzno - two bridges over the Wisłoka river; Rzeszów and Tarnów* - [Russian] *aviation parks;* in the report of December 26, 1914, despite the cloudiness, *the movement of trains in the area of Tarnów and Dębica was stated; a lot of rail transports in Tarnów; in Pilzno - three bridges over the Wisłoka River.*

33 SAW-KA, Auszug aus dem Kreigstagebuch des Oberleutnant Feldpilot Wilhelm Schreiber der k.u.k.

Left: In 1914, KuK Luftschifferabteilung airmen became idols of the society of the Danube Habsburg monarchy…. One of the Flik 8 officers in the photo - Source: Fotoarchiv SAW-KA.

Above: Already on September 12, 1914, when the Austrian army began to retreat, the fear was intensified by Russian planes shelling Rzeszów railway stations.... In the photo taken by the Flik 8 crew in autumn 1914, seen from a height of over 1000 m -Rzeszów and the city's communication routes - source: Fotoarchiv SAW-KA.

Fliegerkompanie nr 15 aus dem Weltkriege 1914-1918, p. 7-9, 10. On December 21, 1914 the crew of Flik 15. (Oblt. W. Schreiber and Zgsf. J. Siegel) made a flight to the area of Dąbrowa Tarnowska - Radomyśl - Mielec - Kolbuszów and back, and on December 23, 1914 to the area of Słomniki - Lipna Wola - Dąbrowa Tarnowska to the artillery command from Radlov.

34 A. Olejko, *Aviation episodes from the c.k. Tarnow*, Tarnow 2013, pp. 77-80; idem - Ł. Chrzanowski, A. Olejko, op. cit. "Berliner Tageblatt" in the issue of 18 January 1915 informed readers about the flights of Flik crews. 15. performed on Lohner B "Pfeilflieger" airplanes, praising the good results of reconnaissance flights, often performed from a height of less than 100 m, and the cooperation near Tarnów with the crews of the "Škoda" M 11. 30.5 cm mortar batteries.

35 SAW-KA, Radio-Artil. Schiesen eingeführt erstmals durch Flik 1. Artilerie vorbereitung der Durchbruchschlacht bei Gorlice. Frühjahr 1915 vom Hptm. Ritter Fiala v. Fernbrug, p. 2-3; A. Olejko, *Aviation...*, op. cit., pp. 55-56. *We fly from Brzesko in 4 airline companies* [Fliegerkompanie]. *With us - the proud bachelor,* [Oblt.] *Benno Fiala, the one who himself led the train with his company's belongings when it was suddenly surrounded by Cossacks. He ran under fire to the machine, replaced the dead mechanic, counterattacked, withdrew the train, firing at himself furiously with a service pistol. He is cheerful and often poses for photos with a cute white spitz named Muzzl. This is also how the "radio" airplane was baptized* [Oblt. M.] *Hesse*- Lt. K. Mroczkowski, K. Wielgus, *Fliegerkompanie 7. Krakow Air Company in the Great War*, Krakow 2014, p. 37.

36 As the pilot of this "radio" plane (Albatros B.I No. 7 "*MUZZL*"), which, having received radio equipment, was renamed the first radio-transmitting plane of the Great War, volunteered Oblt. M. Hesse of Flik 1., and the technical officer Flik was responsible for the efficiency of the device. 1. Lt. B. Fiala von Fernbrug. SAW-KA, Radio-Artil. Schiesen eingeführt erstmals durch Flik 1. Artilerie vorbereitung der Durchbruchschlacht bei Gorlice. Frühjahr 1915 vom Hptm. Ritter Fiala v. Fernbrug, p. 2-3; A. Olejko, *Aviation...*, op. cit., p. 77-80; M. Drożdż, *Małopolska in the fire of the Great War*, Kraków 2015, p. 121. On

Above: In the winter of 1914/1915, the Eastern Front froze ... The photo shows a "frozen" field altar of one of the army's Imperial-Royal troops - source: Fotoarchiv SAW-KA.

the initiative of Obst. M.E. Uzelac and prof. dr. L. Kanna was installed in an exceptionally short time from December 18, 1914 till January 10, 1915, just 16 km from the front line, on the Albatros B.I No. 7 "*MUZZL*" aircraft from Flik 1. at the field airport in Brzesko (Jasień), the first radio station to contact the crews cooperating with the services of the above-mentioned heavy artillery and ground transmitter- por. SAW-KA, Radio-Artil. Schiesen eingeführt erstmals durch Flik 1. Artilerie vorbereitung der Durchbruchschlacht bei Gorlice. Frühjahr 1915 vom Hptm. Fiala v. Fernbrug, p. 4–5, 6–7, 9.

37 SAW-KA, Improvisationen zur Bek⊠mpfung von Luftfahrtzeugen in der Festung Przemyśl 1914/1915 von Generalmajor Hans Schwab, p. 20; J. Magdoń, *Departure field in Biadoliny Radłowskie - on the 85th anniversary of the first combat flight*, „Zeszyty Wojnickie" 2000, no 4 (95), p. 23-25; *idem - Airplanes over Brzesko, Wojnicz and Tarnów (1914-1915) - facts and traditions,* [in:] *Brzesko-Okocim, air base of the Most Bright Lord,* edited by K. Wielgus, R. Panek, *Dębica 2000*, p. 34-36. The effect of the explosion of the "Škoda" M 11 mortar 30.5 cm was described by one of the soldiers of the Imperial-Royal army*: Poles of earth and smoke rose for a few minutes in the place where such a missile deigned to fall* - see M. Drożdż, op. cit., p. 160.

38 For every Flik (six officers, 100 non-commissioned officers and privates) there were six planes, 94 horses and 47 heavy carts for transporting disassembled planes- por. SAW-KA, Die Militärluftfahrt In Öterreich-Ungarn bis zum Beginn des Weltkrieges. Bearbeiter Oberst J. Diakow, Wien, p. 187..

39 SAW-KA, General Uzelac. Uzelac – der schöpner und die seele der Österichisch-Ungarischen Luftstreitkräfte, p. 50, 52.

40 SAW-KA Wien, Auszug aus dem Kreigstagebuch des Oberleutnant Feldpilot Wilhelm Schreiber der k.u.k. Fliegerkompanie nr 15 aus dem Weltkriege 1914-1918. Interesting photographic documentation, including The above-mentioned firing is presented by M. Tomaszewski in the album *Tarnów, The Great War 1914-1918* (Tarnów 2018).

41 SAW/KA, Manuskripte/Luftfahrtruppen, Radio-Artil. Schiesen eingeführt erstmals durch Flik 1. Artilerie vorbereitung der Durchbruchschlacht bei Gorlice. Frühjahr 1915 vom Hptm. Ritter Fiala v. Fernbrug, p. 1.

42 According to the field research (I. Banek, *42 cm fire station Küstenhaubitze M. 14 near* Tarnów, "Odkrywca" 2019, No. 10; https://gazetakrakowska.pl/pasjonat-historii-odkryl-miejsce-z-ktorego-armatagigant-bombardowala-tarnow / ar / c1-14187723) the position of the howitzer was within the boundaries of the village of Łętowice, in the Wierzchosławice commune.

43 Ibid, p. 2-3. W. Brzoskwinia and K. Wielgus (*Fliegerkompanie are going to war*, [in:] *From Galician take-off fields, the beginnings of air mail. Aviation*

Above: Requiem for the dying ... The photo shows the field cemetery of the Imperial-Royal Army in the Carpathian Mountains in the winter of 1914/1915 - source: Fotoarchiv SAW-KA.

traditions and monuments in the landscape of southern Poland, ed. K. Wielgus, R. Panek, Tarnobrzeg-Kraków 1999, p. 73–74) report that on December 27, 1914, an Albatros B.I plane, named "Muzzl", directed to Flik 1.at the airport in Brzesko, arrived at the Krakow-Rakowice fortress airport. According to these authors, the plane was equipped with a radio, the electricity of which was supplied by a 220-volt generator and the entire apparatus weighed over 60 kg, which is not confirmed by the archives. Wojniczer Wald is actually a forest and a small hill next to the railway line in Biadoliny Radłowskie. The M 14 42 cm heavy howitzer-mortar was stationed in the above-mentioned area (the other designation is the 42 cm KHL 15 caliber, 15 m long). There was a heavy coastal and field howitzer - it could fire 800 or 1000 kg projectiles. Before the outbreak of the Great War, only one copy was completed, the second was in production, which was installed in the Pula Fortress located on the Adriatic Sea. After the outbreak of the war, it was decided that the next examples would be used as field guns. The second produced unit was adapted to disassembly (the assemblies were bolted together - the assembly of the guns took about 40 hours. The finished second heavy howitzer was assigned to a specially created battery, which consisted of eight officers and 200 soldiers (including the direct crew of the cannon of 27 people), and entered combat in January 1915, firing at Tarnów, occupied by the Russian army. F. Czernin-Chudenitz (*Erster Einbau eines Funkgerätes in ein Flugzeug der K.u.k Armee* [w:] *Der Dolomitenfreund: Mitteilungen für unsere Mitglieder und Freunde*. Nr 3, Viena 1997) reports that in the area of Wojniczer Wald a battery of 42 cm heavy howitzers was stationed, and on both sides of the railway track two "Škoda" M 11 30.5 cm mortars, through which Tarnów was shelled, and Küstenhaubitze 42 cm M 14, manufactured by "Škoda "Was delivered from the Adriatic Pula. In the literature on the subject, the names of the "Škoda" M 11. 30.5 cm mortars are found as "Skinny Emma".

44 SAW-KA, Manuskripte/Luftfahrtruppen, Radio-Artil. Schiesen eingeführt erstmals durch Flik 1. Artilerie vorbereitung der Durchbruchschlacht bei Gorlice. Frühjahr 1915 vom Hptm. Ritter Fiala v. Fernbrugg, p. 4–5, 6-7, 9. The author of this relation states that the tests were to be carried out in connection with the planned new offensive of the Central Army in 1915 on the Tuchów-Zakliczyn-Gorlice-Nowy Sącz line. Lt. B. Ritter Fiala von Fernbrug, Oblt. L. Dumbacher and Oblt. P. Horwath conducted talks with General Janitschka [Josef Janečka], commander of the artillery during the Battle of Gorlice, to use the experience from the above-mentioned attempts in offensive operations.

Above: On the retreat route, the tsarist army left a large amount of its equipment ... In the photo, a gas apparatus for filling observation balloons captured in Jezierna in 1915 by the Imperial-Royal army - source: Fotoarchiv SAW-KA.

F. Czernin-Chudenitz (op. Cit.) Described the ground devices quite accurately and stated, inter alia, that the cooperation of the aviators with the mortar service was very effective and the first and primary goal was to destroy the branch of the railway line from Tarnów to the south, i.e. to the Biała valley and Tarnów railway station. The second missile hit a branch of the railway line, the fourth and fifth missiles hit the station and one was an unexploded bomb. He also states that after a change of firing direction, the missile hit the "Bankpalais" - the Russian headquarters, and the effects were photographed. This missile, on the night of January 15/16, 1915, made a large hole in the yard of the Bank Zastawniczy, at pl. Casimir the Great, and the command was at the Austro-Hungarian Bank, i.e. at ul. Mickiewicza 2 (today's City Hall).

45 SAW-KA, Manuskripte/Luftfahrtruppen – Improvisationen zur Bek⊠mpfung von Luftfahrtzeugen in der Festung Przemyśl 1914/1915 von Generalmajor Hans Schwab, p. 20. The information given by J. Borzęcki (*Moskale w Tarnowie*, Tarnów 1915, p. 24) is dated January 21, 1915 and it is possible that a "Czech mistake" was made when writing the number. W. Brzoskwinia and K. Wielgus (*Fliegerkompanie ...*, op. Cit.) State that on January 12, 1915, this plane with the crew: blt. M. Hesse and Oblt. L. Dumbacher went into action to destroy Russia's defensive positions and supply lines. A special fire system was set up around the town in the area of Biadolines Radłowskie, for the "Škoda" M 11.30.5 cm mortar battery, consisting of two separate semi-batteries in the mortars, an observation balloon landing field and a small airport for two aircraft, located at the railway station in Biadoliny Radłowskie. The underground shelter-dugout was the center where information obtained by air was provided, from here as well the fire coordinates for artillery were sent by telephone. This place is therefore associated with the first in the history of wars, directing the fire of the heaviest mortars from airplanes via the radio (!). The data on the location of the above-mentioned system is not confirmed by the SAW-KA Wien archives.

46 J. Magdoń, *Airfield in Biadoliny Radłowskie - on the 85th anniversary of the first combat flight,* [in:] "Zeszyty Wojnickie" 2000, No. 4 (95), pp. 23-25; J. Borzęcki, op. cit., p. 34. J. Magdoń, *Airplanes over Brzesko, Wojnicz and Tarnów (1914-1915) - facts and traditions,* [in:] *Brzesko-Okocim, air base of the Blessed Lord*, edited by K. Wielgus, R. Panek, Dębica 2000, p. 34-36. On that day, six more missiles fell on Tarnów. The date of the first use of heavy artillery near Tarnów can be postponed by a few more weeks, because in the memoirs of F. Szewczyk from the collection of J. Magdon there is a record of three missiles that fell next to the building of the Tarnów "Sokół" as early as December 25, 1914 and one missile who fell near the Uhlan barracks on the next day - December 26, 1914. So they were the first shots of the "Škoda" 30.5 cm M 11 mortars from Biadolines Radłowskie, but without observing the fire from the air. Moreover, J. Borzęcki (op. Cit.) And F. Szewczyk (op. Cit.) In their memoirs give the actual beginning of the shelling of the city on

Above: Pre-war lineup of early production Lohner B.I series 11 aircraft. In February 1915 *Blitz* was assigned serial 11.15 and *Barbar* was assigned serial 11.06. Source: Aeronaut.

Below: Turkish allies… In the photo, Turkish airmen at the Flik 32 airport - Source: Fotoarchiv SAW-KA.

Left: On As time passed, better and better flight uniforms found their way to the field Fliegerkompanien…. One of the Flik 32 pilots in the photo in a flight suit - Source: Fotoarchiv SAW-KA.

Above: During the Great War, out of a group of over 2,800 officers, non-commissioned officers and rank-and-file officers in aviation, losses amounted to 1,615 people (57%) .... From this number, 627 members of the flying personnel were killed or missing (including 289 officers), 938 were injured (including 518 officers). The photo shows the Flik 32 flying staff. How many of them survived? - source: Fotoarchiv SAW-KA.

Below: And so it used to be ... During the landing at the field airfield on the Alpine air front, the crew of Flik 32 lacked a bit of luck The photo shows a crashed Hansa Brandenburg C.I. Source: Fotoarchiv SAW-KA.

Above: There was no air happiness ... In the photo the Brandenburg D. I fighter plane from Flik 32 crashed at the field airfield on the Alpine front - source: Fotoarchiv SAW-KA.

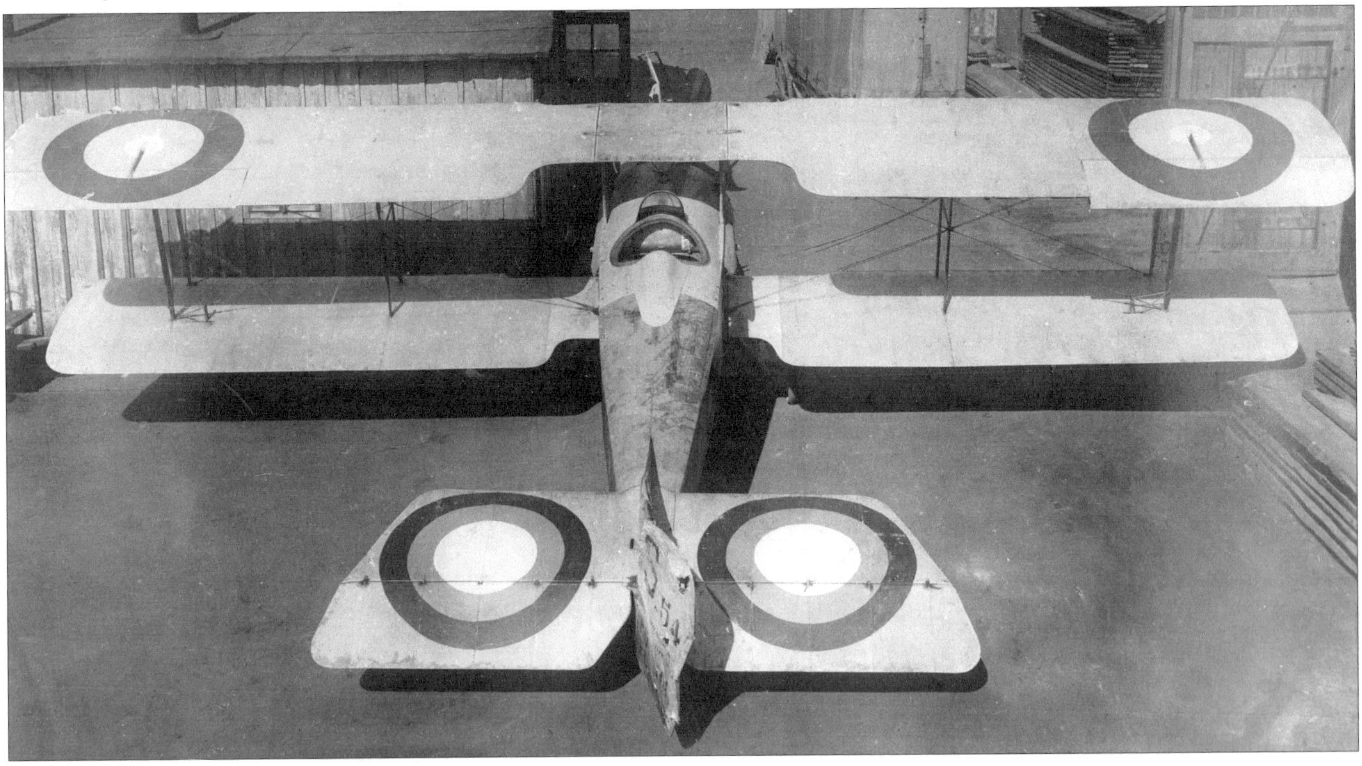

Below: Military prey ... Over time, the Russian air force used better and better French-made flying equipment. The photo shows a Russian type fighter SPAD SA.4 S-54, which on November 27, 1917, after a forced landing near Lviv (visible rudder shot in an air fight), was in the hands of Flik 14 pilots. Different, two-way nationality markings visible on the panels and on the horizontal tail - source: Fotoarchiv SAW-KA.

Above: A war requiem for the dead ... Cemetery of the soldiers of the Great War at Toruńska Pass - status from 2015 - source: author's collection.

January 14, 1915, i.e., two days from the first combat flight of the Albatros B. No. 7 "Muzzl" with radio on board (it is interesting that two years later, during the first Polish aviation courses, this device was called a wireless telegraph) - Lt. M. Sikorski, *Wireless telegraph, and aviation,* [in:] *First aviation courses. Memorial book edited by the technical director of the courses,* Eng. Zygmunt Dekler, Warsaw 1917, p. 55-56. K. Wielgus (op. Cit.) States that because of the first missiles, the Russian command ordered the blackout of Tarnow and the stopping of trains to the east of the city, in Wola Rzędzińska.

47 SAW-KA, Manuskripte/Luftfahrtruppen, Auszug aus dem Kreigstagebuch des Oberleutnant Feldpilot Wilhelm Schreiber der k.u.k. Fliegerkompanie nr 15 aus dem Weltkriege 1914-1918, p. 18–20.

48 SAW-KA, Manuskripte/Luftfahrtruppen, Radio-Artil. Schiesen eingeführt erstmals durch Flik 1. Artilerie vorbereitung der Durchbruchschlacht bei Gorlice. Frühjahr 1915 vom Hptm. Ritter Fiala v. Fernbrug, s. 8; F. Czernin-Chudenitz, op. cit.

49 SAW-KA, Manuskripte/Luftfahrtruppen, Radio-Artil. Schiesen eingeführt erstmals durch Flik 1. Artilerie vorbereitung der Durchbruchschlacht bei Gorlice. Frühjahr 1915 vom Hptm. Ritter Fiala v. Fernbrug, s. 9.

50 Ibidem, p. 8; F. Czernin-Chudenitz, op. cit, pp. 10–11. J. Borzęcki (op. Cit.) Noted in his diary that from January 22 to February 18, 1915, Tarnów was not shelled, but F. Szewczyk (op. Cit.) Recorded several missiles during this period.

51 J. Magdoń, *Airplanes ...,* op. cit. From February 1915, the Flik crew. 10. made flights on airplanes equipped with radio stations, correcting the fire of heavy artillery shelling Tarnów, and carried out direct bombardments of Russian positions (including in Tarnów itself) using special chutes adapted to the discharge of bombs and grenades.

52 SAW-KA, Manuskripte/Luftfahrtruppen, Die durchbruchsschlacht von Gorlice 1915. Bearbeitet von Obst. A. D. Feldp. Erich Kahlen, p. 1.

53 SAW-KA, Manuskripte/Luftfahrtruppen, Auszug aus dem Kreigstagebuch des Oberleutnant Feldpilot Wilhelm Schreiber der k.u.k. Fliegerkompanie nr 15 aus dem Weltkriege 1914–1918, p. 34–38.

54 SAW-KA, Manuskripte/Luftfahrtruppen, Radio-Artil. Schiesen eingeführt erstmals durch Flik 1. Artilerie vorbereitung der Durchbruchschlacht bei Gorlice. Frühjahr

Above: The Frozen Front in the Carpathians in the winter of 1914/1915.... The illustration from the period shows the war cemetery no. 55 in Gładyszów, designed by Dušan Jurkovič- source: R. Frodyma's collection

1915 vom Hptm. Ritter Fiala v. Fernbrug, p. 11–12.
55 SAW-KA, Manuskripte/Luftfahrtruppen, Der feldzug gegen Russland 1915, p. 7.
56 J. Magdoń, *Airplanes ...*, op. cit. pp. 36–37; J. Borzęcki, op. cit. pp. 34–45. After an almost month-long break, on March 25, 1915, shots were fired at ammunition carts standing in front of one of the schools (over 30 people died).
57 SAW-KA, Manuskripte/Luftfahrtruppen, Der feldzug gegen Russland 1915, p. 4, 6.
58 SAW-KA, Manuskripte/Luftfahrtruppen, Auszug aus dem Kreigstagebuch des Oberleutnant Feldpilot Wilhelm Schreiber der k.u.k. Fliegerkompanie nr 15 aus dem Weltkriege 1914–1918, p. 21, 34.
59 M. Kromkayowa, op. cit., p. 20.
60 Fr. K. Wais, *Memoirs of the First Year of World War*, Lviv 1930, p. 65-66.
61 M. Kromkayow, op. cit, p. 23, 27, 29.
62 AMBL in Sanok, *Helena Schramówna ...*, op. cit., p. 63.
63 SAW-KA, Manuskripte / Luftfahrtruppen, Auszug aus dem Kreigstagebuch des Oberleutnant Feldpilot Wilhelm Schreiber der kuk Fliegerkompanie No. 15 aus dem Weltkriege 1914–1918, p. 21–23.
64 Ibid, p. 25.
65 Ibid, p. 26.
66 F. Molnár, *Galicia 1914-1915*, Warszawa 2012, p. 139. On December 30, 1914, the AOK developed the KuK Luftschifferabteilung oleats for pilots for the Przemyśl Fortress, with two spare landing sites in the event of the shutdown of the fortress airport in Żurawica under fire by the Russian artillery. They were in the Lwowskie district, between the town and the Wiar riverbed, and between and along the San Bend, between Buszkowice, Buszkowiczki, Potoczek and Wielki Budy. According to the findings of Franz Czernin von Chudenitz based on the SAW-KA archives, the following flights to the Przemyśl Fortress took place in 1914: On October 1, 1914, the Flik crew made a flight on the Nowy Sącz-Przemyśl route. 14. composed of Oblt. A. Tauszig and Hptm. Raabl; On 12 December 1914, the flight from Košice to Przemyśl was made by the Flik crew. 11.with a package of three kg. materials for Hptm. Fp. Philip Blaschke Ritter von Zwornikkirchen; December 16, 1914 Rtm. von Lehman of Flik 11. made a flight on the Kraków-Przemyśl route, delivering 210 kg of cargo, including mail; On December 16, 1914, Oblt made a failed flight from Kraków to Przemyśl. Fp. Jozef Toth from Flik 8. forcibly landing on the way; On December 19, 1914, the Flik crew. 10. composed of Oblt. Eduard Rzemenowski von Trautenegg and Oblt. Ferdinand Reischer made a reconnaissance flight on the Kraków-Przemyśl route without the mail (!); On December 24, 1914, the Flik crew. 10. composed of Oblt. Father Schmoczer von Mosenzeff and Oblt. Martinek made a reconnaissance flight on the Krakow-Przemyśl route without mail! In the light of the findings of A. Clement, in 1914 the following flights were performed: October 1, 1914, one flight to Przemyśl without post office; On October 6, 1914, one flight from Przemyśl ended with a forced landing in Jasło; 23 November 1914, one flight from Przemyśl to Kraków; December 9, 1914 one flight from Przemyśl ended with a landing in Hungary; December 11, 1914, one flight to Przemyśl; December 16, 1914, one flight to Przemyśl; December 19, 1914, one flight to Przemyśl; December 24, 1914 two flights to Przemyśl; On December 26, 1914, two flights from Przemyśl to Krakow - see G. Schenk, Przemysl 1914-1915. Monographie über ein historisches Ereignis. Flieger und Ballonpost der Festung Przemysl. Handbuch unf Katalog, Wien 2003, p. 105.
67 SAW-KA, Manuskripte – Die durchbruchsschlacht von Gorlice 1915. Bearbeitet von Obst. A. D. Feldp. Erich Kahlen, p. 1–2.
68 Ibidem, p. 27.
69 Ibidem.
70 Росийский государственный военно-исторический архив Москва (РГВИА), Ф. 2067. Оп. 1. Д. 140. Л. 228-229 – Из истории Карпатской…, op. cit., pp. 203-204 - report of 10 (23) March 1915.
71 SAW-KA Wien, Manuskripte/Luftfahrtruppen, R. Grutschnig, Unser letzer Tag und Flug Flug aus Przemyśl, geschriebe in sibirischer Gefangenschaft 1916; J. Zahalka, *Posledni let z Premyślu*, „HPM" 2001, nr. 1; F. Forstner, op. cit., s. 166, 180.
72 SAW-KA, Manuskripte/Luftfahrtruppen, Auszug aus dem Kreigstagebuch des Oberleutnant Feldpilot Wilhelm Schreiber der k.u.k. Fliegerkompanie nr 15 aus dem Weltkriege 1914-1918, p. 45–55.
73 Fr. K. Wais, op. cit., p. 68.
74 More broadly, air operations over the Przemyśl fortress, the author presented in the study Karpacka war ..., op. cit., and in the periodical "Przemyski Yearbook - Military History" -*From air fights ...*, op. cit.Zgsf. R. Meltsch (born on May 10 or 11, 1891 in Loučka near Nový Jičín [German Neutitschen]) joined the Imperial-Royal Air Force from the Festung Artilerie Regiment (FAR) 1. keeping the name of his Fw. For flights in Flik The 8th commander of the defense of Przemyśl awarded him in October 1914 with the Gold Medal for Courage, and later the pilot was awarded the badge of a field pilot (order PVB 1 / Nr.37 of 13 March 1915). From 22 March 1915, he was listed in the statistics of the KuK Luftfschifferabteilung as missing. He was captured on March 22, 1915 near Rymanów, and joined the Czechoslovak legions on March 15, 1916 in Yekaterinburg, and on June 26, 1916, he was assigned to the 1st Rifle Regiment. He graduated from the school of non-commissioned officers in Kiev and officer school in Borispol, he was in the rank of Cpt. commander of air units organized by Czechoslovak troops in Russia.

In the meantime, he changed his surname to a more Slavic-sounding Melč and married a Russian woman. With troops of the Czechoslovak Corps, he withdrew to Siberia and returned to the country via Trieste from Vladivostok in 1919. He served as a captain until February 26, 1920. On April 29, 1920, as Major of Aviation, he joined the Czech 1st Aviation Regiment in Prague, from which he was demobilized on February 16, 1921 - data from the private archive of M. Plavec.

75 РГВИА, Ф. 2067. ОР. 1. Д. 140. Л. 214-215, Д. 2601, Л. 154, Л. 158-160 -Из истории Карпатской ..., op. cit., p. 199–200, 202 - report of 9 (22) March 1915

76 РГВИА, Moscow, Ф. 2067. ОР. 1. Д. 139. Л. 311 - Из истории Карпатской операции 1915 г. Сборник документов, Москва-Санкт-Петербург 2016,p. 122-123 - report of February 15 (28), 1915

77 РГВИА. Ф. 2148. ОР. 2. Д. 63, Л. 56, 73–76.

78 S. Nielipowicz, The Carpathian Operation (Battle of the Carpathians) - January-April 1915: *a brief outline of the activities of the Southwestern Front*, [in:] *Bieszczady Lions 1914/1915. Sketches of military operations in the Beskids during the Great War*, edited by A. Olejko, K. Rejman, Rzeszów 2020 - passim.

79 J. Magdoń, Airplanes ..., op. cit., pp. 38–39. F. Forstner (op. Cit., P. 158) reports that from the airport near Buszkowice the flights were performed by pilots servicing Aviatik [BI] 317 and Albatros [BI] HP 100 aircraft with 100 HP engines.

80 Ibid.

81 Hungarians in the Przemyśl Fortress in 1914-1915, edited by I. Lagzi, Warsaw-Przemyśl 1985, p. 117.

82 "Przemyśl Land" from January 14 and 18, 1915.

83 Fragments of S. Gayczak's diary, compiled by Jan Rożański, were published in "Życie Przemyskie", numbers 8-16 (February-April 1970). Gazeta Podhalańska from January 31, 1915 announced the launch of air communication with the Przemyśl Fortress by airplanes and about the air battle - cf. K. Wielgus, W. Brzoskwinia, op. cit. in which both Imperial-Royal and Russian were killed, which, however, the Viennese archives do not confirm.

84 "Przemyśl Land" from January 24, 1915.

85 H. Jabłońska née Seiferts, *Diary from besieged Przemyśl 1914-1915*, Przemyśl 1994 - passim. This entry and the place where the bombs were dropped are almost identical to the note in "Ziemia Przemyska" regarding the bombing of January 18, 1915, so it may be the same event, and the Russian bombs hit ul. Green twice.

86 Ibid.

87 Ibid.

88 This subject has been described by the author in the studies *1918 - postal airline Vienna-Krakow-Kiev-Vienna*, [in:] Перша світова війна: історія та уроки. 1914-1918 pp., Київ 2015 iPeople writ*e letters ... Postal airline Vienna-Krakow-Kiev-Vienna in the light of the archives of the Osterreichisches Staatsarchiv-Kriegsarhiv Wien*, "Przemyski Yearbook - Military History" 2015, n. 51, Historia z. 1 (13) 2015.

89 G. Schenk, op. cit .; data from the private archive of G. Penz via findings of Franz Czernin von Chudenitz based on SAW-KA archives.

90 Ibid, A. Clement, Handbuch der Militär Luftpost, Graz 1955 - passim. F. Forstner (op. Cit., P. 158) states that on January 5, 1915 the first air mail departed from Przemyśl, but due to the small loading capacity of the planes, only 20 kg of cargo could be taken with four postal flights. K. Wielgus, W. Brzoskwinia (op. Cit., Pp. 108–111) not based on the SAW-KA Wien archives state that: On January 12, 1915, it took off from the Kraków-Rakowice fortress airport to Przemyśl - the Żurawica fortress airport and that day he returned, the plane with the crew of Flik 10. composed of: Oblt. Father Schmoczer von Mosenzeff and Oblt. Emil Kumstat; On January 18, 1915, three planes took off from the fortress airport: Oblt. A. Schindler piloted the Albatros B I plane with its own name "Gabi" from Flik 10. and the crew of the second plane were pilot Oblt. Father Schmoczer von Mosenzeff and Oblt. E. Kumstat and Oblt. Fp. Eng. R. Stanger of Flik 8. he probably piloted a Lohner C "Gebirgspfeilflieger" airplane (these were officially recognized postal flights on the Przemyśl route as 6, 7 and 8); On February 1, 1915, the Albatros B I plane, piloted by Oblt, arrived in Przemyśl. A. Schindler from Flik 10. these flights are not confirmed by F. Czernin von Chudenitz in his list based on the SAW-KA Wien archives. Schindler from Flik 10. these flights are not confirmed by F. Czernin von Chudenitz in his list based on SAW-KA Wien archives. Schindler from Flik 10. these flights are not confirmed by F. Czernin von Chudenitz in his list based on SAW-KA Wien archives.

91 "Przemyśl Land" from January 24, 1915.

92 SAW-KA, Als Flieger In Przemysl von Oberst Feldpilot Nikolaus Wagner Edler von Florheim, p. 20.

93 J. Zahalka, op. cit. J. Adamczyk, M. Chodurek (op. cit.), not based on the SAW-KA Wien archives, state that: on February 7, 1915 the ninth postal flight took place on the route Przemyśl-Kraków-Rakowice fortress airport and the flight crew, which carried 130 kg of mail were Oblt. Georg Edler von Lehman from Flik 11. and Hptm. Rankenberger; the tenth flight on the route from Przemyśl to Jasień was made on 10 February 1915 by the crew of the Oblt. A. Schindler from Flik 10 and Oblt. Ignatz Kozma of Flik 11; 11 postal flight on the Przemyśl-Brzesko route was made on February 12, 1915 by Oblt. G. Edler von Lehman from Flik The 11th and the twelfth flight from besieged Przemyśl was made on February 18, 1915 by the crew of Lt. B. Fiala von Fernbrug from Flik 1. and Zgsf. R. Meltsch of Flik 8. who previously had to bring an Aviatik B.I plane to Przemyśl;

94 SAW-KA, Fliegerkompanie 10 - Okt. 1914-2 May 1915. Studies. Bearb. Obst. Fp. AD Erich Kahlen, p. 9.

95 SAW-KA, Manuskripte / Luftfahrtruppen, Auszug aus dem Kreigstagebuch des Oberleutnant Feldpilot Wilhelm Schreiber der kuk Fliegerkompanie No. 15 aus dem Weltkriege 1914-1918, p. 18-20.

96 Ibid.

97 Ibid. This flight is not included in F. Czernin von Chudenitz's list.

98 This flight is included in the list by F. Czernin von Chudenitz, however, information in the Polish literature on the subject says that after taking off for the return flight, the plane was shelled by Russian troops besieging the Fortress of Przemyśl - a hit in the fuel tank meant that the pilot had to forcibly land, and the plane burned down after the crash and the crew managed to return to the fortress - this author is not included in the list.

99 SAW-KA Wien, Manuskripte / Luftfahrtruppen, R. Grutschnig, Unser letzer Tag und Flug Flug aus Przemyśl, geschriebe in sibirischer Gefangenschaft 1916. J. Zahalka (op. Cit.); F. Forstner (op. Cit., Pp. 166, 180). Departure of six planes from Flik The 11th from Przemyśl is dated in the literature on March 18 or 19, 1915. According to W. Brzoskwini and K. Wielgus (op. Cit.), Four planes took off: Aviatik B.I named "Baba", piloted by Oblt. Sesergo and Albatros B.I planes with crews - Oblt. Father Schmoczer von Mosenzeff and Oblt. A. Schindler and Oblt. F. Reischer and Lt. F. Junker of Flik 10, and five Perseval-Sigsfeld Drachenballon M 98 balloons, prepared for free flight, were launched into the air. F. Stuckheil (Second siege of the fortress Przemyśl. III. Koniec, Przemyśl 2010, pp. 47–52) states, that the airship "*Przemyśl*" landed near Nowogrod Wołyński, the airship "*Austria*" and the airship "*Erz.* [herzog] *Josef Ferdinand*" - near Brześć nad Bugiem, the airship "*Schicht*" near Sokal - the crews were captured by the Russians. The airship "*Steiermarkt*" was destroyed on take off. The reason for the rumors in the fortress was the mistress of the commander of the 23rd Honveda DP and two ladies from the underworld, "pilot princesses", whose conduct caused a general scandal, and were supported by pilots stationed in Przemyśl. There was a rumor among the soldiers that the fortress commander escaped from Przemyśl by plane - see F. Forstner, op. cit., p. 180.

100 F. Stuckheil, op. cit., pp. 47–52. The airship "*Przemyśl*" landed near Nowogród Wołyński, the "*Austria*" airship and the "*Erz. Josef Ferdinand*" - near Brest on the Bug, the airship "*Schicht*" at Sokal - the crews were captured by the Russians. The airship "*Steiermarkt*" was destroyed on take-off.

101 SAW-KA, Manuskripte Luftfahrtruppen, Als Flieger In Przemysl von Oberst Feldpilot Nikolaus Wagner Edler von Florheim, p. 20.

102 Ibid.

103 SAW-KA, Manuskripte Luftfahrtruppen, Auszug aus dem Kreigstagebuch des Oberleutnant Feldpilot Wilhelm Schreiber der kuk Fliegerkompanie No. 15 aus dem Weltkriege 1914–1918, p. 42, 43. There is no information about the Przemyśl flights by F. czernin von Chudenitz. Oblt. Fp. W. Schreiber.

104 SAW-KA, Als Flieger In Przemysl von Oberst Feldpilot Nikolaus Wagner Edler von Florheim, op. cit.

105 F. Molnár, op. cit., p. 146.

106 Памятная записка Свиты его величества генерал-майору графу Граббе. Взятие крепости Перемышль, [in:] Великая война 1914–1918, Выпуск 7, Москва 2018, p. 123–124.

107 Fr. K. Wais, op. cit., p. 68.

108 SAW-KA, Manuskripte Luftfahrtruppen, Auszug aus dem Kreigstagebuch des Oberleutnant Feldpilot Wilhelm Schreiber der kuk Fliegerkompanie No. 15 aus dem Weltkriege 1914-1918, p. 42, 43. There is no information about the Przemyśl flights by F. czernin von Chudenitz. Oblt. Fp. W. Schreiber.

109 J. Zahalka, op. cit.

110 J. Zahalka, op. cit.

111 SAW-KA, Auszug aus dem Kreigstagebuch des Oberleutnant Feldpilot Wilhelm Schreiber der KuK Fliegerkompanie No. 15 aus dem Weltkriege 1914-1918, p. 40, 41–42. These are the words of the last letter from D. Wagner, serving in Festungballonabteilung 1/3. R. to living in Viennawives: Dearest Lotto on March 22, 1915 at 12:50 pm. I lay down ten hours to get some sleep, but I couldn't sleep. The guns thundered and flashed so that the windows of our camp were smashed. It has never been like today, all day, all night, but soon there will be a change. We must get up at half past three in the morning to send the plane to Krakow. *Most of us stay awake due to strong thunder and flashes. You can't write to me anymore because we are going far to Przemyśl. Your D.* see J. Zahalka, op. cit

112 J. Magdoń, Airplanes…, op. cit. p. 150. F. Forstner (op. cit., p. 159) states that technical faults and weather conditions caused min. loss of 12 planes, while seven pilots and seven observers were captured or killed. The literature on the subject contains information that in the period of November 7, 1914–22 March 1915, 33 courier flights to and from the besieged Przemyśl Fortress were carried out, carrying many valuable military documents and civil correspondence.

113 SAW-KA, Manuskripte Luftfahrtruppen, Als Flieger In Przemysl von Oberst Feldpilot Nikolaus Wagner Edler von Florheim, op. cit. Oblt. Fp. ing. R. Stanger in the later period of the Great War was commanded by Flik 18th and the Flik hunting lodge. 14. J.

114 J. Magdoń, Airplanes …, op. cit., p. 38–39.

115 F. Forstner, op. cit., p. 129.

116 S. Pilecki, B. Skąpski, On the fronts of World War I. Memoirs, Warsaw 2015, p. 65–70.

Above: Fokker B.I 03.19, a type M10, in flight. Iron cross national insignia have been applied over white backgrounds on the wings and rudder; it retains the red and white elevators. It was assigned to *Flek* 6 at Wiener Neustadt. (Peter M. Grosz collection/STDB)

Below: Another view of Fokker B.I 03.19. (Peter M. Grosz collection/STDB)

Above: Fokker B.I 03.19, a type M10, in winter flight.It has wingtip skids (Peter M. Grosz collection/STDB)

Below: Fokker B.I 03.3X has a 7 cylinder engine and wingtip skids. (Peter M. Grosz collection/STDB)

Gorlice Breakthrough. Aviation in the Tarnów-Gorlice Operation in 1915

April 29, 1915. We've been watching for three days [in Biecz] *areoplanes* [airplanes - planes - spelling consistent with the original throughout the quoted text] *flying over us. The Saldat* [soldiers of the tsarist army] *shoot with handguns and polymotys* [HMGs] *because they are Austrian* [Austro-Hungarian] *and Germanic. Sensation! Areoplanes* [airplanes - planes] *throw bombs. In Górne Przedmieście* [in Biecz] *one fell and plunged into the ground. This displeased the soldiers, and they ran away with the horses. But we also look at areoplanes* [airplanes - airplanes] *with concern*[1].

This is what M. Kromayowa, living in Biecz, wrote in her diary. The fights in the Carpathians aimed at the deblocking of the Przemyśl Fortress by the field 2nd and 3rd armies, in the winter of 1914/1915, took a huge number of victims, in which the Austro-Hungarian, German and Russian armies lost over a million soldiers killed or wounded, frozen and missing (only during the so-called Easter Battle, the Russian army - Southwestern Front - lost about 40,000 soldiers). Due to the critical situation on the Nord-Ostfront, the help of a German ally turned out to be indispensable for the AOK. The head of the German General Staff, General E. von Falkenhayn, at the end of March 1915 ordered to develop a plan for a possible transfer from the Western Front of three Army Corps (six Infantry Corps) possibly to the Nowy Sącz-Tarnów line, and on April 1, 1915 a German liaison officer at the AOK informed OHL that half of the 2nd c.k army must be withdrawn from the front due to its difficult situation. Gen. F. Conrad von Hötzendorf urgently needed the support of German forces subordinate to AOK to strengthen it. Initially, it was considered to use about four German Infantry Corps to attack the Gorlice-Tarnów section to break the Carpathian front of the Russian army, but General E. von Falkenhayn was ready to send four army corps of eight divisions to Galicia. A joint attack in the above-mentioned section was to permanently change the situation on the Eastern Front. On April 14, 1915 in Berlin, during an official conversation between the heads of the AOK and OHL, a decision was made on how to launch the offensive and force the Russian army to withdraw from Western Galicia. The newly established German field 11th army was to be transported by rail to the area of the Austro-Hungarian field defense of the 4th Army, south-east of Krakow, to break the front line of the Russian 3rd Field Army. The head of AOK assured that he would strengthen both armies with strong infantry divisions (two for each army), the allied 11th field army, additionally supplying at least one DK, heavy artillery, and light transport equipment... Preparations for the Gorlice Breakthrough began, in which ... aviation became the supplier of information about the enemy!

At the end of July 1914, the status of the Imperial-Royal Air Force amounted to 39 aircraft and 85 airmen, which were divided into 13 part-time Fliegerkompanien, seven of which went to the Nord-Ostfront and two to the Süd-Ostfront. As a result of significant purchases of aircraft in allied Germany, the amount of flying equipment (in view of the considerable losses suffered) increased gradually, and moreover, while in the summer of 1914, the Lohner factory was the only company in the Austro-Hungarian Empire that was able to mass-produce aircraft of by 1915 the number of factories had increased from one to seven (Lohner, Albatros, Aviatik, Lloyd, Ufag, Oeffag and Fischamend). At the turn of 1914/1915, there were 14 Fliegerkompanien at the front, including ten on the Nord-Ostfront. A total of 147 aircraft - including training aircraft - were ready for combat use, including: 64 Lohner B, C "Pfeilflieger", 41 Aviatik B.I, 22 Albatros B.I, 13 D.F.W. B.I, one Lloyd type and another six different types of machines[2].

The Memorable Aviation Spring 1915

After three Carpathian battles at the turn of January, February, and March 1915 and the so-called In the Battle of Easter 1915, pas the Carpathians turned into a gigantic "winter cemetery". As part of the long-distance reconnaissance, from the beginning of April 1915, the Flik crew. 15, having considerable experience in conducting remote reconnaissance, wandered deeper and deeper into the Russian hinterland. On April 2, 1915, a radio flight from the Jasień airport was performed by a crew of Hptm. Hautzmayer and Zgsf. J. Siegel and on 3 April 1915 the same flight was performed by the crews of the Oblt. B. von Losonczy and Oblt. Cserich and Hptm. Robert Eyb and Zgsf. J. Siegel. The next reconnaissance flights were carried out on April 4, 1915 and April 9, 1915. Flik 15 made a long flight to the Jasło-Sieniawa area, directing the crew of the Hptm. Hautzmayer and Zgsf. J. Siegel, which took off despite being completely cloudy, but after a 1½ hour flight in a snowstorm, the pilots returned to the field base in Jasień3. On April 7, 1915, in the southern section of the Russian 3rd Army, heavy shelling of enemy heavy artillery was recorded, and in the following days, a large activity of enemy aviation was observed around its stationing, along with the bombing of railway stations4. The intensity of April flights from the airport in Jasień was gradually stabilizing, although it depended on the weather conditions. On April 14, 1915, the Flik crew. 15 composed of Oblt. B. von Losonczy and Oblt. J. Wierzejski was heavily shelled over Tarnów, and on April

17, 1915, Oblt. Fp. W. Schreiber from Flik 15 made a flight to the area of Pilzno-Dębica-Frysztak-Krosno-Jasło, in Jasło, detecting seven tent hangars at the Russian field airport in Lipie. Unfortunately, the weather conditions in this period were changeable and the cloud cover made reconnaissance flights impossible on 18 and 19 April 19155.

Oblt. Fp. W. Schreiber from Flik 15 has recalled the April reconnaissance flights from the period preceding the Tarnów-Gorlice operation:

April 6 [1915] *(Tuesday).* [Oblt. A. von.*] Mandelsloh got a letter today from Lt. Lord's sister, in which she reported that Lord was also wounded and was in the same hospital as* [Lt.] *Bratmann* [March 21, 1915, Lt. Bratmann and Lt. Lord of Flik 15 did not return from a combat flight, forcibly landing near Dębica and being captured by the Russians]. *Since my Albatros* [BI] *is under repair, I flew today in the Lohner* [B] *Pfeilflieger C-48 making a flight* [to the region of] *Miechów-Wola Rzędzińska. I took two photos and returned to the base* [in Jasień*]. Suddenly I noticed that I was attacked by an enemy plane following me. I didn't have any weapons with me. We passed Dunajec and the chase continued with me. A Russian or a French* [Russian AO and KAO used French-made planes], *or whoever it was - did not give up. Eventually I was hit by a burst of his machine gun, because suddenly I heard a bang, the engine stopped, and I started to fall. However, I did not lose consciousness of mind. With the help of the elevator, I smoothed the fall and turned it into a diving flight. I also managed to land on a farmland near the German airmen* [FFA 57.] *near Biadolines* [Szlacheckie]. *There was, of course, no question of landing in accordance with the "art of landing". I hit the ground hard, even though I picked up the machine at the last moment. I touched the ground first with my right wing, then with my left wing, which had completely torn off me, and then "drilled" with the propeller into the ground. I sat in a daze for a moment, and then in the first move I reached for my cigarette case. I lit a cigarette to calm down. I sat like this for about 15 minutes. Suddenly I heard a buzzing noise above me. Go to hell, is this Russian still not wanting to leave me alone? - no, it's Austrian. At first, he circled over me for a while, I waved him, and he started to approach for landing. To my great joy,* [Oblt. J.] *Wiezejski* [Wierzejski]. *I told him what had happened and how lucky I was. He shook my hand and congratulated me once again on surviving the first air fight (or being shot down).* [Zgsf. F.] *Benda, who was with me, took a photo of all of us, which is attached below* (this photo is missing from the archival material - AO), *and flew towards the base* [in Jasień*], promising that he would arrange everything. I also told him to send only a crewed technical rescue car; I, however, will go to the German airmen* [FFA 57.] *to Biadolines* [Szlacheckie] *and I should get a car there. Then we shook hands again and he flew away. Of course, the Germans welcomed me joyfully and congratulated me. I was fed, and then captain (Hptm.) Heyder gave me a car to drive me to my base. This is how that memorable day ended for me ... From then on, I took a machine gun with me on the plane. After all, I didn't want to get shot down, unable to defend myself!*

April 9 [1915] *(Friday). Quite a fruitful reconnaissance flight on the Jasło-Sieniawa line. The Russians built a fort near Sieniawa* [field fortifications], *which, it seems, should be taken seriously. April 9th. It's totally cloudy.* [Oblt.] *Hautzmayer and* [Zgsf. J.] *Siegel started anyway, but after 1.5 hours they had to turn back because of a snow storm. At 4 o'clock in the afternoon the engineer Schiske* [Schicke] *from the company Werner und Pfleiderer in Vienna, the inventor of the Schicke carburetor in Lohnerach* [B] *Pfeilfliegerach, came to us.*

April 17 (Saturday) [1915]. *Long flight through Pilzno, Dębica, Frysztak, Krosno and Jasło. Air quite foggy, visibility up to about 20 km. Slight shifts of troops from Pilzno to Tarnow and from the east towards Pilzno. In Frysztak, a train going* south (100 axes), in Jasło a lot of rolling stock and 7 hangar-tents [field airport]. *Today's flight was in the gustiest wind I have experienced so far. It was a very strong headwind, so I thought I would run out of gas before I could get back to the base* [in Jasień]. *Besides, the trash was rising miserably; over Tarnów it was only at 1300 m. I was under heavy fire there, both during the flight to and from. So far, I have covered 15,254 km in 138 hours*[6].

From the Western Front to Galicia, at the same time by rail transports led to confuse the Allied intelligence in various directions through the territory of Germany, the field troops of the German 11th Army were transferred. It was to be the main strike force in the Tarnów-Gorlice operation of May 1915, when the front of the Russian 3rd Army was interrupted, and then the entire Russian South-Western Front retreated to the east In connection with the situation on the Nord-Ostfront, when the intensity of air reconnaissance at the beginning of April 1915 in the area of the field fights of the Imperial-Royal The 4th and 3rd Army grew significantly, deployed on the left flank of the field front of the Imperial-Royal The 4th Army was subordinate to Flik 8, 10, and 15 while righteous neighbor, c.k. 3rd Army, had only one Flik 13. In accordance with the AOK order, the entire area from the Dunajec estuary to Zborov was entirely photographed from the air. In April and May 1915, the average condition of the planes and airmen at Fliegerkompanien on the Nord-Ostfront was as follows: Flik 10- planes - 3-4, pilots - 3, Flik 15 - planes - 2, pilots - 2, Flik 8 - planes - 2, pilots - 2, Flik 13 - planes - 2, pilots - 3. The average readiness for use was therefore about 37% of the planes! The following types of aircraft were available at that time: Flik 10 - Albatros B.I, Flik 15 - Lohner C "Gebirgspfeilflieger", then Aviatik B.I, Flik 8 - Aviatik B.I and Albatros B.I, Flik 13 - Aviatik B.I. At the beginning of hostilities, they used airplanes which were not equipped with on-board weapons, because the light and low-power engine types Lohner B "Pfeilflieger" and Lohner C "Gebirgspfeilflieger" available at that time were not suitable for mounting a machine gun in them. Despite this, the first attempts to equip the aircraft with machine guns were made in August 1914, but the problem of water cooling the Army's Standard-MG machine gun, i.e., 8 mm M7/12 Schwarzlose System, designed for completely different purposes, appeared. The attempt to use the 6.5 mm Schwarzlose-MG system machine gun, originally produced for the Greek army, also failed and it was only in the summer of 1915 that the problem could be solved. In the planes, behind the observer sitting in the rear, a rotating ring was mounted, used to attach the much more efficient 8 mm MG 7/12 rifle, without water cooling, from which it was possible to fire at air and ground targets. However, in the period before the Gorlice Breakthrough, this solution had not been applied yet, as there were virtually no air fights taking place over the Nord-Ostfront and only the aircraft crew brought their personal weapons on board. The personal equipment of the aircraft crew was the pilot's Mauser multi-shot pistol and the observer's 8mm Mannlicher multi-shot rifle for self-defense in case it was necessary to land behind enemy lines. However, this weapon was not suitable for air combat. There were also bombs of a smaller caliber, which were dropped in case needed, but they caused little losses to the enemy. The fuel was sufficient for a flight of about four, four and a half hours on average, and with the maximum refueling of the aircraft and in combination with very economical control, it was possible to extend the flight to six, six and a half hours. This remarkable achievement was possible only with the appropriate use of air currents; however, weather conditions often drastically limited the capabilities of the aircraft. For example, the Fliegerkompanie of the field corporation of the 4th Army in the period between April 18 and May 8, 1915 could not take off only on April 18, 1915, while the neighboring, field corporation of the 3rd Army, had to deal with a completely different weather conditioned by the terrain mountain. For 21 days, there were only 12 during which it was possible to start. The photographic equipment of airplanes at that time was still very modest and bulky and included ordinary ground cameras with 9x12 cm film[7].

By mid-April 1915, thanks to aerial reconnaissance, important information about the location of enemy artillery batteries was obtained, and artillery fire on important sections of the front began with the use of airplanes equipped with radio stations, especially in the area around Gorlice[8]. However, it was clear to AOK that the field armies bled out after a series of four winter battles would not be able to hold back the Russian Southwest Front on their own. The Eastern Front, previously treated by the allied OHL as secondary, began to change its status in the spring of 1915. Alarming reports from the allied AOK contributed to the change in the attitude of the German General Staff, which, with the then OdeB, noticed a real threat from Stawka. The field armies of the Southwestern Front in April 1915 had the opportunity to cross the main ridge of the Carpathians and enter the Great Hungarian Plain, which, together with the threat to Budapest and Vienna, could eliminate Austria-Hungary from the war. According to the directive of the commander of the Russian Southwestern Front of April 15, 1915 on the continuation of the attack in the Carpathians, with the earlier strengthening of the occupied positions and the lifting of human reserves and supplies, the commander of this, in accordance with the new plan, ordered to prepare for the imminent, decisive attack on the territory of the Kingdom of Hungary, in order to "take control of them once and for all". In the area conquered so far in the Beskidy Mountains, it was necessary to strengthen and draw strong reserves, which, in the strength of a two-division corps, should have been collected in the Stanisławów region, the second one in the Sambor region and the third one in the Krosno region, within two weeks, starting the entire maneuver not earlier than in the evening 22 April 1915. Movement of troops was to be continued only at night and every effort had to be made so that this night maneuver was not noticed by the enemy air force[9]!

Above: Postal Albatros B.I ... At the Przemyśl-Żurawica fortress airport, the Albatros B.I plane from Flik 1 - in the cabin Oblt. Jozef Smetana. SAW / KA Wien collection

Endnotes

1 M. Kromkayow, op. cit., p. 34.

2 G. Artl, *Air reconnaissance KuK Luftschifferabteilung before the start of the offensive in the vicinity of Gorlice in 1915,* [in:] *Aviation in the Gorlice Breakthrough 1915. A look after 100 years*, ed. A. Olejko, A. Kharuk, Gorlice-Rzeszów 2015, p. . 145-147; R. Keimel, *Österreichs Luftfahrzeuge. Geschichte der Luftfahrt von den Anfängen bis Ende 1918*, Graz 1981, p. 231; E. Peter, *Die kuk Luftschiffer- und Fliegertruppe Österreich-Ungarns 1794-*

Above: Przemyśl 1914 - the sting of Aviatik B.I ... Machine gun mounted in the observer's cabin: source: SAW/KA Wien

1919, Stuttgart 1981, p. 115.

3 SAW-KA, Manuskripte Luftfahrtruppen, Auszug aus dem Kreigstagebuch des Oberleutnant Feldpilot Wilhelm Schreiber der kuk Fliegerkompanie No. 15 aus dem Weltkriege 1914–1918, p. 47–49.

4 РГВИА. Ф. 2067. ОР. 1. Д. 142. Л. 68 - Из истории Карпатской ..., op. cit., pp. 262-263 - report of March 26 (April 8) 1915.

5 SAW-KA, Manuskripte Luftfahrtruppen, Auszug aus dem Kreigstagebuch des Oberleutnant Feldpilot Wilhelm Schreiber der kuk Fliegerkompanie No. 15 aus dem Weltkriege 1914-1918, p. 54–55, 56; G. Artl, op. cit., p. 149; R. Keimel, op. cit., p. 231.

6 SAW-KA, Manuskripte Luftfahrtruppen, Auszug aus dem Kreigstagebuch des Oberleutnant Feldpilot Wilhelm Schreiber der kuk Fliegerkompanie No. 15 aus dem Weltkriege 1914–1918, p. 45–55.

7 G. Artl, op. cit., p. 149–150; R. Keimel, op. cit.

8 G. Artl, op. cit .; R. Keimel, op. cit.

9 РГВИА. Ф. 2067. ОР. 1. Д. 142. Л. 189–191, 195–196 -Из истории Карпатской ..., op. cit., p. 276–277, 278 - reports from 2 (15) April 1915.

Above: Miechów 1915 ... At the Flik 8 field airfield Aviatik B.I aircraft no. 19. KVH Beskydy collection

Five to 12 ... Gorlice on a Combat Course

Beginning of May [1915] *was warm and cheerful. Bright spring days, blued, favored the activities of air troops, which - although under fire - could nevertheless take precise photos of enemy positions; the roads and paths, dry in this beautiful weather, facilitated both the marches of troops and the transports of guns and all kinds of war material; - so everything was prepared with haste to achieve the intended goal - that is, to break the Russian lines* [spelling in accordance with the original in the entire quoted text], *- the terror of the coalition and the achievement of peace ... In the evening of April 30* [1915] *I and III battalion of our regiment* [ck 57. pp. 12th Infantry Division] *to Wilczyska, IV. battalion and the 9th Marching* [Marszbatalion] *to Jeżewo. With these ordinances, we learned that a general attack is scheduled for May 5* [1915]; *this was confirmed by the transports of German troops, which went night after night with trains and an enormous amount of war material, cannons, and ammunition, while the German airmen carried out an enormously lively activity during the day. Despite all precautionary measures, the enemy saw through the intentions of the combined German-Austrian armies; - maybe a bit late, but he understood them exactly, which was since he had brought in significant infantry reinforcements, and his heavy cannons, for which no ammunition had arrived anyway, began to dismantle, and remove from the position. This circumstance, checked by the airmen, caused the general attack to be postponed from May 5 to May 2, 1915*[1].

The above mention of one of the officers of the Imperial-Royal 57th Infantry Regiment 12 DP, from May 1915 shows how the eye of an infantry soldier perceived air operations in the period preceding the Gorlice Breakthrough. The launch of the victorious offensive on the Eastern Front, because of the defeat of Russia, opened not only to the OHL but also to the German Empire, the prospect of subjugating Central and Eastern Europe and gaining permanent influence in the Balkans at the expense of the weakening ally - Austro-Hungarian[2]. Already on April 15, 1915, OHL issued recommendations regarding the transport to Galicia of the forces of the 11th field army, and on April 16, 1915 its Chief of Staff, Obst. Hans von Seeckt was orally instructed by General E. von Falkenhayn about the German attack plan (on April 19, 1915, Obst. H. von Seeckt gave orders to quartermaster officers regarding the conference to be held between the 4th and 3rd Imperial-Royal field commands Army on April 21, 1915)[3]. Along with the transfer of the German 11th Army's field division to the Eastern Front, German air units, intended to participate in the offensive, also arrived in Galicia, but they needed time to acclimatize and start combat flights. As of May 1, 1915, the OdeB of the joint Austro-Hungarian-German forces on the Nord-Ostfront section south of the Dunajec estuary to the Vistula River amounted to 471 thousand bayonets, 28 thousand sabers, 2,468 guns and 102 heavy mortars[4].

The air reconnaissance necessary for the offensive could only be carried out without the use of the crews of the Air Force. This action was planned to start on April 18, 1915. The gradual increase in flights over Galicia by the crews of the Imperial-Royal Air Force was noticed behind the front line, on its Russian side, not only by military units. Feliks Grząglewski described the spring of 1915 in Gorlice in his diary:

On Easter holidays [2-5 April 1915] *very often aeroplan's* [airplanes-planes - spelling consistent with the original in the entire quoted text] *Austrian and the following days - there was no more day without an Aeroplan* [aeroplan-airplane] *not being noticed - when Muscovites saw them, they shot rifles and cannons, but there was no accident that they had hit the Aeroplan* [aeroplan-plane] *at least once - and civilians were hiding in the cellars for fear that they could throw bombs on the city from the Aeroplan* [airplane][5].

In the "Diary of Mieczysław Pattek" written in Rzeszów, the situation in the back of the Russian front was described as follows:

April 2, 1915 []. Good Friday.... An airplane [plane] flew over the city, which could not be seen, only after the buzz about it was known[6].

April 7, 1915 [r.]. Wednesday... morning an airplane appeared over the city[plane] 1 coat. [yznowy - monoplane] high up, it seems hostile. Behind him flew 2 2-plane [biplanes] as if chasing him. All of them were to be Russian [yjskie], there were 7 of them, they were to return from Przemyśl [the monoplane was most likely of the Nieuport IV or Morane Parasol type, used in Russian aviation - at that time, monoplane planes were not used in the Imperial the services on the Western Front entered fighter planes of the Fokker E. I type].

April 8, 1915 [] ... Russian [Yi] army, infantry goes to Tarnów in the morning, fresh new troops arrived in the afternoon. About 3,000 prisoners of war were brought from the Carpathians. They say that there are enough troops there.

April 9 [1915]. Friday ... Yesterday the gathered prisoners this morning was taken deep into Russia. They were gathered in the marketplace, where they were like a whole regiment, and they went with the four. From the morning you can hear the roar of cannons from the west. In the afternoon, they were taking the badly wounded from Jasło. They say that hospitals are being transported from there, because Austrians are to be too close ... They say that a terrible fight is going on on the Dunajec and Gorlice. Muscovites are to surrender (...), and yet they go further deep into Galicia ... In the afternoon, new prisoners from the Carpathians were brought[7].

April 10, 1915 []. Saturday ... They say that trains run only from Dębica, because the bridge over the Wisłoka River is to be smashed with [air] bombs [ts]. They say further that the entire Austrian [Iacki] team was taken prisoner in the Carpathians [ck of the 28th IR of "Praga's children"]. The news was brought by an officer, one of the many who were brought with prisoners today, and there were about 1,500 of them ... There is to be severe havoc with Austrian [air] bombs in Jasło ... In the morning, a new Russian army goes to the Carpathians. The Russian [yi] circles claim that they are doing poorly on the fronts in Gal. [icia], and that the

Above: Przemyśl windows to the free world ... Aviatik B.I plane at the Przemyśl-Żurawica fortress airport - collection of the SAW/KA Wien

Right: Everyday aviation ... Aviatik B.I postal plane at the Przemyśl-Żurawica fortress airport - collections of the SAW/KA Wien

Above & Above Right: Przemyśl stronghold under the bombs of 1915 ... The effects of bombing by Russian aviation. collections of the SAW/KA Wien

Carpathians have buried thousands of Mosk. [al][8]
April 11, 1915 []. Sunday ... About 1,500 Austrian prisoners were brought, including 21 Prussian ones. They were brought from the [?] war square from Mezö Laborcz [today Medzilaborce] to Tarnów. Aust. [rian] was to break up the entire Russian division and take all prisoners. 6 soldiers returned from the division. New troops go to the Carpathians. It is doubtful whether Austria will defeat them ... Four Prussian corps and from Tirol were to come to the Carpathians ... All hospitals from Krosno are in Rz. [eszow], they must have been hot there.
April 12, 1915 [r.]. Monday...About 200 prisoners came today. Muscovites had to withdraw from Bardejów and left the Dukla Pass. The fights there are supposed to be gigantic.
April 13, 1915 []. Tuesday ... Mos. [kals] are to retreat from the Carpathians and are already looking for accommodation in Bukowsko [?]. Hospitals from Tarnów have moved back to Czarna [Tarnowska] and are in tents there. It is going to be move back all the way, we'll see it in a few days. They almost daily lead fugitives, Russian deserters, to the army headquarters[9].
April 14, 1915 []. Wednesday ... In the morning the infantry with music went to the railroad - it was probably going to Tarnów. Mos. [Kal] were to fail at Uzhok and retreated to some area in the Carpathians.
April 15, 1915 [r.]. Thursday...Normal traffic in the city. In the morning, the troops and infantry go to Jasło[10].

In preparation for an extensive air reconnaissance operation, on April 18, 1915, the command of the field corps of the 4th Army from among its three Fliegerkompanies withdrew from the neighboring field Imperial-Royal 3rd Army Flik 8, which was stationed beyond the Carpathians, at the field airport in Örmezö (today Strážske in Slovakia). Flik 10 was transferred from the airfield in Jasień, where it had been stationed so far with Flik . 15, to the field airport in Zakliczyn, so that the Gorlice region was closer to the airmen. Oral reconnaissance orders for Fliek. Field Imperial-Royal of the 4th and 3rd Army (flight reports were passed to the staff officers only orally!) included the following tasks: observation of the movements of enemy troops from the east to the area of Pilzno-Tarnów and Biecz-Jasło-Nowy Żmigród; controlling the existing and damaged bridges and fords, in particular the bridges over the Vistula; observation of railway transports, including railway stations; control of the status of increasing or changing positions of enemy artillery; control of changes in positions of Russian units; observation of changes in the locations of hostile airfields-landing sites, the number of hangars and possibly airplanes; observation of cases of seizure of places near the frontier by supply units. The command of the field corps of the 3rd Army reported to AOK on April 20, 1915 at 01.20 am:

The airmen reported that by noon in the area from Gorlice to Rymanów, on the roads leading to the south, there were no traffic except a slight movement of the rolling stock. No aviator took off in the afternoon. Due to unfavorable winds, the pilots did not conduct long-range reconnaissance today. 13. Flik carried out a detailed survey of the Šandal *area* [in Ondavskej vrchovine], *Stropkov, Breznička,* Chotča, Duplín - *8 Flik* [Flik 8] *it could not carry out any flight because its planes were unreliable* [unreliable engines]. *Repeatedly summoned technician from Moravska Ostrava has not yet arrived*[11].

The first flights of Flik crews. 13. in the period from 17 to 23 April 1915, in the explored area, they initially did not show anything worth mentioning (no movement of Russian troops and no increased movement of their supply units). The first reconnaissance reports also showed that the frequency of railway transports did not differ much from their normal traffic. It was only noted that the Russian troops were constantly expanding their field positions[12]. Command of the field corps of the 3rd Army in the report to the AOK Op. No. 3883 of April 23, 1915 reported:
Air-report: poor rolling stock traffic on the Stropkov road- Svidnik, strong on the Stropkov – Bukovce road. At 9 am on the Svidnik road-Nižný Orlík 2 km long column of infantry marching towards Nižný Orlikfollowed by a two-kilometer rolling stock. It was not possible to continue the long-distance reconnaissance due to the cloud cover and fog and the increasing south-east wind[13].
The mountainous terrain and bad weather conditions made it impossible from the very beginning to conduct intensive and cyclical observations from the air on the field front of the 3rd Army, where the observations were initially to be carried out by Flik 13. Flik 8 was already dislocated from the field facilities of the 4th Army to the area of its neighbor's front, but its crews could not start flights so far, because the planes were not fully operational and had to be subjected to a thorough technical inspection. The moment the first Flik 8 approved for use, on April 23, 1915, a need arose to use it to radio-secure the position of the German Beskidenkorps in the Łupków-Csabahaza area (today the Slovak Čabalovce, part of the village of Sterkovce in the Medzilaborce region). It was not the only case when air support for artillery was recorded in the Western Bieszczady Mountains[14]. In the field report of the Imperial-Royal 3rd Army to the AOK of April 21, 1915 a 06.30 pm, the command of the army reported to the AOK: *Kaschau [today Košice] Op. no. 3270/19. Defeating enemy heavy artillery right next to Zemplenoroszi [today Ruské], which is urgently needed to recapture Kozielata [today Kozialat], is possible only with the use of aerial reconnaissance. However, the planes assigned to Burgflink [front-line units] are not equipped with [on-board] radios, which makes the reporting of the effects of the shelling and communication between the air scout and the artillery commander difficult and slow. Because Burgesche [command of the field Imperial-Royal 3rd Army] in Strážskem has the Flik 8 at its disposal. [Flik 8] equipped with [on-board] radios, please provide Burgflink [code name of first-line units] with an aircraft with a transmitting / receiving station so that you can observe the artillery fire in the following days. In exchange for the Burgflink plane [codenamed first line units] they could take over the reconnaissance in the eastern part of Burgesche [the area under the command of the field command of the 3rd Army]. It is also requested to completely take over the observation of the fire [34th Artillery Brigade] while fighting*

Above: Fokker B.I 03.22, a type M10, was worknumber 243. Iron cross national insignia has been applied to the fuselage in addition to the wings and rudder. (Peter M. Grosz collection/STDB)

Above: Fokker B.I with engine removed for maintenance. (Peter M. Grosz collection/STDB)

Above: Fokker B.I 03.24, a type M10. Iron cross national insignia have been applied to the wings and rudder. (Peter M. Grosz collection/STDB)

Above: Lineup of aircraft at *Fliegerersatzkompagnie* 6 at Fischamend with Fokker B.I 03.30 at left and B.I 03.31 to its right. (Peter M. Grosz collection/STDB)

the enemy artillery in the region of Łupków-Čabalovcewhich is also in the interest of Burgesche [command of the field-command of the 3rd Army] and the appointment of the 8 Flik for this purpose. [Flik 8][15].

In the area of the aforementioned Kozialat peak, the fights fought there earlier and on April 23, 24 and 25, 1915, fought with unprecedented intensity and the troops storming the summit of the Imperial-Royal, received powerful fire support from howitzers and the 17th "Škoda" M 11 30,5 cm mortar battery from the area of Velka Polana, but the enemy artillery did not give up the field here despite the precise recognition of its new positions by air reconnaissance KuK Luftschifferabteilung[16]. In the report, Op. 3883 / III of the command of the 3rd Army in the field of April 23, 1915 was writen:

8 Fliegerkompanie [Flik 8]. *Make the machine available in the coming days*[plane] *with a transmitting and receiving station to observe the fire* [Kozialata's own artillery]. *Detailed orders should be collected immediately from Rapsflink in Uzhgorod*[17].

On April 26, 1915, despite the exceptionally strong artillery fire, with the use of the 17th "Škoda" M 11 mortar battery, 30.5 cm, the fire of which was carried out based on accurate air reconnaissance data (shots were fired at the target with the accuracy of 500 m), but Kozialata was not conquered[18].

On April 24, 1915, the weather conditions in the Gorlice direction again made it impossible to perform flights – in report Op. No. 3883 / VI of April 24, 1915 of the 3rd Army field command contained the following content: *Due to strong wind, it was impossible to carry out a long reconnaissance on April 24.* [IV]. *The Flik 13 plane fell* [damaged] *the crew was not injured*[19]- only on April 25, 1915, two planes from Flik 8 could be used. On April 24, 1915, the Flik 10 and Flik 15 of the field Imperial-Royal 4th Army during the observation of the Russian airfields in Jasło and Tarnów-Lipie, stated that although nothing had changed in Jasło, tent hangars had been removed in Tarnów-Lipie. The long flight over Przemyśl showed that two new bridges were built in place of the bridges on the San river that were blown up by the defenders of the fortress on March 22, 1915. The first dynamic shifts of the units of the Russian 3rd field army were noticed on April 25, 1915 by Flik 10 and Flik 15. (an approximately 4 km long supply column was observed on the Dąbrowa Tarnowska-Tarnów road and a heavily loaded rolling stock in the march near Jasło and Krosno and two 4 km long columns during the march from Krosno to Dukla; a two-kilometer column was also noticed on the march from Dębica to Pilzno). On April 25, 1915, five Russian infantry battalions were discovered, marching towards Pilzno along with supply and artillery units, and on the return flight the observer stated that these troops had set up a camp in the village of Bielowy south-east of Pilzno at noon. Moreover, four enemy infantry battalions were observed along with the rolling stock on the road from Kołaczyce towards Jasło, where they stopped; another, three km long column, was observed walking from south to north towards Nowy Żmigród and during the return flight, the observer stated that at noon these units had set up a camp in the village of Bielowy, south-east of Pilzno. In addition, four enemy infantry battalions were observed along with the rolling stock on the road from Kołaczyce towards Jasło, where they stopped; another, three km long column, was observed walking from south to north towards Nowy Żmigród and during the return flight, the observer stated that at noon these units had set up a camp in the village of Bielowy, south-east of Pilzno. Moreover, four enemy infantry battalions were observed along with the rolling stock on the road from Kołaczyce towards Jasło, where they stopped; another, three km long column, was observed walking from south to north towards Nowy Żmigród[20].

Oblt. Fp. W. Schreiber from Flik 15 recalled his April reconnaissance flights:

April 18 (Sunday) [1915]. *The sky is completely obscured, the clouds are low and there is no visibility.*

April 24 (Saturday) [1915]. *Moderately good weather, very windy.* [Oblt. B. von] *Losonczy flew out of* [Oblt. J. von] *Wierzejski and was heavily fired from Tarnów. It seems there is a solid battery of anti-aircraft artillery there because the peasants shoot quite accurately. We will not be flying over Tarnów for a while if it can be avoided. The Russians seem to be very well informed. For example, the airport in Tarnów is deserted!*

April 25 (Sunday) [1915]. *Again, quite a long and interesting flight from* [Zgsf. F.] *Benda: Jasło- Dukielska Pass- Svidnik-Bardejov- and back. Initially, the weather was very nice. In the Carpathians, however, we fell into a very strong snowstorm twice, from which our faces felt very painfully. When we got back, we were very swollen and cold, it was so cold. When I got out, I fell flat and couldn't stand; I had no feeling in my feet at all*[21].

In the operational area of the 3rd Army, air reconnaissance initially did not bring the desired results, and on April 25, 1915, one plane did not return from the flight and it was declared missing, as well as the crew. In report Op. No. 3883 / VI of the field command of the 3rd Imperial-Royal Army of April 25, 1915, written: *Plane* [Albatros BI type No.] *2128 with corporal* [Corp.] *pilot Stolba and observer Oblt.* [E.] *Grycz on board, flew today at 7 am towards Krosnofor long-distance reconnaissance and has not yet returned. (Flik 8) - if they land on our premises, please let me know.* In the report, Op. No. 3883 / VII of April 25, 1915, the command of the field corps of the 3rd Army reported to AOK: *Distant reconnaissance of the 25th 4* [1915] *without results. The plane, which took off at 7 am, with Zgsf.* [Corp.] *pilot Stolba and Oblt.* [E.] *Grycz has not returned yet. Rapsflink* [in Uzhgorod] and *Rapsamel have been notified*[22]. However, in the light of Russian archives, on April 25, 1915, one enemy plane was shot down by artillery fire and landed in the Rymanów area - the crew (two officers of the Imperial-Royal Air Force) were taken prisoner[23]. On the Gorlice section on April 26, 1915, due to bad weather conditions, another flight could not be carried out - report Op. No. 3883 / VIII of April 26, 1915 from the command of the above-mentioned field Imperial-Royal 3rd Army stated: *Due to unfavorable weather, it was impossible to perform a long reconnaissance by the airmen. Despite two departures, no observations were made*[24]- and only from April 27, 1915, the plane's crew of Flik 10 and Flik 15 subordinates field commanders of the 4th Army could continue to observe the movement of Russian troops – report NO. Op. 3883 / IX of 27 April 1915 from the above command reported: *Aerial reconnaissance: slight rolling stock traffic on the Svidnik road-Dukla. In Tylawa, Trzciana and Dukla lively traffic of rolling stock, large stopping place for rolling stock in Jaśliska. Slight traffic on the* Čertižné *road-RabbitPolish. Purely on the Zborov road-* [New] *Żmigród and Zborov-Gorlice to Małastów (Malastow), only slight rolling stock traffic*[25]. In addition, increased traffic was observed on the railway lines towards Jasło, which proved that Stawka had shifted the 63rd Reserve Infantry Division, 7th DK and part of the units of the 3rd Caucasus Corps from south to west. Due to the large distance from the front line, it was impossible for the staffers to conclude whether these maneuvers were to relieve the front troops or strengthen them. On that day, the crews of the Fliegerkompanies of the 13th and 8th field Imperial-Royal 3rd Army could finally make reconnaissance flights, but no shifts of enemy forces were noticed during them. A lively supply traffic was observed on the Tylawa-Trzciana-Dukla road, and a large rolling stock was also detected in Jaśliska. On April 27, 1915, in the late evening hours, the command of the field corps of the 3rd Army received an order from the command of the German 11th Army field:

The command of the 11th Army orders the following air operations to be performed for its benefit: 1) Air operations begin in the morning of April 28 [1915]. 2) *The tactical reconnaissance was planned from the Małastów-Tuchów line east to Wisłok.* 3) *Long reconnaissance - from the Gorlice-Tarnów line to the Sanok-Rzeszów line. The roads leading from Bartfa* [today Bardejov] *to* [Nowe] *Żmigród and from the Dukla Pass and the Laborca Valley to the north are also monitored. The 3rd Austro-Hungarian army was asked to observe the road leading from Hommon* [today Humenne] *to Sanok and the area to the east of it. The Army Headquarters will keep the Eleventh Army Headquarters informed of the results of the reconnaissance*[26].

The long reconnaissance on 28 April 1915 was not made again due to bad weather conditions[27] and therefore the command of the field commander of the 3rd Army decided to apply defensive measures against the noticed, more and more active flights of Russian aviation planes: *Recently, enemy planes flying low over the positions have started to appear frequently. As soon as it is determined beyond doubt that it is an enemy machine, it should be fired with artillery fire*[28]. On April 28, 1915, however, Flik 10 reported to the field command of the 4th Army that the Frysztak-Jarosław section and the area from Jarosław to the Dukla Pass were free from caravans and large Russian troops. Jarosław was heavily manned by enemy forces, three new infantry battalions and about 100 transport vehicles were discovered near Biecz, moreover, a 2 km long infantry column was noticed marching from the south towards Krosno[29]. In the light of subsequent air reports, the areas of Jasło-Nowy Żmigród-Bednarka and Dukla-Dąbrowa Tarnowska were heavily manned by Russian supply columns and military units; several infantry battalions marched from Jasło towards the north and west; there was little traffic on the railway sections Tarnów-Rzeszów and Gorlice-Krosno; in the Brzostek area, the presence of delivery columns moving in different directions was found; around 1000 transport vehicles (wagons) have been spotted in the area of Dukla; a very large vehicle camp was discovered in the area of Krępna and Nowy Żmigród; from the side of Dukla towards Tylawa, a military column with a rolling stock of 5 km was moving; on the road between Krępna and Nowy Żmigród there was lively traffic of transport vehicles in both directions[30]. Order of the command of the German field command of the 11th Army 1a No. 118 Op. 3883 / XI of 27 April 1915, assigned to the field command of the 3rd Army, spoke about recommended aviation activities for the allied 11th Army: 1) *the activities will begin on the morning of April 28, 1915;* 2) *tactical reconnaissance was designated from the Małastów line-Tuchówto the east as far as Wisłoka; 3) long reconnaissance was set from the Gorlice-Tarnów line (Tarnov) to the Sanok line–Rzeszów; moreover, the roads from Bardejov to* Żmigród *should be under surveillance , as well as those that lead from the Dukla Pass and the Laborca Valley to the north. The 3rd Army is requested to observe the road from Humenné to Sanokand the area to the east of it. Let* [ck] *The 3rd Army hands over the results of air reconnaissance to the command of the 11th Army*[31]. Report Op. 3883 / XII from the command of the Imperial-Royal 3rd Army from April 28, 1915 to AOK in Cieszyn and Rapstaube in Nowy Sącz said: *Long aerial reconnaissance on April 28 [1915] impossible due to strong wind*[32]. Nevertheless, quite interesting thing was detected by Flik 15 in the area between Sękowa and Zborov, south of Gorlice, where a total of 19 field artillery batteries were found. Due to its extreme importance, the report was handed over in person, orally, during a stopover at the command of the German 11th Army in Nowy Sącz (the field airport was situated around the present Błonie housing estate). On that day, in the afternoon, it was planned to send for the first-time reconnaissance flights the crews of the German FFA, which were to be taken over by further, close reconnaissance directly around Gorlice[33]. On April 29, 1915, the crews of Flik 13 and Flik 8 Field Imperial-Royal of the 3rd Army were again unable to perform reconnaissance from the air due to bad weather conditions, and only one of the Flik 13 crews was only able to report that the road Stropko (today Stropkov) -Felsövizköz (today Svidnik) -Ladermervagasa (today Ladomirová) between 7.30am to 8.30am and road Mezölaborcz (today Medzilaborce) -Zagórz between hours. 9.00 a.m. and 10.00 a.m. were free of any Russian forces[34]. The April 29, 1915, passed in the field of the Imperial-Royal 4th Army without any special events. Report Op. No.

Right: In the former territories of the Austro-Hungarian Empire, one can now find a lot of evidence of a sentimental return to the past…. In the photo, Emperor Franz Josef I Habsburg invites you to a chair in Ivano-Frankivsk (Stanisławów) - source: author's collection

3883/13 of the field command of the Imperial-Royal 3rd Army on April 29, 1915, assigned to the AOK in Cieszyn and Rapstaube in Nowy Sącz said: *Air report: road Stropkov-Svidnik-Ladomirová* [Hungarian Ladomervagassa] *from 7:30 to 8:30 am clean, Medzilaborce road- Go downbetween 9 and 10 am clean. Strong north wind made it impossible to carry out long-distance reconnaissance*[35]. In the area from Tarnów to the area of Dębica-Mielec to the Vistula line and to the area of Dąbrowa Tarnowska, no enemy forces were noticed. Also, south of the Tarnów-Dębica road through Jasło-Biecz-Tuchów, no major displacements of the enemy's forces were found. At the train station in Jarosław, five trains ready for departure were observed, and the city itself was empty. The bridges on the Wisłoka near Krajowice, northwest of Jasło, in Wróblowa, near Buków and near Bielowice, were detected and photographed[36]. Only in the vicinity of Jasło and Biecz a lively movement of the Russian forces was recorded on that day. Exceptionally modest staffing of Flik 10 and Flik 15 (both had only five pilots), prompted their command on April 29, 1915 to submit a request to the field command of the 4th Army to increase the number of pilots to at least eight. Due to the constant departures on reconnaissance flights of the same crews, as well as due to the prevailing bad weather conditions and the few days of rest, the commanders were afraid that these few "exhausted pilots" would not mentally withstand the tension during the upcoming offensive. In response to this request, the army command ruled out the assignment of additional pilots (!) within the next two weeks[37] while the commander-in-chief of the Imperial-Royal army, Archduke Józef Ferdinand Habsburg, expressed his wishes to the exhausted flying personnel[38]. Air reconnaissance carried out on April 30, 1915 by the field aviation of the 4th Army confirmed the observations from the previous day. OP report. 3883/15 of 30 April 1915 from 15.35 assigned to the field command of the 3rd Army from the field command of the German 11th Army said:

Lt. Barlen and Oblt. Becker, pilot of the Flik 8, did not return on time. Asking for a message to the AOK if they did not accidentally land in that area[39].

In a report from the army command in Nowy Sączto AOK Op. No. 3883/15 of April 30, 1915 it reads:

Regarding Op. No. 3883/15 mentioned plane [Flik 8] *landed in an emergency today at 11 near Koškovce. Both officers, safe and sound, were picked up by the plane of the 8 company* [Flik 8] *and are now in Strazskem.*

Central Command in Košicereported on April 28, 1915 between 10.30 a.m. and 12.00 noon reported:

Our plane was constantly under heavy shelling by Russian artillery in the Smerekowiec region (Smerekowice), Regietowa (Regetow), Hańczowa(Hanczowa). About 160 shrapnel were fired against them. Both [gen.?] *von Schneider and Oberst Korzer emphasize the technical skill of aviators*[40].

In the report of the OP. 620/10, sent on April 30, 1915 to the AOK in Cieszyn it is noted that

none of the Burgesche airmen [the command of the 3rd Army] *was in this area on 28 April* [1915], *it could have been an aviator from Rapstaube* [in Nowy Sącz] *or Rapsamsel.*

In the report from *Burgesche* [command of the 3rd Army] to the AOK in Cieszyn, to *Burgflink* in Uzhgorod and to *Rapstaube* in Nowy Sącz Op. No. 3883/16 of April 30, 1915 it reads:

Air report: On the Sanok road–Medzilaborce (Mezőlaborcz) in the afternoon, increased movement of the rolling stock itself, no troop movements. On the Rymanów road-Jaśliska–Medzilaborce only slight rolling stock movement. On the Sanok – Medzilaborc railway routee, and later Sanok – Krosno rail traffic in both directions[41].

On the road section from Dąbrowa Tarnowska through Radomyśl Wielki-Kolbuszowa-Sokołów-Łańcut-Rzeszów-Jasło to Pilzno, slight troop movements and light rail traffic were observed. A considerable movement of Russian troops and a large camp with rolling stock were observed in Jasło and the surrounding area, as well as *huge clouds of dust rising after the march of troops from Krosno to Piastowe Place*[42]. The crews of the Fliegerkompanien of the 3rd Army fieldwork noted a lively transport movement with little movement of Russian troops[43].

On May 1, 1915, 24 hours before the start of the Tarnów-Gorlice offensive, the crews of the Fliegerkompanien field of the 3rd Army continued reconnaissance flights, reporting that on the roads of Sztropko (today Stropkov) -Alsovizköz (today Nižný Svidník - now Svidnik) -Ladomervagasa (today Ladomirová), Felsövizköz (today Svidnik) -Alsopagony (today Nižna Pol'anka), Zboro (today Zborov) -Ożenna and Zboro (today Zborov) -Kieczna ,in the early morning no movement of any enemy troops was observed, only a slight movement of rolling stock was noted. On that day, the action of observing the forces and permanent positions of the tsarist army from the air ended, and the increasing violent north-west wind disrupted further reconnaissance flights. The crews of Flik 10 and Flik 15, from the morning of May 2, 1915, have continued the reconnaissance flights, in the area assigned to them: Tarnów-Dębica-Rzeszów-Sokołów-Kolbuszowa-Radomyśl Wielki-Szczucin, where no troop movements have been observed so far, a 1 km long infantry column was noticed and a half-kilometer-long supply column (the plane sent to the flight in the afternoon, due to bad weather conditions, could not reach the Nida region)[44].

Results of aerial reconnaissance conducted by the Flik the KuK Lufyschifferabteilung carried out on April 28 - May 2, 1915, enabled the command of the German 11th Army field to be fully aware of all maneuvers of the Russian field forces of the 3rd and 8th Army, and provided information on the arrival of the III Caucasian Corps to the front zone observed by the Flik field army of the 4th Army just in time in the area of Brzostek-Jasło and Chyrów-Krosno. All these successes were quite a feat, considering that the reconnaissance flights of the crews of two field companies of the 3rd and 4th Army were carried out with airplanes with very modest technical equipment, as well as with an extremely small and exhausted personnel crew of the 3rd and 4th field Fliegerkompaniens. During these flights, only one plane was lost (the crew of Flik 8 shot down on April 25, 1915)![45]. The threat rate in the field section of the Russian 3rd Army was neglected by STAWKA, although the concentration of troops about to start the offensive in the Tarnów-Gorlice section was observed. The scientific point of view that this threat has not been observed by the Russian aviation due to the insufficient number of aircraft[46] in the light of recent research, it is not true. In the period preceding the Tarnów-Gorlice operation, the Russian 3rd Army field aviation included five KAO / AO - four KAO were subordinated to the field commander of the 3rd Army, and one to the command of the XII Corps, an additional AO was located at the Army commander[47]. In total, these forces consisted of 17 aircraft and 17 pilots, which accounted for more than half of the tsarist aviation and 26% of the number of aircraft in all five field armies of the Southwestern Front[48]. In fact, however, the 9th, 11th, 12th and 24th ACA had only technically efficient airplanes (in April 1915, in fact, the Russian 3rd Army had 10 aircraft, of which only ... three made combat flights)[49]. The Russian air reconnaissance did not overlook the preparations made for the great offensive on the enemy side, despite the low intensity of air reconnaissance (already on March 24, 1915, changes in the position of the enemy's forces in Galicia were detected). Report Op. 3883 / IX of the field command of the Imperial-Royal 3rd Army of April 28, 1915, addressed to all commands subordinate to him, said:

Recently, enemy planes, very low flying, often appear above our positions. As soon as the plane belongs to hostile forces, it can be attacked by our artillery[50].

According to the Stawka report sent on April 30, 1915 to the headquarters of Emperor Nicholas II Romanov, two days before the offensive Gorlice Breakthrough (!), the situation on this section of the Eastern Front was as follows:
17 (30) April 1915 the concentration of German-Austrian forces on the front in Western Galicia continued. The enemy conducts vigorous aerial reconnaissance in the section of our 3rd Army[51]*!*

This little-known entry is only a pure quotation from one of the documents, which is only an introduction to the thesis introduced into historiography: in the case of the Gorlice Breakthrough 1915, which was considered a landmark event on the Eastern Front in Great War, it should be unequivocally stated that *Without an aviator, there would be no success near Gorlice* (Ohne Flieger - kein Gorlice)! Without close and distant air reconnaissance and direct air support for the land forces, it would be impossible to achieve Gorlice success by the combined forces of the Central Powers in 1915! The mentioned above- April 17 (30), 1915, a Polish aviator of the 11th KAO, Capt. Antoni Emerykowicz Mroczkowski, was the first to detect the concentration of AOK forces and the German 11th Army around Nowy Sącz-Grybów-Bobowa-Ropa, passing this information on to the field command of the Russian 3rd Army and the Southwestern Front. He was awarded the Order of St. Jerzy, awarded to him on January 24, 1917[52]. STAWKA incorrectly assessed the real threat and, with great delay, decided to counteract the concentrated, overwhelming forces of the Central Powers. Reconnaissance flights of the crews of the Imperial-Royal Air Force, in the period of the second half of April 1915, were also observed by several Galician diarists and others. Father Bronislaw Świeykowski wrote in his diary in Gorlice:
19 IV [1915] *Airplane* [plane] *over the city - thunderous and long-lasting rifle and cannon volleys* [of Russian troops] *aimed at it, but in vain*[53].

Father K. Wais wrote in his memoirs:

From the direction of Dukla and Jaśliska, we were sometimes heard by clear, silent echoes of cannon shots. After April 20 [1915] *Austrian aeroplanes* [airplanes] *began to appear more often than before. One of them threw two bombs in Rymanów, which set the house on fire, injured two soldiers, and killed two Russian horses. Since then, aeroplanes* [airplanes] *have been shelled, but to no avail. Besides, the Cossacks rested freely and happily*[54].

Arato Jenö, one of the soldiers of the tsarist army, wrote in Cirocha's valley in his diary:

April 24, 1915 Airplanes fly over our heads. These are ours. They fly over enemy lines. Of course, the Russians shoot them. You can see how Russian artillery shells are flying past our planes. Fortunately, they didn't manage to shoot down one[55].

The chronicle of the village of Targowiska says:
30/4 [April 1915] *you could hear a bang from the bang* [spelling consistent with the original in the entire quoted text] *thrown by the Germans from aeroplanes* [planes], *which killed horses and people*[56].

The atmosphere of the April days was presented in this way in the "Diary of Mieczysław Pattek" written in Rzeszów[57]:
April 17, 1915 [] ... The infantry goes to Jasło. You can see that there must be a demand there.
April 18, 1915 [r.]. Sunday... In the morning the cavalry went to Jasło ... near Jasło or in the vicinity of Gorlice they were to pound the Mos. [Kals]
April 19, 1915 [r.]. Monday ... The troops passing through Rzeszów go north through Bratkowice. Prussians and Aus. [trians] were to break through the Nida River and are standing near the left bank of the Vistula. The right bank of the Vistula in the territory of Galicia will strengthen Mos [kals] and will defend access to Galicia ... They are talking about a defeat and the capture of 5,000. Aust. [tians] taken captive near Tarnów when they crossed the Dunajec to the other side[58].
April 20, 1915 []. *Tuesday ... In Dębica, an airplane* [plane] *crashed a train - the crushed carriages arrived ... Today 50 prisoners of war were brought from the Carpathians.*
April 21, 1915 [r.]. *Wednesday... They fortify Mosk.* [als] *Sandomierz and Tarnobrzeg. Some hospitals are fed up. Damn*

is supposed to be on the prowl in Jasło. In the afternoon, a rumor spread that the staff of the 3rd [field 3] *Army was to go to Old Town for lodging ... Yesterday evening around 10 an airplane* [plane] *was driving over the city in the moonlight. Same this morning at 5 and 10. he soared very high, so he had to be Austrian or Prussian* [German].
April 29, 1915 []. *Thursday ... The Circassians are robbing the Old Town. An airplane* [plane] *crashed a train in Jasło. About 30 bombs were to be thrown on the train and the station ... There is to be an offensive all along the line. So murderous battles are expected*[59].
April 30, 1915 [] ... *Today,* [air] *bombs were to be thrown again, and at Boguchwała, on a train bound for Jasło, in Dębica and Ropczyce. Dębica is to be on fire. An airplane* [plane], *I think Prussian* [German], *was going over the city today. He circled the railroad, city, and river. Yesterday, an airplane was fired heavily in the evening.* [Russian] *troops go to the Carpathians at night*[60].

Dr Karol Zaleski described the aerial episodes from Sanok from this period:
26.04.[IV] *1915 Monday ... My first exit from the house led me through Sujec for Posada Olchowskato to a sick child. The plots of Wojciech Drwięga were already plowed in the Błonie near the San. I wholeheartedly wished them "God bless"! I looked towards the city. The Castle Hill - in the steepest place under the central windows - covered with white spots in a wide strip from the bottom up, you would think with primroses [primroses] dotted with, and these are just discarded papers, files [c.k.] of the Sanok district office... Rudnicki kept his word: two white birches were brought to me by Patała from the Royal Well to decorate the Krasicki Rock. When, after returning from the city at noon (+22 [0] C), I started with my boy to plant them, Jan's family from the castle showed me a kite* [plane] *extremely high under the sun, and soon there was a huge roar towards the railway station and a ball dark smoke rose up, followed by a strong flame* [that day, recognition in the Sanok areawere led by the Flik crews. 10, Flik 15, and Flik 13] ... *I thought that the cannon was fired into the kite, but it was he who "spat" the bomb - with fire on the ground, on our poor land ... They still have too little blood and destruction! As of this writing, seven buildings and this picturesque hill of Posada Olchowiecka are already burning opposite the station it looks like an active volcano - without lava ... The northeastern wind ignited this terrible fire to destroy our poor people ... and moreover, thick rifle shots for the kite are heard all around ... On the bridge below there is an entire infantry unit with an officer, ready to shoot ... The trumpet called them - "to the church." The kite* [plane] *disappeared among the clouds ... Kuba went to the fire and brought the news that the first house of Mazurek was on fire, where the Cossack horse was charred, that she was injured there, as I was told, some young Miss Lewandowska and that something burned down 25 huts, and some people fell to the ground in fear during the explosion, and now they walked like sleepwalkers ... They said that one bomb fell into the San and did not explode - probably on a targeted bridge if it was so. Kubuś said that after the explosion the kite was heading towards our house.*
27.04.[IV] 1915 *Tuesday. A wonderful morning, but I woke up after 6, when the sun was already working ... Błachowski told Kuba that at 6 there was already one kite* [plane] *over Sanok ... I haven't seen the sick well yet, until a breathless woman runs in with the news that a misfortune is ready again, because there is an airplane above us* [on that day, reconnaissance in the Sanok areawere led by the Flik crews; Flik 10, Flik 15, and Flik 13.] ... *When, after giving what I needed to do, I went out on the road, people looked at the sky and indicated the direction of flight, but because of quite significant clouds I could not see or hear anything. And these clouds filled me with a little fear when I was walking near the railway warehouses and rolling stock by the dam and then on Sujac ... Late in the evening I was called to the owner of the house on which the bomb fell yesterday ... her 3-year-old son did not have an accident, for, looking for an airplane, he ran over a stream far from home.*
28.04.[IV] *1915 Wednesday. Cool but lovely and sunny and without an airplane* [plane] *day. At Posada Olchowiecka the mother of the sick Piotr Mielecki told me that yesterday, as the village leader somehow said - a bomb in Lisko fell on a car park and killed 60 people by tearing up carts and horses ... the whole war is wicked - like a meager humanity that must resort to war ... If at least there is a convention to come, at least regarding kite bombs* [plane].
29.04.[IV] *1915 Thursday. Cold night - in the morning everything shines with white frost in the bright sun, People, however, do not tremble from the cold ... They are looking to see if a new airplane is flying] - the personification of the "dove - love" ... Yesterday they were supposed to be at Sanokas two, according to the assurance of one boy, but they flew on ... At noon I was riding my Moscow horses to Zagórze. Soldat was driving. The daughter of a sick woman abandoned 4 years ago by her drunkard husband, an elderly lady told me that during the stay of the Russian chief staff in Zagórze thrown from a kite* [plane] *a bomb on a wireless telegraph station, but it fell into a pile of dung near the roadside near the church and did not explode. The second one, thrown, exploded near the forest but did not damage anything. She showed me relevant places. Crazy clouds of dust on the highway, I also swallowed the dust well and coughed well afterwards.*
1.05.[May] *1915 Saturday ... Masses of infantry, ammunition, cannons, hidden branches, cavalry go west* [the Third Caucasian Corps] *and go, and over them, in front of and behind them, clouds of dust. I could barely walk-through Posada from home all by myself. In the evening, the procession continues through the city. The market has turned into a camping site. They eat, smoke, read newspapers and one of them wins the harmony ... People look up to see if the airplane is not flying. Yesterday it was said that a Sabbath house had brought the reliable news that the railway station in Lviv had been bombed*[61]!

On the other hand, the commander of the Russian 13th Infantry Regiment, a Pole, General Eugeniusz de Henning-Michaelis, who fought in the Bieszczady Mountains wrote: *Often "birds" began to fly over us* [planes] *but didn't throw confetti*

Above & Below: The harsh winter of 1915 did not stop the postal line to the Przemyśl Fortress ... Aviatik B.I planes in the "polar" scenery of the Przemyśl-Żurawica airport. SAW/KA Wien collection

[leaflets] *they just watched. It was necessary to improvise anti-aircraft artillery from field guns placed on bases made of logs; He could not, of course, give tangible results such a primitive, but he forced airplanes*[planes] *to high flight*[62].

Just before the commencement of the Tarnów-Gorlice operation, on the night of April 30 / May 1, 1915, as many as 32 of the heaviest shells from the heavy M 11 "Škoda" mortars, 30.5 cm caliber and the M 14 42 cm heavy howitzer, fell on Tarnów and its vicinity, and the next day, 15 of them fell near the city, which resulted in the rapid evacuation of Russian forces. The fire of the heaviest artillery, directed by the Imperial-Royal Air Force, lasted for the next few days, until May 4, 1915 (a 42 cm missile hit the railway station in Tarnów; the radio was also used to hit the monastery located 8-10 km from Tarnów in Zbylitowska Góra)[63]. In total, 48 30.5 cm and 42 cm rounds fell on Tarnów, which resulted in the death of over 150 people - soldiers and civilians. F. Molnár, a Hungarian war correspondent, who watched the effects of the fire from Tarnów, recalled:

We're coming from where forty-two stood [M 14. 42 cm], aimed at Tarnów. Here on the Dunajec, soldiers talk about this mortar [M 14. 42 cm] as about some god. This cannon first stood right next to the embankment. When they were set up, the soldiers entered the barrel with joy - I saw a photo that captured the scene. The "deity" appeared in this area three times, twice at the embankment and once in the forest, where a special railway line was built for him. When the first shots were fired one night, officers photographed a pillar of fire that fired from the cannon's throat. This photo is also kept by the local railroad officers. In the photo, you can accurately measure the height of the fire by comparing it with the size of the people and trees standing around the cannon. This comparison shows that the pillar of fire is forty meters. The missile was also photographed, placed in a row with the soldiers - its height corresponds to the average height of a soldier. When the first shot was fired, they all walked away to a decent distance. But the officers were surprised to find that the noise made by the cannon was not annoying. It is fundamentally different from the hard and unbearable thunder of other heavy guns - it is a very deep but hollow sound. Then they dared to approach the monster so that they finally stood eight or nine meters away when the gunner jerked off the long rope. The forty-two [M 14. 42 cm] rides a train consisting of twelve carriages. The total weight of this train is 235 tons. The long-range missile itself, when it reaches its peak, flies higher in the clouds than any artillery shell known so far. The missiles weighing over 800 kilograms were fired very carefully to secure Tarnów. One missile can destroy half a street. The aiming, as I could say later, was extremely precise. Far beyond the Dunajec, hidden among green trees, stands alone the huge building of the theological seminary - it occupies the same space as the Parliament building in Budapest. He was given one missile because it housed important Russian warehouses. One part of the building just fell off the other and collapsed. It also got - for the same reason - a real school in a villa district. For a long time, classes in higher classes will not be held ... Our soldiers are most proud of the large pile of bricks lying next to the railroad - there was a storey house in this place, but forty-two [M 14. 42 cm] were interested in it. This pile of bricks is famous for being only a few hundred pieces. You can't see the rest. They say that the "deity" [M 14.42 cm] hit the very center and before the cloud disappeared after the explosion, the house simply disappeared, until today it is not known where the furniture and other larger objects are. It is not difficult to imagine the physical effects of the forty-two [M 14 42 cm] missiles hitting the Russian troops. As for the moral consequences, the captives told one captain that they had seen many cases of madness. In addition, the outbreak of many soldiers stunned, but it happened that they blinded them, literally knocking out both eyes. The very center of Tarnów forty-two [M 14. 42 cm] saved, practically not causing any serious losses. It "honored" one more school and destroyed the cinema[64].

Endnotes

1 *War memoirs of the Tarnów Region Regiment (now 16 p.p.) from 1914 and 1915 (Compiled by officers of the 16th infantry regiment)*, Warsaw 1919, p. 35–36.

2 Z. Palski, op. cit., p. 247-249; M. Klimecki, *Gorlice 1915*, Warsaw 1991, p. 65, 78, 80.

3 *Der Weltkrieg 1914–1918, hg. vom Reichskriegsministerium, Bd.* 7, Berlin 1931, p. 350, 361; August von Cramon, *Unser Österreichisch – Ungarischer Bundesgenosse im Weltkrieg*, Berlin 1920, p. 11.

4 Из истории Карпатской..., op. cit., p. 9. The literature on the subject also includes other data that in the period from the second half of April to May 1, 1915, AOK and OHL concentrated 558,000 people south of the Dunajec estuary to the Vistula River. infantry soldiers, 32 thousand Cavalry, 1506 heavy machine guns, 2911 guns and 178 mortars, on the main strike section against the Russian 3rd Army, generating numerical superiority in manpower and artillery. At that time, there were 564,000 in the field armies of the Russian Southwestern Front. infantry, 60 thousand cavalry, 1,522 heavy machine guns, 1,809 guns and 30 aircraft.

5 *Stanisław Feliks Grząglewski's diary from the Russian invasion of Gorlice in 1914-1915*, p. 41–43 - a copy in the private archive of R. Frodyma.

6 B. Gąsiorowska, *Rzeszów during the Russian occupation of 1915 in the "Diary of Mieczysław Pattek"*, From the past of Rzeszów (III), Rzeszów 1998, p. 102.

7 Ibid, p. 104.

8 Ibid, p. 104–105.

9 Ibid, p. 105–106.

10 Ibid, p. 106.

11 SAW-KA, Die durchbruchsschlacht von Gorlice1915. Bearbeitet von Obst. AD Feldp. Erich Kahlen, p. 18–22.

12 G. Artl, op. cit., pp. 150–151; SAW-KA, Manuskripte Luftfahrtruppen (MS / Lu) No. 11, Die entscheidende Schlacht bei Gorlice 1915. Entwickelt von Oblt. Fp. Erich Kahlen, p. 5–7.

13 SAW-KA, Die durchbruchsschlacht von Gorlice 1915.

Bearbeitet von Obst. AD Feldp. Erich Kahlen, p. 18-22.

14 Honv⊠delmi Miniszt⊠rium Hadtört⊠neti Int⊠zet ⊠s Muzeum (HMHIM) Budapest, Hadtört⊠nelmi Lev⊠ltar ⊠s Iratttar, *Der letzte grosse rusische Durchbruchversuch 20 March–20 April 1915*, part IV, p. 965.

15 SAW-KA, Die durchbruchsschlacht von Gorlice1915. Bearbeitet von Obst. AD Feldp. Erich Kahlen, p. 18-22.

16 HMHIM, *Der letzte…*, part IV, op. cit., p. 959-960.

17 SAW-KA, Die durchbruchsschlacht von Gorlice1915. Bearbeitet von Obst. AD Feldp. Erich Kahlen, p. 18-22.

18 HMHIM, *Der letzte…*, part IV, op. cit., p. 961, 963.

19 SAW-KA, Die durchbruchsschlacht von Gorlice 1915. Bearbeitet von Obst. AD Feldp. Erich Kahlen, p. 18-22.

20 G. Artl, op. cit., p. 156; SAW-KA, Manuskripte Luftfahrtruppen Nr. 11, Die entscheidende Schlacht bei Gorlice 1915. Entwickelt von Oblt. Fp. Erich Kahlen, p. 7.

21 SAW-KA, Manuskripte Luftfahrtruppen, Auszug aus dem Kreigstagebuch des Oberleutnant Feldpilot Wilhelm Schreiber der kuk Fliegerkompanie No. 15 aus dem Weltkriege 1914-1918, p. 45-55.

22 SAW-KA, Manuskripte Luftfahrtruppen, Die durchbruchsschlacht von Gorlice 1915. Bearbeitet von Obst. AD Feldp. Erich Kahlen, p. 21.

23 РГВИА. Ф. 2067. ОР. 1. Д. 142. Л. 384 - Из истории Карпатской…, op. cit., p. 307 - reports from April 13 (26), 1915. The photographic documentation from the epoch proves the shooting down of the Imperial-Royal Air Force plane in the Daliowa area and most probably it was an airplane from that flight.

24 SAW-KA, Die durchbruchsschlacht von Gorlice1915. Bearbeitet von Obst. AD Feldp. Erich Kahlen, p. 18–22.

25 Ibid.

26 G. Artl, op. cit., p. 157; SAW-KA, Neuefeldakten 3. AK Ktn. 262, Op. No. 3883/11, 27 April 1915, at 22.20.

27 G. Artl. Op. cit., pp. 157-158; SAW-KA, Manuskripte Luftfahrtruppen Nr. 11, Die entscheidende Schlacht bei Gorlice 1915. Entwickelt von Oblt. Fp. Erich Kahlen, p. 13.

28 G. Artl, op. cit .; SAW-KA, Neuefeldakten 3. AK Ktn. 272, 3. AK Op. No. 3883/9, 28 April 1915.

29 G. Artl, op. cit., p. 158; SAW-KA, Manuskripte Luftfahrtruppen Nr. 11, Die entscheidende Schlacht bei Gorlice 1915. Entwickelt von Oblt. Fp. Erich Kahlen, p. 8.

30 G. Artl, op. cit .; SAW-KA, Neuefeldakten 4. AK Ktn. 11, Op. No. 3330/6 and 3330/21, 28 April 1915.

31 SAW-KA, Die durchbruchsschlacht von Gorlice1915. Bearbeitet von Obst. AD Feldp. Erich Kahlen, p. 18–22.

32 Ibid.

33 From now on, on April 28, 1915, the reconnaissance in the Gorlice region will be carried out by German airmen. On that day, four German crews took part in a group flight and bombing targets in the far hinterland of the front in Starzawa. They bombed the Russian transport at the railway station together with the adjacent yards. 25 bombs were dropped on ground targets, but no major movements of units or delivery columns were observed - see S. Rosenboom, op. cit. p. 61.

34 G. Artl, op. cit .; SAW-KA, Manuskripte Luftfahrtruppen Nr. 11, Die entscheidende Schlacht bei Gorlice 1915. Entwickelt von Oblt. Fp. Erich Kahlen, p.13.

35 SAW-KA, Die durchbruchsschlacht von Gorlice1915. Bearbeitet von Obst. AD Feldp. Erich Kahlen, p. 18-22.

36 G. Artl, op. cit .; SAW-KA, Neuefeldakten 4. AK Ktn. 11, Op. No. 3350, 29 April 1915 at 20.50.

37 G. Artl, op. cit ,; SAW-KA, Neuefeldakten 4. AK Ktn. 11, Op. No. 3355 and 3355 / II, April 29, 1915.

38 *For a long time, I have been observing with particular interest the successes of air crews, which, even in the most unfavorable conditions, are able to perform the tasks entrusted to them. I express my appreciation to each of them for their great achievements and I expect that also in the next important days my pilots will not weaken and will do their best* - cf. G. Artl, op. cit.; SAW-KA, Neuefeldakten 4. AK Ktn. 11, Op. No. 3356, 29 April 1915.

39 SAW-KA, Die durchbruchsschlacht von Gorlice1915. Bearbeitet von Obst. AD Feldp. Erich Kahlen, p. 18–22.

40 Ibid.

41 G. Artl, op. cit., p. 159; SAW-KA, Manuskripte / Luftfahrtruppen Nr. 11, Die entscheidende Schlacht bei Gorlice 1915. Entwickelt von Oblt. Fp. Erich Kahlen, p.9.

42 G. Artl, op. cit., p. 159; SAW-KA, Manuskripte / Luftfahrtruppen Nr. 11, Die entscheidende Schlacht bei Gorlice 1915. Entwickelt von Oblt. Fp. Erich Kahlen, p.9.

43 G. Artl, op. cit .; SAW-KA, Manuskripte / Luftfahrtruppen Nr. 11, Die entscheidende Schlacht bei Gorlice 1915. Entwickelt von Oblt. Fp. Erich Kahlen, p.14.

44 G. Artl, op. cit .; SAW-KA, Neuefeldakten 4. AK Ktn. 11, Op. No. 3370/1, May 1, 1915.

45 The results of the reconnaissance concerning the positioning of Russian forces, carried out mainly by airplanes, were in fact completely in line with the actual state of affairs - see G. Artl, op. cit.

46 M. Ronge, Kriegs- und Industriespionage. Zwölf Jahre Kundschaftsdiens, Zürich-Leipzig-Wien 1930, p. 155.

47 SG Nielipowicz, Operation *gorlicka in the documents of the Russian State Military and Historical Archives,* [in:] Military and political significance of the Gorlice operation in the warfare of World War I, Gorlice 1995, p. 21.

48 S. Błażewicz, *Russian defense of Gorlice on May 2, 1915,*[in:] Materials from the international conference Signs of Remembrance - in the footsteps of World War I, vol. II, Gorlice 2009, p. 64; II Rostunov, Russkij front, first mirowoj war, Moscow 1976, p. 99.

49 W. Brzoskwinia, K. Wielgus, *Fliegerkompanie...*, op. cit. p. 79.
50 SAW-KA, Die durchbruchsschlacht von Gorlice1915. Bearbeitet von Obst. AD Feldp. Erich Kahlen, p. 21–22.
51 Велика війна 1914–1918 pp. and Україна. Мовою документів і свідченъ, Київ 2015, p. 83.
52 *Wings of victors, wings of the defeated, wings of freedom, wings of independence* 1918, edited by A. Olejko, P. Mrozowski, D. Mechin, B. Ciglić, Zabrze-Tarnowskie Góry 2019, pp. 514-515; data from private archives of prof. S. Januszewski and prof. A. Kozycky.
53 B. Świeykowski, *From the days of horror in Gorlice*, Krakow 1919, p. 77–78.
54 Fr. K. Wais, op. cit., p. 81–82.
55 Extract from the war journal - www.csongor6.fw.hu/Neus12/index3.htm - entry from March 16, 2021.
56 *Liber Memorabilum (Memorial Book). Chronicle of the village of Targowiska near Krosno, 1846-1926,* p. 77, via M. Klara.
57 B. Gąsiorowska, op. cit., p. 106.
58 Ibid, p. 108.
59 Ibid, p. 109.
60 Ibid, p. 109–110.
61 *Diary of Dr. K. Zaleski,* collections of the Historical Museum in Sanok- a copy of the fragments in the author's collection, no page pagination. On May 1, 1915, the airplanes of the Imperial-Royal Air Force bombarded Krosno- one Russian soldier was killed and five were wounded - see the school chronicle of School No. 1 in Krosno - a copy from the collection of Dr. E. Bereś; Central State Historical Archives of Ukraine in Lviv (Centralnyj Derżawnyj Istoricznyj Archiv of Ukraine at Lvov - CPAHU), Fond 178, op. 1, spr. 313, p. 7.
62 E. de Henning-Michaelis, The Storm of History. Diary in the World War 1914-1917, Volume I, Warsaw 1928, p. 166.
63 SAW-KA, Manuskripte Luftfahrtruppen, Fliegerkompanie 10 - Okt. 1914–May 2, 1915.Studies. Bearb. Obst. Fp. AD Erich Kahlen, pp. 12–14. S. Chrapczyński, *Muscovites in Brzesko,* [in:] "Zeszyty Wojnickie" 1995, No. 4 (36), pp. 8–9; Z. Jordan, *Memoirs of the Old Granaciarz* (II), "Zeszyty Wojnickie" No. 5 (37), pp. 9–11; J. Magdoń, *Aeroplan*..., op. cit., pp. 34–35; J. Borzęcki, op. cit. p. 37. W. Brzoskwinia and K. Wielgus (op. cit.) report that one of the missiles destroyed a wing of the monastery in Zbylitowska Góra, half an hour after the departure of the commander of the Russian 3rd Army from the briefing taking place there General R. Dimitriev. In the light of the archival information from KA Wien, this event was different. On the basis of Stefan Bielawski's photographic collection - photographs taken in the area of the monastery of PP Sercanek (Sacrecceur monastery) in Zbylitowska Góra - it can be established.
64 F. Molnár, op. cit., p. 170–171.

Below: The harsh winter of 1915 did not stop the postal line to the Przemyśl Fortress ... Aviatik B.I planes in the "polar" scenery of the Przemyśl-Żurawica airport. - SAW/KA Wien collection

Above & Above Right: The time of capitulation ... March 22, 1915 brought the capitulation of "Przemyskie Lions" - collection of the SAW/KA Wien

Above, Above Right, & Below: The battlefield of Gorlice 1915 ... Battle images of the Gorlice Break recorded on photographic plates. collections of R. Frodyma and the author

East direction ... Przemyśl-Lviv 1915

There would not be such a great triumph of the Central Powers' troops in the Tarnów-Gorlice operation if not the aerial reconnaissance, which was performed by Flik 10, Flik 15, Flik 13, and Flik 8 crews at the turn of April and May 1915 who provided AOK and OHL with important information about the opponent. And it was an exceptionally cold, foggy, and windy time, in which the crews of the Imperial-Royal Air Force performed miracles, recognizing the area of the future breakthrough near Gorlice. For these flights, the crew of the KuK Luftschifferabteilung - 24 airmen performing combat flights - deserved the highest recognition, being co-contractors of the later Gorlice success. The list of air operations of the crews of the Fliegerkompanien subordinate to the field commands of the 4th and 3rd Army, in the zone planned by OHL and AOK to break the Russian front between Tarnów and Gorlice from 19 April - 8 May 1915 was as follows:

Flik 10 - Zakliczyn airport; commander Hptm. E. Kahlen - flights: April 19, 1915 - 2; April 22, 25, 27 and 29, 1915 - 1 each; May 3, 1915 - 1 /total seven flights.
Oblt. O. Schmoczer von Mechenzeff - flights: April 1 - May 8, 1915 - 15 flights, including in the period April 17-May 8, 1915 - 9 flights (April 19, 1915 - 2 flights, 22, 25, 27, 29 April 1915 - one flight each, 1, 4, 7 and 8 May 1915 - one flight each)
Oblt. E. Rzemenowsky von Trautenegg - flights: 20, 22, 26, 27, 29 April 1915 - 1 flight, 1, 4, 7, 8 May 1915 - 1 flight - nine flights in total.
Oblt. Hans von Reidlinger - 0 flights
Oblt. Emil Kumstat - flights: 20, 22, 26, 27, 29 April 1915 - 1 flight, 1, 5, 7 May 1915 - 1 flight - total - seven flights
Lt. Cik - flights: April 20, 21, 26, 30 April 1915 - 1 flight each, May 2, 1915 - 2 flights, May 5, 1915 - 1 flight - seven flights in total
Lt. Ischpold - flights: 24, 28 April 1915 - 1 flight each; May 7, 1915 - 1 flight - three flights in total
Lt. Csicseri - 4, 7, 8 May 1915 - 1 flight each - three flights in total
Lt. Konschel - flights: 17, 21, 24, 26, 28, 30 April 1915 - 1 flight, 2 May 1915 - 2 flights, 5, 7 May 1915 - 1 flight - 10 flights in total.
Daily Flik 10. operated with a force of 3-4 Albatros B aircraft[1].

Flik 15 - Jasień airport: Commander of Hptm A. von Mandelsloh - flights: April 22, 1915 - May 1, 3, 4, 6, 1915 - 1 flight each - four flights in total
Oblt. B. Losonczy von Losoncz - April 1 - May 8, 1915 - 21 flights, including in the period April 17-May 8, 1915 - 15 flights (17, 19, 22, 24, 26, 28, 29 April 1915 r. 1 flight, 1 May - 1 flight, 2 May - 2 flights, 3, 4 May 1915 - 1 flight, 6 May - 2 flights, 8 May - 1 flight)
res. Lt. Hautzmayer - flights: 19, 24, 29 April - 1915 - 1 flight each, May 2, 1915 - 1 flight - 4 flights in total; Lt. Rath - flights: 21, 26, 27 April - 1915 - 1 flight, 1, 2, 8 May 1915 - 1 flight - six flights in total
Oblt. J. Wierzejski - flights: 24, 28 April 1915 - 1 flight, 2, 6 May 1915 - 1 flight - four flights in total
Lt./Hptm. Fp. W. Schreiber - flights: April 17, 25, 1915 - 1 flight each, May 6, 1915 - 1 flight - three flights in total
Oblt. Cserich - flights: April 19, 26, 1915 - 1 flight each, May 4, 6 and 8, 1915 - 1 flight each - total - five flights
Feldw. J. Siegel - flights: April 1-May 8, 1915 - 21 flights, including in the period April 17-May 8, 1915 - 13 flights (17, 19, 20, 21, 24, 25, 26, 27, 29 April 1915 - 1 flight, 2, 4, 6, 8 May 1915 - 1 flight).
Daily Flik 15 operated two aircraft, initially Lohner C "Gebirgspfeilflieger" and then Albatros B.I[2]

Flik 8 - Strážske airport - today Slovakia
commander - Hptm. Studeny - flights: - 0
Lt./Oblt. Eng. R. Stanger - on leave
Oblt. Maythenyi - 5 May 1915 - missing
Oblt. J. Kostrba - 0 flights
Oblt. Hartzer - 0 flights
res. Oblt. Langl - 0 flights
Oblt. Emanuel Grycz - April 25, 1915 - one flight
Zgsf. Siara - 5 May 1915 - missing,
Zgsf. Stucs -?
Zgsf. Stolba - April 25, May 7, 1915 - two flights in total
In Flik 8. On April 25, 1915, only two planes were ready for flights - the unit used Aviatik B.I and Albatros B.I[3]

Flik 13 - Bardejov airport - today Slovakia
Commander: Hptm von Wilhelm - 0 flights
Hptm. J. Cejnek - flights: April 2 - May 8, 1915 - 17 flights, including in the period April 17 - May 8, 1915.12 flights (19, 20, 21, 22, 25, 25, 27, 29 April 1915 - after 1 flight, on 1, 2, 7 and 8 May 1915 - after 1 flight)
Oblt. von Nowy - 0 lots
Hptm. Steiner - flights: 19, 20, 21, 22, 25, 26, 27, 29 April 1915 - 1 flight, 1, 2, 7, 8 May 1915 - 1 flight - 12 flights in total
Oblt. Pfeifer - 19, 20 April 1915 - 1 flight, 2, 3, 8 May 1915 - 1 flight - total - five flights
Lt. Otto Seidl - April 29, 1915 - 1st flight
Zgsf. Till - 19, 20 April 1915, 2, 3, 8 May 1915 - 1 flight each - five flights in total
Zgsf. Horwath - April 29, 1915 - one flight.
Daily Flik 13 operated 2-3 Aviatik B.I plane[4]

From May 1915, not only the Austro-Hungarian air force operated actively over Galicia, thanks to which preparation of the directions of the attack of the German 11th Army field and of the field armies subordinate to AOK[5]. As of the morning of May 2, 1915, the following units of German aviation performing systematic photo recognition of the enemy's position in the "Gorlice Stellung" area were stationed at Galician field airfields:

• FFA 1st commanded by Hptm. Jasper von Oertzen (six planes), who was to support the Guard Corps, stationed at the field airfield in Czchów, separated from the Brief Abteilung Ostende (BAO - unit after being transferred by rail from the Western Front - Metz-Frescaty area - in the period 10 March-22 April 1915 r. stationed at the airport Deuthen [Dajtki] near Olsztyn)

• FFA 69th (formerly FFA "Müller" - six aircraft) commanded by Hptm. Kurt Müller (4 planes) to support Imperial-Royal VI Corps from the field airport in Falkowa

• FFA 21. (six aircraft) to support the X Corps from the field airport in Nowy Sącz

• FFA 8th (six aircraft) to support the XXXXI Reserve Corps of the German 11th Army field from the field airport in Nowy Sącz

• BAO (about 16 aircraft) assigned to the field command of the 11th Army was located from April 26, 1915 at the airport in Kraków-Czyżyny (Rakowice airport)

• FFA 57. (six aircraft) assigned as support to the 47th Reserve Field Infantry Division of the 4th Army at the field airport in Biadoliny Szlacheckie[6].

During the Gorlice Breakthrough, the crews of the KuK Luftschifferabteilung and the German FFA became the "eyes of the AOK" and of the German 11th Army field command, because of which its commander constantly received air reports about the retreat of the Russian 3rd Army's forces from the area of Jasło and Kołaczyce, as a result of which the main forces of his army did not act in the dark, forcing Wisłok between Haczów and Frysztak and strengthening east of the riverbed. Little losses were incurred during the reconnaissance flights (Flik 10 lost in June 1915 the Albatros B I 24 planes, forcibly landing near Grybów)[7].

On the day of the beginning of the Gorlice Breakthrough on May 2, 1915, the actions of the Imperial-Royal and German aviation were aimed at obtaining the latest information from the battlefield. On the first day of fighting at 15.45 in the field section of the Imperial-Royal 4th Army, subordinate to Fliek. carried out reconnaissance of the front line: in the Biecz section, a 2 km-long infantry column was detected, and a further 14 enemy infantry companies were detected in the section from Rzeszów to the west, as well as one artillery battery and a 2-km-long transport column. The Tuchów-Jasło-Frysztak-Rzeszów-Łańcut-Sokołów-Kolbuszowa-Mielec-Dąbrowa Tarnowska region was free from new enemy forces[8]. On May 3, 1915, the Flik crew. 15. provided reports on the retreat of Russian forces in the area of Biała and Dunajec; the bridges in the entire region over the Vistula and Dunajec River and from Tarnów to the north were on fire; the bridges over the Dunajec River were destroyed; the large transport column was in the retreat in the Kowalowy section - east of Tuchów, and the small column was detected in the reverse movement near Brzostek; in the retreat there were columns to Pilzno, while in the Tarnów-Pilzno-Dębica-Ropczyce-Brzostek-Jasło section no new enemy forces were reported. The reconnaissance of the Nowy Korczyn-Dąbrowa Tarnowska-Tarnów region and the Vistula region showed the enemy's forces in the march near Szczucin. Air reports from the period up to 5 May 1915 showed that the field units of the Russian 3rd Army were retreating towards the Wisłok line; reconnaissance on 4 and 5 May 1915 showed the opponent's retreat from the region of Jasło and Dębica towards Rzeszów, and from the region of Tarnów to Radomyśl Wielki-Mielec towards Sandomierz. On 5 May 1915, only one infantry battalion was detected in Radomyśl Wielki, and another battalion in the Tarnów-Pilzno section; bridges and modern field fortifications were reported near Sandomierz, and further on the section along the Vistula to Dęblin[9]. The crews of Flik .10 and Flik 15 in the period of May 2-5, 1915, they recognized the area of Ołpiny-Biecz and Tuchów-Tarnów from the field airfields of Jasień and Zakliczyn, and on May 6, 7 and 8, 1915, they recognized the retreat of Russian units in the strength of 1 1/2 DP on north of the Tarnów-Rzeszów road to Szczucin and Połaniec to Sandomierz; south of the Tarnów-Rzeszów route, enemy units in the strength of about the corps in the Jasło-Frysztak area in the retreat to Rzeszów-Łańcut and fighting along the entire Wisłoka line - burning villages and forests near Mielec. On May 6, 1915 heavy fighting was observed in the area to the north-east of Jasło and to the west of Pilzno[10]. On 5 May 1915, the Flik 15 composed of Oblt. B. von Losonczy and the commander of Flik 15 Hptm. A. von Mandelsloh made a long-distance reconnaissance of the Sandomierz and Dęblin regions, and Lt. Cserich during the next flight in the Russian Lipie field airport near Tarnów, where he detected five Farman-Doppeldecker planes and the lack of tent hangars, which indicated that his evacuation had started, a bomb caused an explosion right next to a landing Russian plane[11]. On May 6, 1915, Oblt. Fp. W. Schreiber from Flik 15 made one flight from the Jasień airfield to the area of Tarnów-Dąbrowa-Tarnowska-Szczucin-Pacanów-Stopnica-Polanice-Ostrówek-Mielec-Kolbuszowa-Sędziszów Małopolski-Dębica-Pilzno-Radomyśl Wielki-Dąbrowa Tarnowska-Wał Ruda where a forced landing was made, so that, after 2 ½ hours in the area of Jadowniki Mokre, he would drop a report for the General Stöger-Steiner group and return to his home airport. Now of the rupture of the southern section of the Eastern Front, on May 9, 1915, Flik 15 recognized the line Wisła-Dąbrowa Tarnowska-Dębica-Krosno and further east and south-east, and in Mielec Oblt. B. von Losonczy and Oblt. Tsserich reported the observed carriage of Russian troops[12]. On the same day, in Jasło, the Flik 15 composed of Oblt. Father Schmoczer von Mosenzeff and Oblt. E. Kumstat had to land forcibly and on 10 May 1915 in the morning the Flik 15 composed of Oblt. B. von Losonczy and Lt. Hautzmayer recognized Russian positions - five lines of trenches - in the Czudec area. It was observed that all detected Russian transports were on the march to the east, which indicated the retreat of the Russian army. As part of the observation of the retreat of the tsarist army to the east on May 11, 1915 Oblt. Fp. W. Schreiber from Flik 15 made a long reconnaissance flight from the Jasień field airfield to the area of Tarnów-Frysztak-Jawornik Polski-Przeworsk-Jarosław-Sieniawa-Sokołow-Mielec-Wisła-Szczucin-Dąbrowa Tarnowska-Tarnów, landing in Tarnów where it was stationed from 10 May 1915, at the former Russian airport in Lipie, his mother Flik 15 was transferred from the Jasień region (Hptm. E. Kahlen in his study gives 10 May 1915 as the date of ferrying Flik 15 to Tarnów, and Oblt. Fp. W. Schreiber, 12 May 1915 - AO's note). Report on the observed Russian columns in the march near Jawornik Polski and in the Przeworsk-Jarosław-Błażów-Dynów region, as well as on burning villages east of Rzeszów and the observation at 08.00 in Rzeszów, many Russian troops in the retreating march to the east, and over the Wisłok, Wisłoka and Vistula burning bridges were sent to the AOK[13].

On the day of the Gorlice Breakthrough on May 2, 1915, the crews of the 8th and 13th Fliegerkompanie subordinated to the field command of the 3rd Army, carried out reconnaissance of the routes Zborov-Krempna, Nižná Pol'anka-Svidnik, Krempna-Nowy Żmigród, Nowy Żmigród-Dukla, Nowy Żmigród-Jasło, Jasło-Biecz, Gorlice-Nowy Żmigród and Zborov-Gorlice[14]. An interesting fact was the use of the 4th Army's Imperial-Royal Field Corps in the area west of the Tuchów-Tarnów line in the period just before the beginning of Operation Ballonabteilung 2/2. R. and Ballokompanie 5. described as Ballonabteilung 1/5. B. soon joined with Ballonkompanie 12 (commander Hptm. Hermann Kühn, number - five officers, nine non-commissioned officers, 71 soldiers, 44 carts, 12 horses and four reserve horses, one Drachenballon M. 98 balloon, one Kugelballon M. 96th and three gas filling units). On May 3, 1915, both units carried out reconnaissance for the Imperial-Royal artillery (Ballonkompanie 5., which was assigned to Krakow on August 20, 1914, had one Drachenballon M. 98 with a capacity of 750 m3 and one Kugelballon M. 96). The balloons were used for observation in favor of the 15 cm heavy howitzers, 24 cm, and heavy mortars "Škoda" M 11. 30.5 cm on the impact section of the 14th Corps. Then they were used for the benefit of the Artillery Group, including one M 11 30.5 cm heavy mortar "Škoda" and one M 14 42 cm heavy howitzer installed at the IX Corps, and for a battery of 24 cm mortars and heavy mortars "Škoda" M 11. 30.5 cm. After May 3, 1915, the Ballonkompanie 12th was moved back to Kraków. In the days of the Gorlice Breakthrough, no balloon company worked for the field corporation of the 3rd Army[15].

As part of reconnaissance flights in the Gorlice region, the crews of the German BAO made between landings at the airport in Nowy Sącz, where the headquarters of the German command was located. The next BAO airports were: 11-14 May 1915 Tarnów, 14 May-12 June 1915 Rzeszów, from where air reconnaissance for the Bavarian Corps was carried out, 12-20 June 1915 Munina near Jarosław, from where the above-mentioned Bavarian Corps, June 21-July 27, 1915 Wróblaczyn from where the unit, after being transferred to the Western Front in mid-July, was stationed at the Ghistelles airport until the end of August, and at the end of September it was at the Rathel airport[16]. The Central Powers' air force systematically bombed the front lines - in the "Mieczysław Pattek's Diary" written in Rzeszów in those days, it was noted:

May 2, 1915 []. Sunday... In the afternoon, around [h] 4 [4 pm], the airplane [plane] arrived over the city from the Carpathians. The soldiers hid themselves, fled the bridges. He circled over the station and dropped 2 bombs. Until now, it has not been established where he had thrown and what kind of havoc he had wreaked. They say that he hit the Circassian camp in Staromiasto(Old town). Others say he crashed 2 train carriages and destroyed a railroad track[17].

May 3, 1915 []. Monday ... An airplane [plane] threw one bomb yesterday, which was intended for the railway bridge. He hit away from him ... making a pit [funnel] and breaking trees. Yesterday afternoon, the Circassians went to Tarnów, because they passed Austria through the Dunajec... The witness talked about the bombing of Tarnów. The shore occupied by Austrians [s] is all burning and rumbling with the roar of cannons. The headlamps work all night long. The Aero. [Plans - planes] circle the city and attack the Rus [Russian] position and ammunition depots.

May 4, 1915 [r.]. Tuesday ... They say Sunday's airplane [plane] accompanied the Cossacks coming from here and threw bombs at them.

5 May [1915]. Wednesday ... The army's movement is intensified. In the evening news came that Tarnów was taken by the Austr. [ian] and Prussian [German] armies. People who had contact with the Russian army are even fleeing from Dębica, where there is supposed to be such a roar of cannons that it is hard to hear anything else. The troops lost their heads. It all looks like panic. In the afternoon, the submarine movement began. They ride a long rope to Łańcut, others to Sokołów [Małopolski]. They drove from Tarnów and Jasło all night. Tarnów was supposed to fall into the hands-on Wednesday morning[18].

May 6, 1915 []. Thursday ... Today a rumor spread that Jasło was taken at night yesterday. Austria [s] do not yet dispose of Tarnów. Roads from Tarnów are packed with submarines going north. All day long they haul submarines from Jasło, and in the evening they ride pontoons. These are, as convoying soldiers say, transports of two armies.

May 7 [1915]. Friday ... In the morning over the city around 5 'oclock, 5 Austr. [ian] and Prussian [German] airplanes arrived. They circled over the city, and 6 o'clock they started throwing bombs. They threw about 40 bombs on the city, they say. 3 of them fell on the market, near the Stawisko 9 house, airplane hangars, etc. A bomb crashed into Laub's house [?] On Mickiewicz Square, broke the roof and started a fire, another one next to the Trybowicz building, smashed a piece of cornice, torn off the facade and smashed windows same in the whole neighborhood. killed and wounded over a dozen soldiers and officers. Of the civilians,

Above: Aviatik B.II 32.15 at the front was powered by a 120 hp Austro-Daimler. Source: Aeronaut.

Below: Aviatik B.II 32.37 after crashing into a building on the airfield. Source: Aeronaut.

Above: German-built Aviatik B.I 31.03 was powered by a 100 hp Mercedes and purchased by Austria before the war. Source: Aeronaut.

Left: A crew with their two-seater at the front. Source: Aeronaut.

16 are wounded and 2 have already died ... The staff of the 3rd [3rd] army with [general] Radko Dymitriewicz [Dmitriev] is in Rzeszów. All hospitals are already packing, and some have already left. They are to go all the way to Jarosław. Prisoners brought from the Carpathians claim that Austrian troops will be here within 3 days
May 8 [1915]. *Saturday day* ...About ½ 9 [hr. 8.30] *2 airplanes* [planes] *appeared over the city* [Rzeszów]. *Cannons fired at him from all sides. One circled over the city and threw a bomb next to Janikowa's house, the other hovered above the railway threw a bomb on Stefan Czarniecki street, from which the cottage caught fire and 3 wooden houses burned down, he threw bombs again at Stawisko, in total 3, but he did not cause any damage. The nervousness in the city is great, and when the airplane* [plane] *flies over, everything escapes in terrible panic, civil.* [ilians]*, the army, everything so that in a few moments there is no one on the street. Everything is hidden in tenement houses. Everyone who goes looks up, and when he hears the roar of the cannons - without even seeing the airplane* [plane], *he hides into the tenement houses. I had to do the same today while being on Dr. Jabłoński street. In the afternoon, around 2 PM, the airplane* [plane] *showed up again. At that moment, the cannons began striking. The airplane was circling far and high, and smoke from bursting shrapnel hovered around it. He made full circle and fled the city without dropping a single bomb. The retreat continues. Inflatable boats, artillery, ammunition, everything retracts. The infantry is to pull the roads north. At home, I ordered a vat of water to be prepared so that the fire could be extinguished if it was started by a bomb ... A Russian officer was to bring the news that there are Austrians in Dębica. They brought prisoners from the Carpathians ... You can see nervousness among the army. It was announced that all non-school maps were to be submitted to the military for review. There is a fear of spies. Austr.* [ians] *are expecting in a few days*[19].

Fr. K. Wais wrote in his memoirs:
Third of May [1915], *the day of the indulgence in the Klimkowska parish, was, however, a sad day. In the village there was a tumult, constantly made by several thousand soldiers, who, worn down and covered with dust, washed in the river, and cleaned their clothes.* The aeroplanes [airplanes] *flying in the sky every now and then seemed to announce with their muffled rumble some important events. From the Dukielska Pass and* [Nowe] *Żmigród, the roar of cannons could be heard, much louder than in the previous days. Moreover, the voices of rifles firing from aeroplanes* [planes] *were heard from all sides ... Suddenly, at half past noon, the infantry left behind the rest of the Cossacks, because the news that "German is following" appeared... aeroplanes* [planes] *flew from west to east and on the contrary: cannon shots came only from the direction of Krosno*[20].

Whereas Dr. K. Zaleski in his diary written in Sanok wrote:
2.05.[V] 1915[r.]. Sunday ... Muscovites were to say, that they were expecting a bomb at the train station, they worried about the barracks where there were supposed to be large supplies of ammunition, so that in the event of an explosion, half of Posada would be left and therefore cities.
9.05.[V] 1915[r.]. Sunday ... Kites. There were no [planes], thank God, and the cannons rang, not as close as yesterday, only towards evening, and thundered until late, the sky was red after sunset. In the Sanoczek side I saw a lot of smoke and people said they saw an even larger area in the smoke.
10.05.[V] 1915 [r.]. Monday ... When the sun was shining and the clouds parted, a kite was hovering high above us [plane - on that day reconnaissance in the Sanok area were led by the Flik 10, Flik 15, and Flik 13], despite the sometimes quite strong wind. Fr. Witoszyński, son-in-law of Fr. Mykiełyty from Sanok that I should go to his sick 10-year-old brother-in-law with typhus, promising to drive me back himself and ensuring that the roads are free and that the position is still well away from him. For a moment, as if an evil fly buzzed in the head: ae-ro-plan-ny ...[planes]. And the ride really took place without an accident and without even a single kite showing up [plane].
16.05.[V] 1915 [r.]. Sunday ... In the morning, he flew right next to us over the castle trees, gliding really like a bird, a kite [plane] and descended on the station, which was arranged in front of the Castle beyond the San River in the fields of Nowakowa (Olchowce- manor house). On the other hand, there is a radiographic station in Sujac again. What is the most interesting in the cultural world is coming or going to our Sanoczek... The half-burnt bridge makes an unpleasant impression (6 water spans on this side are left).
17.05.[V] 1915 [r.]. Monday. A lovely morning, but I got up after 6. During the morning exercise, he flew up [from the field airport in the present "pines"] Kite [aircraft of the Imperial-Royal Air Force] and flew towards Hungary (?). Then they snarled again in air[21].

From the auxiliary direction of fighting in 1914, the Carpathians became in 1915 the main, strategic direction of the Russian army's offensive on the Eastern Front. Until May 1915, Stawka concentrated here the maximally increased forces of the South-Western Front, which during the Carpathian Battle captured 662 guns (including 600 in the Przemyśl Fortress), 455 machine guns, two planes and many other supplies of military equipment and ammunition. Unfortunately, in the spring of 1915, the front forces were practically exhausted and during the Gorlice Breakthrough in May1915, the line of defense of the Southwestern Front was broken. The field forces of the 3rd and 2nd Army and the German Southern Army in mid-May 1915 completely cleared the Bieszczady Mountains of the Russian 8th Army's field forces, and by the end of May 1915 forced them and the 3rd field army to withdraw beyond the San and until it is temporarily closed in the ruins of the Przemyśl Fortress. The Carpathian passes, which were to become the road to the victory of Stawka over the Central Powers bloc, in the Beskids and Bieszczady Mountains became the scene of the most serious defeat of the Russian army in the Great War[22].

During the fighting in the Carpathians, from January

Above: The reality of the air service on the Serbian Front ... Crashed Flik 4 plane - Source: Fotoarchiv SAW-KA Wien

Above: From the Aviano and Comina airports on February 18, 1916, nine bomber planes, Italian "flying giants" of the Caproni Ca. 3 type, took off for the Laibach raid. (six flew over the target, the rest returned to their home airports due to mechanical problems). Approx. time. At 9.00 the expedition was attacked by Flik 4 pilots piloting Fokker A.III (E.III) Eindecker fighter planes of German production ... The photo shows a shot down Italian "giant" type Caproni Ca 3 No. 703. Source: Fotoarchiv SAW-KA Wien

1915, the crews of the 30th FFA worked for the German South Army, in February and March 1915 supported by the crews of the 54th and 59th FFA, taking photos of reconnaissance of enemy positions in the Carpathians, which were particularly active in early June 1915. In early spring 1915, the army received as its own air support FFA 26.a (six aircraft), which was transferred by rail from the Western Front after arriving in Laon, field airport in Mukachevo. Operating from here over the Western and Eastern Bieszczady Mountains, the crews of the unit renamed to FFA 59, operated as assigned to the Corps of General Feliks von Bothmer, and recognized enemy artillery positions in Zwinin, Klewa, Tuchla and Jarowiszcz, performing their photo recognition as well as hostile positions near Libochora and Hołowczanka. The second task of the 59th FFA crews was to recognize General Szandor Szurmay for the Imperial-Royal Group, with particular emphasis on the position of the enemy artillery in the area south of Wołosaty and in Wołosaty itself (church area), Tarnica region and Wyżniańska Pass. On April 6, 1915, one of the crews of the unit recognizing the Użocka Pass and Sambor, due to an engine failure, forcibly landed on the Użocka Pass just behind its lines, crashing the plane. With the approaching date of the offensive in the Carpathians, the German Southern Army received another air unit as support, which was the FFA 54. As soon as the front line began to move east, the FFA 59. was at the field airport in Synowecko, performing aerial reconnaissance Drohobycz region and the Stryj valley[23].

The German aerial reconnaissance was especially important during the May breakthrough of the front at Gorlice and on May 14, 1915 in the Nadwórna-Dolina-Halicz area and during the fights on June 4, 1915 on the Dniester and during the subsequent fights of the Southern German Army on both sides of the Bug towards to Brześć Litewski, which turned out to be particularly helpful during the fights at Grabowice, Hrubieszów, Wojsławice, Chełm, Włodawa and in Brześć Litewski in the further march to Kobryn. The German 11th Army field support in the spring offensive received 54 aircraft belonging to the 1st, 8th and 21st FFA and one Kampfgeschwader (KG), which originally performed photo reconnaissance and during the offensive, correcting artillery fire[24]. During the fights over the middle San, the FFA 8 crews operated over the Przemyśl Fortress, while the FFA 1st and 21st crews and the above-mentioned KG operated for the artillery of the "Müller" Group in the Lubaczów region. After the fighting near Lviv, the crews of the FFA 66th and 69th supported the field troops of the 11th Army in their offensive actions in the east, and then in the second half of September 1915 the FFA. The 1st, 66th, 69th and 57th were directed to the Serbian Front (Süd-Osttfront)[25].

Flik 15 after ferrying to the Lipie airport near Tarnów, after a few days, it was prepared for another, imminent ferry to the field airport in Rzeszów, where it was located on May 14, 1915. At that time, the German FFA 57. Hptm. Heyder, with whom Flik 15 was stationed throughout the winter in the region of Jasień-Biadolina Szlacheckie, was assigned to the newly created Fliegergruppe Hptm. R. Eyba and on May 14, 1915, the German FFA 57. planes left the Brzesko area, crossing via Tarnów further east (on May 15, 1915, strong winds prevented combat flights and erecting two tent hangars, and damaged the reserve Flik 15). With the change of the front line and with the prepared attack on the Przemyśl fortress, on May 17, 1915, the Flik 15 and the above-mentioned FFA 57 (?) carried out a bombing raid of 14 planes on Przemyśl from the airport in Rzeszów with heavy bombs weighing 150 kg[26]. Along with the change of the front line, also Flik 10 was transferred from Zakliczyn to Tarnów, to the former Russian airport in Lipie, two days later to the field airport in Dębica, and its next airport was Rzeszów and Przeworsk. The retreating Russian units were bombarded with carbonite bombs -they hit the railway station in Rzeszów - and the air reconnaissance reached the area of Niemirów-Jaworów-Gródek Jagielloński and led along the railway line to Przeworsk and the Jaworów-Sądowa Wisznia road, where in one flight two battalions of Russian infantry with artillery on the march were detected. A large concentration of Russian troops was also detected in the railway station in Sądowa Vishnia and in the Melniki area[27]. At that time, the Flik 10 airport (in Dębica?) was bombed for the first time by a Russian plane that dropped two bombs that hit the airfield without causing any damage. Before ferrying to Rzeszów, the Flik 10 made 200 hours of flights, but the exhausted planes needed repairs and in Leżajsk Flik 10 received newer Albatros type B.II planes, faster than version B.I by 30 km / h. The flights were carried out gradually to the area of Tanwia in order to explore the area of the local forest complexes, but due to the cloud cover, they did not provide any specific information, although in one of them, carried out in the area of Lubaczów-Cieszanów-Ruda Różaniecka-Aleksandrowo-Biłgoraj, Russian forces were identified and in the area of Rozwadów (in May 1915 a Russian plane dropped a bomb on a field airfield near Rozwadow, but with no damage)[28]- Tanew-Kraśnik and in the area of the Vistula and San forks. This area was recognized as particularly important twice every day. Gradually the crew of Flik 10. made reconnaissance flights in the area of Lublin-Chełm-Dęblin-Kazimierz-the mouth of the San near Sandomierz, going up to 100 km into enemy territory, e.g., near Krasnystaw[29]

Correspondent of the Imperial-Royal Army F. Molnár wrote in his memoirs as follows:

May twelfth [1915] *appeared in Przemyśl the first Austro-Hungarian aeroplane* [plane]. *The people and our soldiers lying in the hospital knew that it was not a Russian plane from the fact that it was heavily fired from the Tatar Hill* [Mound]. *But to no avail. A week later, a squadron of five machines bombed the Russians while being loaded at the station, causing heavy losses. Russian caravans were also bombed. On the fourteenth* [June 1915] *the guns thundered for the first time, causing enormous nervousness in the hospital. Now our people knew everything. From then on, the guns thundered every day, until at about eleven o'clock on the night of June 2–3* [1915], *the Russian troops left the city in a hurry. Exactly at midnight, bridges built on the ruins of the old ones blew up. At half past four in the morning they entered Przemyśl from the north Bavarians* [from the royal 11th Infantry Division], *and from the south-west soldiers of our* [Imperial] *Ninth Division* [Infantry]. *People poured out onto the streets and helped build the bridge. By four in the morning, the city was full of soldiers, and women were throwing flowers at officers*[30].

A resident of Przemyśl Helena Jabłońska wrote in her diary:

May 11 [1915]. *Tuesday ... There was supposed to be a* [airship] *Zeppelin in Brzozów* [in fact, such an event did not take place], *he threw a bundle of hay which, when the Muscovites approached, exploded* [military facilities in Przemyślthey became permanent targets for air strikes [CC and German aviation]: *May 12* [1915]. *Wednesday ... In the morning there is a panic in the city - this is a German aeroplane* [plane] *throwing bombs. They shoot shrapnel at him. Around 10 am he threw a bomb on the Market Square near Szancer. 3 Russian soldiers and 2 women killed, several wounded. It is said that he also threw a bomb on* [street.] *Grodzka and burned a woman. They say that these bombs are completely different, they do not burst or punch holes - they burn. I don't know what to make of it. Could they be gases? Far more people fear of them. A terrible panic arises when they see the kite* [plane]. *Everything escapes, abandoning horses, carts, and goods. This wasn't there before. May 13* [1915]. *Thursday. In the morning, around 5 o'clock, the air raids started again. It had to throw a lot of bombs and I don't know with what result. It showed up several times a day, it was shot with shrapnel ... on May 14* [1915]. *Friday. At 5 o'clock in the morning there is such a crash, they are hitting the cannons this way, they are firing rifles from the roofs; they hit aeroplanes* [airplanes] *from mashinagers* [machine guns]. *Two of them fly high, they throw bombs densely ... The aviator was also throwing bombs in the evening. Hear numerous cannon shots. We live under such a threat that we do not know what to fear anymore, whether military revolt, slaughter and attacks, aeroplanes* [planes], *bombing, imprisonment, Siberia or knuts. May 17* [1915]. *Monday. What fear engulfed us with the dawn ... If anyone could get dressed, everything went out into the courtyard, sheltered in the basement and hall. Crazy crash, bang, hell has raged. It is an aviator circling over our heads, throwing bombs densely, and pounding at him from the hoops* [of the forts] *repeatedly. Shrapnel is flying and whistling on all sides ... Ears are bursting. He has already hit the mill through Dawid's tenement house on Dworski street, the roof is already torn off, the windows around* [spelling consistent with the original report] *broken, and a fire. Now in the courtyard of the Landwerkasarnia* [national defense barracks]. *6 horses, 9 soldiers and 2 officers fell victim ... In Zasanie, around 20 bombs were thrown on shafts, and they were effective. Panic in the city. One fell and killed several people in Błonia, damaged houses and casualties in Parkowa Street. There were five aeroplanes* [airplanes] *... on May 19* [1915]. *Wednesday. Everyone so scared only thinks about themselves ... Already at 5 in the morning the kites* [planes] *throw bombs from the frames* [forts] *and shoot at them from everywhere ... The kites* [planes] *raged. At 7 o'clock such a crash, thunder, lightning, the earth shakes, thunder after thunder, howl, the whistle of shrapnel only, the game of maschinengewers* [machine guns], *the roar of cannons; 40 bombs were thrown. Countless damage. Many human lives fell victim, many injured, many horses ... Everyone froze with terror. The terror is such that people walk like mannequins. I don't know what I fear more, bombs or an emergency sentence for the smallest thing. May 28* [1915]. *Friday ...*

A few hours of sleep, from which we woke up early in the morning, terrified by the bang of bombs. A whole flotilla of kites [airplanes], *I believe 5, circled over the city, throwing bombs very damaging densely. Muscovites hit them with cannons, from all forts ... on May 29* [1915]. *Saturday ... It seemed that the whole day would pass quietly, then in the evening, despite the unclean sky, the kite* [plane] *hovered over the station, threw 3 bombs, and escaped, but again 9 people were killed and many injured ... on May 30* [1915]. *Sunday. Ah God! Save! God, have mercy on us! Today it felt like the end of the world. At 5 o'clock in the morning a kite* [plane] *with bombs, threw a few bombs at once, and it was shot so much that we thought at times that it was already somewhere on the streets of Przemyśl fight. Right in underwear, everything escaped from the beds to the hall and cellars. I was shaking with terror, certain that our tenement houses were already being destroyed ... June 2* [1915]. *Wednesday. At two in the morning, who lived in Przemyśl had to jump up. After a few hours of relative silence, three terrible shots. What was it, or big walruses*[mortars], *or bridges blown up ...? And in the morning, first aeroplane* [planes] *with bombs, shots at them, the roar of cannons on roller frames* [forts] *... In Mościska where the Muscovites settled so comfortably, our kites* [the planes] *reportedly threw so many bombs that everything fell apart. They fled to* Lviv *and to Przemyśl... It is already approaching the evening, and here the cannons are roaring relentlessly, close, very close. Bullets hissing through the city fly from one side to the other. 7 versions* [forts] *ours have won. From the mass that marched yesterday to the shafts, not even one came back, and the leg was broken. The shafts of human and horse corpses lie ... Around 6 o'clock there is a terrible panic. They run away in a rush along the Lviv route: army, train* [carriages], *cannons, horses, cattle, swordsmen, merchants - with the greatest haste. And behind them there are grenades flying. A fire in the city from grenades in several places. Increasingly denser arrows. Panic. Everything rumbles from a bang; you can go crazy from the whistle of bullets ... It gets worse, more and more scary. The whole orgy. 11 o'clock, when I am writing into the night, and the whistling and roaring above us do not stop ... I would try to describe the impressions of that night in vain, it was terrible, it lasted endlessly ... on June 3* [1915]. *Thursday ... is the first brighter day for me. At 7 o'clock in the morning our and the Germans entered. They walked enriched with greenery and flowers, joyfully greeted, and blessed. It seemed as if millions were walking, infantry and cavalry and so many regiments, so many shades. I will say that they are not tired and skinny, all joyful, because they were still winning with almost no difficulty*[31].

On June 22, 1915, the troops of the Central Powers captured Lviv - an inhabitant of the capital of Galicia Józef Białynia Chołodecki wrote in his memoirs:

Every day the anxiety grew, and the nervousness among the inhabitants of Lviv grew at the sight of flying planes [aircraft] *Austrian and German forces, to throw bombs on the city, in the face of sporadic injuries or deaths of civilians passing through the streets. At the same time, they transported Russians from government and city buildings, as well as from the offices of banks and institutions* [spelling consistent with the original throughout the quoted text], *as well as from private apartments, which only fell into their hands ... population. About an hour* [0] *9* [00] *in the morning, accompanied by a Cossack, a Russian officer went to the town hall tower, took off a flag with the Russian colors, rolled it up and walked away in silence, and at the same time the banners disappeared from the spar on the Union of Lublin Mound and at the Governorship building. 293 days, the ongoing Russian rule in Lviv has ended*[32].

With the retreat of the Russian Southwestern Front, Russian air units were also retreating to the east. 7. KAO, during the 1915 campaign made 52 flights in a total of 74 hours and 37 minutes, as part of the retreat, he found himself at the Lewandówka airport in Lviv, reaching there via Sambor, Nowosiółki and Pidhaitsiand then he moved to

Above: Aviatik B.II on its nose. Source: Aeronaut.

Below: Aviatik B.II *Elsa*. Sources: Aeronaut.

Above: Aviatik B.II *Rosa* after a rough landing. Source: Aeronaut.

Below: Aviatik B.II *Annie*. Sources: Aeronaut.

Złoczów. During a heavy night march along the route of 40 versts (approx. 416 km), and then along a further distance of 50 versts (approx. 530 km), due to the difficult road conditions, planes transported by land transport suffered several damages (broken springs, bent engines, etc.). In July 1915, the 7th KAO made 34 combat flights, and on July 13 (26), 1915, the airport was changed, flying to an airfield near the village of Kadłubiska. In August 1915, the unit was transferred to the Gaj Starobrodski airfield, then to the Dunajów airfield, Kremenets, the village of Nowostawto be in Jampol on August 28 (September 10) 1915. On November 24 (December 7), 1915, the 7th KAO was stationed at a field airfield near the village of Matwiejowce in the region of Szumsk, north-east of Tarnopol[33]. In the fall of 1915, the capital of Bukovina - Chernivtsi, were the target of systematic air raids by the Russian air force: on 22, 23, 25 and 28 September 1915 the city was bombed from the air, and in response to this, the Imperial-Royal Air Force bombed Kamieniec Podolski on 29, 30 and 1 October 1915 . On October 7, 1915, five Russian planes dropped 23 bombs on Chernivtsi - seven of them were incendiary, and on November 19, 1915, the Soviet Air Force bombed targets in Lipkanach[34].

Gradually, the Galician offensive of OHL and AOK on Nord-Ostfront was losing momentum. Forces of the right wing of the allied armies of the Central Powers (field 3rd, 2nd, 7th Armies and the German Southern Army) with 243,392 infantry soldiers, 19,742 cavalry soldiers, 759 heavy machine guns, 196 mortars and 1,308 guns took up fights to take over the bridgeheads on the rivers Bug, Lipa and Dniester faced stiff field resistance from the 8th, 11th and 9th Armies of the South-Western Front (247,743 infantry soldiers, 44,213 cavalry soldiers, 886 heavy machine guns, 1,198 guns and 16 aircraft)[35] only in the area of Sokal and Krystynopol, displacing the enemy's forces beyond the old state border of Austria-Hungary. Only 10,000 km2 remained in the hands of Stawka in the Galician area. At the end of August 1915, the AOK planned a great offensive towards Równe with the field forces of the 1st and 4th Armies, and at the same time the field forces of the 2nd, 7th and the German Southern Army consisting of 239,863 infantry soldiers, 19,804 cavalry soldiers, 711 heavy machine guns, 978 guns and 232 mortars were to finally drive out the Russian troops with 205,085 infantry soldiers, 25,478 cavalry soldiers, 906 heavy machine guns, 997 guns and 22 planes from the borders of Austria-Hungary and, if possible, occupy Podolia[36].

From then on, the numerous rivers that crossed Eastern Galicia from north to south became the natural lines of defense, the conquest of which was fought until the breakthrough for tsarist Russia in 1917, fierce positional battles. Losses incurred in the period of January-April 1915, if not higher, certainly equaled the "bloody shambles" on the Western Front in 1916 fought at Verdun and on the Somme. The Carpathian Winter War, in terms of the losses suffered by both sides, deserved a dubious glory title introduced into historiography by prof. GA Tunstalla - Stalingrad of the First World War[37] while the Low Beskids and the Western and Eastern Bieszczady deserved to be called "cursed mountains", which is confirmed by the following judgment by a Russian officer of the 9th Rifle Regiment of the 3rd Rifle Brigade:

Carpathians! Wild and charming nature. Steep slopes, winding streams, soggy mountain valleys. In winter, everything is covered with deep snow. Carpathians! Wooded mountains, over a thousand meters high, seen by hundreds of fights. These hills, which the heroes of our heroic army conquered, cost the lives of thousands of Russian soldiers. Ominous Carpathians! They took from us the best cadre officers and soldiers whom the Russian army carefully trained in times of peace. As a result, the fighting lasted several years. These struggles created the conditions for the unfortunate February revolution and the defeat of the Bolshevik coup. In addition, the Carpathians gave us the opportunity to show the whole world the glory and heroism of the officers and soldiers of the tsarist army. The slopes of the Carpathian Mountains are flooded with the blood of these soldiers, while the Carpathian forests hide numerous war cemeteries[38].

This court was confirmed by the opinion of the commander of the Russian 13th Infantry Division from the field 8 Army:

So, the army lost [spelling consistent with the original throughout the quoted text] *Russian first-rate strategic operation besides, it became disorganized for a long time. Lesser Malopolska* [Galicia] *and Kingdom* [Polish - Priwislienski Kraj] *were executed*[39].

Endnotes

1 SAW-KA, Manuskripte Luftfahrtruppen, Die durchbruchsschlacht von Gorlice 1915. Bearbeitet von Obst. AD Feldp. Erich Kahlen, p. 18.

2 Ibid,p. 19; SAW-KA, Manuskripte / Luftfahrtruppen, Die durchbruchsschlacht von Gorlice 1915. Bearbeitet von Obst. AD Feldp. Erich Kahlen, pp. 1-2. On May 10-14, 1915, the Flik 15 airfield was Tarnów, and on May 14- June 18, 1915 - Rzeszów. From June 18, 1915, Flik 15. was transferred to the Balkans (Obst. M. E. Uzelac informed by a telegram sent from Rzeszów airport to Nowy Sącz and Aspern airports that they would be prepared to receive planes flying to Ujvidek) and stationed there at the following airports; June 18-October 18, 1915 - Ujvidek; October 18-November 4, 1915 - Belgrade; 4-18 November 1915 - Aranglelovac; November 18-December 2, 1915 - Čačak; December 2, 1915 - January 29, 1916 - Mitrovica. During the 1916/1917 campaign, Flik 15. stationed at the following Alpine airports: January 29-March 6, 1916 - Igalo; 6 March-6 August 1916 - Perigne; August 6, 1916 - January 14, 1917 - Bruneck; 14 January -16 June 1917 - Brixen; June 16-August 30, 1917 - Gardolo; August 30, 1917 - Dato - Bruneck. Flik 15. As of 20 September 1915, it had only three non-commissioned

pilots, of which Zgsf. Brath, he had yet to reach the unit and needed officers. Therefore, in connection with the forwarding of the unit to the Balkan region, commander Flik 15 asked the command of K.u.K. Luftfahrtruppen that the Duke of Oblt was assigned to the unit, which was waiting for short and long reconnaissance flights with the use of radio and cameras. Leon Sapieha (pilot diploma No. 160 received on April 14, 1914) as an observer.

3 Ibid, p. 20.

4 Ibid, p. 21.

5 These issues are comprehensively presented in the study edited by A. Olejko and A. Kharuk (op. cit.), in which an international team of authors presented the hitherto unknown contribution of aviation to the success of the field armies of the Central Powers in Durchbruchschlacht Gorlice. It is hard to find such a point of view in the not very successful studies of J. Centek (Korpus Gwardii w bitwie pod Gorlice, Warsaw 2015) and edited by J. Centek, S. Kułacz and K. Ruszała (Bitwa pod Gorlicami. Studies from the perspective of a century, Warsaw 2015)).

6 BMA Freiburg, PH 19/118, BAO - Brieftabt. O; *Handbuch der neuezeitlichen Militärwissenschaften Franke*, Hermann (Hg.) 3. Band, 2. Die Luftwaffe, Berlin 1939, p. 163; data from the private archive of H. Jäger. According to other sources, the FFA 54 was stationed at the field airport in Biadoliny Szlacheckie.

7 *Österreich-Ungarns Letzter Krieg 1914–1918*, vol. II, Wien 1929–1930, p. 337–338.

8 SAW-KA, Manuskripte / Luftfahrtruppen, Die durchbruchsschlacht von Gorlice 1915. Bearbeitet von Obst. AD Feldp. Erich Kahlen, p. 8.

9 Ibid, p. 10.

10 In the report of May 2, 1915 of the Flik 10 crew reported the detection of movements of the Russian forces in the Jasło-Biecz section, and after a few days, a wide retreat of Russian troops to the east was observed along the routes: Tarnów-Pilzno-Dębica-Rzeszów and Jasło-Strzyżów via Wisłok-Rzeszów. On May 2, 1915 at 06. 00 Oblt. Fp. W. Schreiber from Flik 15. made a reconnaissance flight in the area of Biała, observing the beginning of the offensive of the Central Powers' troops, and at At 12.30 he made another flight, observing the impact of the German Guard Corps in the area of Rzepiennik, reporting on his return seeing a particularly interesting sight of artillery duels. On the second day of the Battle of Gorlice, on May 3, 1915, in the afternoon, Oblt. Fp. W. Schreiber from Flik 15. made a reconnaissance flight near the front line. Activation of flights of the crews of the Imperial-Royal aviation resulted in the days of the Gorlice breakthrough from the need to provide AOK with the latest information about changes on the front line. On May 4, 1915, Zgsf. J. Siegel and Lt. Cserich made a flight to the area of koty 419 south of Tarnów - seeIbid,p. 11; SAW-KA, Fliegerkompanie 10 - Okt. 1914-2 May 1915. Studies. Bearb. Obst. Fp. AD Erich Kahlen, pp. 21, 23-25; SAW-KA, Manuskripte Luftfahrtruppen, Auszug aus dem Kreigstagebuch des Oberleutnant Feldpilot Wilhelm Schreiber der kuk Fliegerkompanie No. 15 aus dem Weltkriege 1914-1918, p. 58–60.

11 SAW-KA, Manuskripte / Luftfahrtruppen, Auszug aus dem Kreigstagebuch des Oberleutnant Feldpilot Wilhelm Schreiber der kuk Fliegerkompanie No. 15 aus dem Weltkriege 1914-1918, p. 58–60.

12 Ibid, p. 60–61.

13 SAW-KA, Manuskripte Luftfahrtruppen, Auszug aus dem Kreigstagebuch des Oberleutnant Feldpilot Wilhelm Schreiber der kuk Fliegerkompanie No. 15 aus dem Weltkriege 1914–1918, p. 62–63.

14 SAW-KA, Manuskripte Luftfahrtruppen, Die durchbruchsschlacht von Gorlice 1915. Bearbeitet von Obst. AD Feldp. Erich Kahlen, p. 15.

15 Ibid.

16 BMA Freiburg, PH 19/118, BAO - Briefabt. O; *Handbuch der neuezeitlichen Militärwissenschaften Franke*, Hermann (Hg.) 3. Band, 2. Die Luftwaffe, Berlin 1939, p. 163; data from the private archive of H. Jäger.

17 B. Gąsiorowska, op. cit., p. 110.

18 Ibid, p. 111.

19 Ibid, p. 112-113

20 Fr. K. Wais, op. cit., p. 83-93.

21 *Diary ...*, op. cit.

22 S. Nielipowicz, *The Carpathian Operation...*, op. cit.

23 During the fights for Stryi, the Russian troops used an observation balloon, which was also fought by FFA 59 airmen. During one of the flights, the crew of the Oblt. Michel and Oblt. Görhardt, died while attacking his position (most likely from the explosion of his own 0.5 kg bombs with a bad reputation among the crews) and was buried in the park in Stryj. After the Russian troops left the city, the local airport located on the outskirts of the city was seized, and the next FFA 59 field airports were Chodorów, Rohatyn, Brzeżany, Mikulińce and Zborów - see data from the private archives of H. Jäger.

24 HMHIM, *Der letzte...*, op. cit., p. 464-465.

25 Ibid, p. 465.

26 AW-KA, Manuskripte Luftfahrtruppen, Auszug aus dem Kreigstagebuch des Oberleutnant Feldpilot Wilhelm Schreiber der kuk Fliegerkompanie No. 15 aus dem Weltkriege 1914-1918, p. 64-65.

27 SAW-KA, Fliegerkompanie 10 - Okt. 1914 - May 2, 1915. Studies. Bearb. Obst. Fp. AD Erich Kahlen, p. 26. In the Imperial-Royal Air Force, improvisation in the manual dropping of bombs ended in 1915, when special bomb ejectors were used in place of special drop pipes for this purpose. In March 1915 the Bomben visier V 1-Fenstersenkel was tested in Fliegerarsenal in Pola, and in 1917 the more perfect model Bombenvisier V

2-Fenstersenkel was created. In February 1918, 500 units of the model with electrical installation were ordered from CP Goerz Wien (in 1918, the Bombenvisier V 3-Fenstersenkel and Bombenvisier V 4-Fenstersenkel models were also tested) - see SAW-KA, Bomben Ziel- und Abwurf- Vorrichtungen Kriegsjahre 1914-1918. Bearbeitr Obst. AD Erich Kahlen, p. 1–2, 3.

28 SAW-KA, Manuskripte / Luftfahrttruppen, Die Entwicklung d. Russ. Milit☒rluftfahrt vom August 1914-1916. Diakow-Kahlen, p. 19.

29 SAW-KA, Fliegerkompanie 10 - Okt. 1914-2 May 1915. Studies. Bearb. Obst. Fp. AD Erich Kahlen, p. 4, 27-29, 30-32, 36-46. The new Flik airport. The 10th was the airport in Strzeszkowice - 12 km south of Lublin - from where they operated in Ulanów on the San River, Domostawa, the Tanew Valley, Modliborzyce, as well as Lubaczów, Dęblin and Chełm. On one of the flights, the crew of Oblt. O. von Schmoczer and Oblt. E. Kumstat did not return from reconnaissance, forcibly landing near Kurów and being captured (the plane was set on fire by the airmen - they were sent to Siberia). Another Flik airports. 10 were Lublin and Kock from where reconnaissance was carried out around Włodawa and Brest-Litovsk, throwing 10 kg carbonite bombs at the local citadel, and another at Flik airport. The 10th was Biała Podlaska. Flik 10 since the Gorlice breakthrough, operating together with Flik 8 and Flik 15 from the region of Brzesko-Okocim-Jasień was constantly in motion all the way to the region of Lublin, Bug, and Brest-Litovsk. From the airport in Biała Podlaska, the fortress in Brest-Litovsk was photographed from a height of 2600 m, gradually forcing into the area of Kobryn, Drohiczyn and Lutsk. The "Kurier Lwowski" published in Lviv occupied by the Russian army described the retreat of the tsarist army on Galician soil in 1915: On 1 and 2 [May 14 and 15, 1915] fights broke out in the region of Jarosław ... With our strong artillery fire, the enemy troops are aiming to fight on the right bank of the [San] River, regardless of the countless losses. During the day, we knocked down several enemy airplanes [planes] correcting the fire of numerous artilleries. The crew of one of the German planes shot down in these fights' rests in the cemetery in Jarosław.

30 F. Molnár, op. cit., p. 195. An artilleryman from the 10th Corps, returning to Przemyśl, described in his memoirs an air event from Galicia: May 31 [1915]. There was also one comical incident with a Russian airplane. The Russian aviator, probably due to a defect in the motorbike, was forced to land - he flew over the forest near the [artillery] battery ... on the other side of the San he descended [landed]. Fortunately, his infantry was not there - and our people were so surprised that before they decided on something - the aviator could jump off, remove the defect in the motorbike and fly away, leaving behind a gasoline stench - see S. Pilecki, B. Skąpski, op.. cit.

31 H. Jabłońska, op. cit.

32 https://www.kuriergalicyjski.com/historia/smaczki-z-archiwum/4284-koniec-okupacji-1915 - entry from 2 February 2019.

33 M. Hajrulin, *7 Korpusnoj aviacionnyj atriad*, Moscow Saint Petersburg 2008, p. 138, 146, 148. During the campaign of 1915 7. Aviaatriad used the following airfields: 1 (14) I - 18 II (3 III) - Sambor; 18 II (3 III) - 23 III (5 IV) - Straszewicze near Sambor; March 23 (April 5) - April 28 (May 11) 1915 - Rozłucz [ukr. Розлуч]; 28-30 April (11-13 May) - Sambor; 30 April (13 May) - 2 (15) May - Nowosiółki; May 2 (15) -24 (6 June) 1915 - Podhajczyki; May 24 (June 6) - June 3 (16) - LubieńWielki (Lubień Mały) [ukr. Великий Любінь], 3 (16) - 6 (19) June 1915 - Dobrosin [ukr. Добросин]; 7 (20) - 8 (21) VI - Dmitrovka; 8 (21) VI - 3 (16) VII - Złoczów; 30 VII (12 VIII) - 17 (30) VIII - Kadłubiska; in the period 17 (30) - 21 August (3 September) 1915 - Gaj Starobrodski; 21-26 VIII (2-8 IX) - Kremenets[ukr. Кременець]; 26-28 VIII (8-10 IX) - Nowostaw; August 28 (September 10) - November 24 (December 7) - Jampol; November 24 (December 7) - December 31 (January 13, 1916) - Matwiejowce. The losses of the unit during the campaign were as follows: On February 10 (23), 1915, during a reconnaissance flight, the engine of one of the Nieuport IV aircraft crashed - during landing in a forested, mountainous area, in a river valley near Sambor, the plane crashed. Crew - lieutenant pil. M. Sabielnikow and Chief Captain A. von Rentel, the observer, left from an accident with injuries, and the plane was repaired to the 6th Rota Lotnicza. April 12 (25), 1915, during the flight of the staff captain pil. I. Stepanova from Sambor to the Loops because of the engine failure of the Nieuport IV aircraft, a forced landing took place, because of which the plane was crashed. 3 (16) May 1915 during the take-off of the Nieuport IV No. 370 aircraft piloted by the staff captain pil. I. Stepanova with a mechanic non-commissioned officer Gerasimov on board, the plane crashed.

34 W. Zapołowskij, *Bukovina in the last prison of Awstro-ugorszczini 1914–1918*, June 2003,p. 81, 92.(On April 12, 1916, Russian planes bombed targets in Chernivtsi and Żaniec, and on April 14, 1916, a nearly two-hour air battle took place over Chernivtsi, in which eight Russian single and biplane planes participated, and on the Austro-Hungarian side - two biplane planes.

35 SAW-KA, AOK Quartiermeister-Abteilung, Ktn. 2680, Materielle Situation 1. Armee. July 15, 1915. Op. 25800; Ktn. 2688, Materielle Situation 2. Armee, July 15, 1915, Op. 23616; Kt. 2718, Materielle Situation 7. Armee, July 15, 1915, Gstb. No. 220 (July 28, 1915); Ktn. 2729, Etappen-Gruppenkommando No. 9, Materielle Situation am August 1, 1915, Op. 808/48 via Dr. S. Nielipowicz; РГВИА, Ф. 2003, О. 2, Д 377,

Сведения о списочном, наличном и боевом составе Действующей армии на 15 июля 1915 г. Л 29–30, 334–339, 415–420 via Dr. S. Nielipowicz.

36 Armee, Ktn. 2718, Materielle Situation 7. Armee, Kt. 2729, Etappen-Gruppenkommando No. 9 via Dr. S. Nielipowicz; РГВИА, Ф. 2134, О. 2, Д. 424, Таблица потерь 8-й армии, Л. 127; Ф. 2139, О. 2, Д. 86, Л. 4-19; Ф. 2148, О. 2, Д. 90, Л. 141, 144, 148; Д. 91, Л. 8, 32–33 via Dr. S. Nielipowicz. SAW-KA, AOK, Quartiermeister-Abteilung, Ktn. 2688, Materielle Situation 2. Armee, August 15, 1915, Op. 25279; Kt. 2718, Materielle Situation 7. Armee, Gstb.290 (September 6, 1915); Ktn. 2729, Etappen-Gruppenkommando No. 9, Materielle Situation am 1 September 1915, Op. 902/69 via Dr. S. Nielipowicz; РГВИА. Ф. 2003, Оп. 2. Д. 379, Сведения о штатном, списочном и боевом составе Действующей армии 15 августа 1915 г., Л.117–121, 141–145, 287 via S. Nielipowicz, PhD.

37 GA Tunstall, op. cit., p. 1; H. Stratchan, op. cit., p. 29; МВ Оскин, Нападение Карпат - зима 1915, Москва 2007, p. 46.

38 M. Drobňák, M. Korba, R. Turik, Mementá prvej svetovej vojny, Humenne 2010, p. 33; "Časovoj" No. 178 (November 15, 1936), Paris: Russkoje vojinstvo i nacionaľnoje dviženije za rubežom, 1936 - translated by Dr. D. Hanik.

39 E. de Henning-Michaelis, op. cit., p. 216.

Above & Facing Page, Above: The victorious march over San ... Pictures presenting the fights of Austro-Hungarian and German troops with the tsarist army.

Below: Across the San to Lviv ... Construction of the bridge over the San River in the Przemyśl region in the spring of 1915 during the pursuit of the tsarist army retreating to the east. collection of the KVH Beskydy

Above: Jarosław 1915 ... Air graves of the German aviation crew at the cemetery. R. Frodyma collection

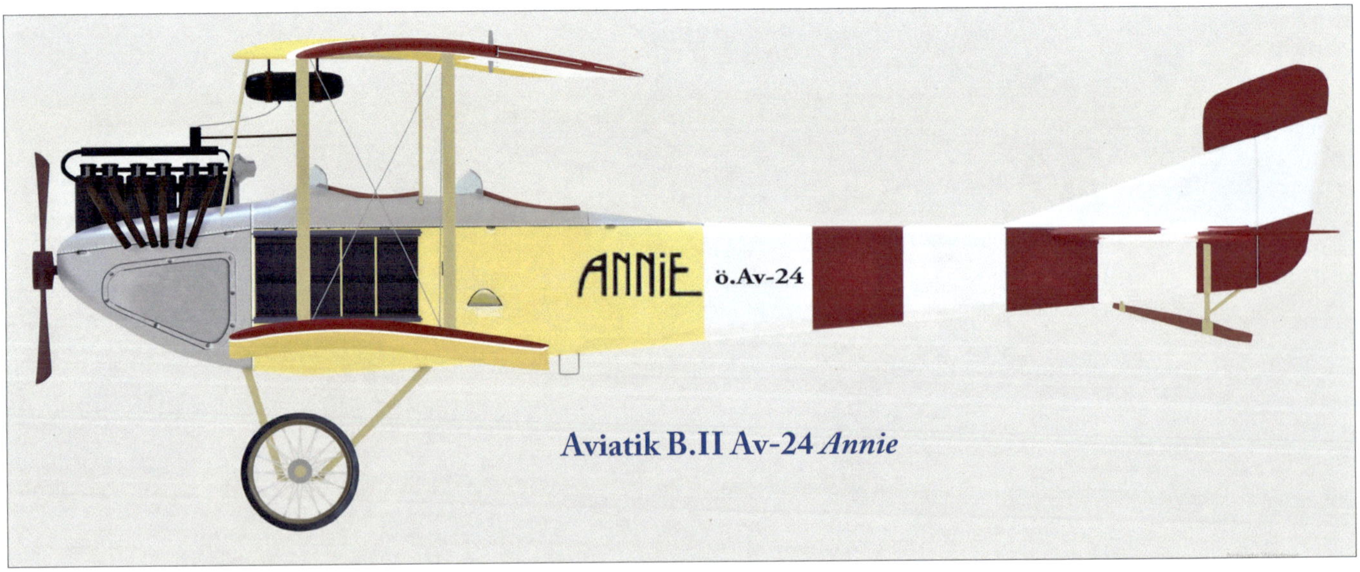

Aviatik B.II Av-24 *Annie*

Punitive Expedition Against Serbia ... Aviation Activities in the Balkan Campaign 1915–1916

The failure of the central states on all fronts In France, Serbia and Poland, the accompanying armies, despite all their efforts from the beginning of the war and throughout 1914, defeated the Berlin and Vienna governments with confidence that defeating the coalition is not as easy as it was before the beginning of the war. war steps[1].

The above opinion of one of the officers of the Imperial-Royal of the 57th Infantry Regiment, who fought in 1914 in Galicia and in "Kongresówka", reflects not only the "staff" situation. During the planning by OHL and AOK of the 1915 campaign against the Kingdom of Serbia, as before the beginning of last year's campaign of 1914, the use of the air force was assumed. The experiences of both Supreme Commands from the 1914 campaign in Galicia, on the Eastern Front and after the "Serbian spanking" showed that "Ohne Flieger kein Sieg". As air support, OHL assumed the allocation to the Serbian front of seven German FFAs for the support of the "third dimension" of the German field 11th Army (during the 1914 campaign, the crews of only one FFA - FFA 36 on Nord-Ostfront, Süd- Ostfront none) and four Imperial-Royal Fliegerkompanien KuK Luftfahrtruppen[2].

C.k. naval aviation (Flotten Flug Abteilung Kotor KuK Seeflugwesen) using flying boats of the Lohner E, M and Mn type, was to operate as before from the Kotor region over Dalmatia (Croatian Dalmacija, Italian Dalmazia, Serbian Далмација) and Herzegovina (Serb.., croatia, bosnia, montenegro Hercegovina / Херцеговина)[3]. In May 1915, KuK Seeflugwessen had 64 seaplanes at its disposal at the time of joining the Great War on the side of the Ententa of the Kingdom of Italy – six of them were transferred to Seeflugstation Kumbor and assigned them to the Flotten Flug Abteilung Kotor. Their crews, apart from the flights so far performed over the Adriatic coast of Montenegro and Albanian, were to stand face to face with a new enemy, the Italian Navy (it. Regia Marina) and the Italian naval aviation (it. Servizio Aeronautico Della Regia Marina)[4]. On the night of May 31 / June 1, 1915 the crews of Lohner L 32 type flying boats (Warnemünde type, 150 HP Rapp engine - the crew consisting of Linien Schiffsleutnant Hugo Ockermüller and Linien Schiffsleutnant Alois Poljanec), Lohner ML 41 (Rapp 150 HP engine - crew consisting of Linien Schiffsleutnant Hugo von Wiktorin and Linien Schiffsleutnant Arpad Mindszenty), Lohner LL 43 (Hiero 140 HP engine, crew consisting of Fregattenleutnant Konstantin Maglic, Linien Schiffsleutnant Stephan Wolleman Schiffs) and Lohner LM 39 (Mercedes-Benz Linnienant engine 100 HP - crew Franz Mikuleczky [Mikuletzky]) attacked targets on Italian territory in the Brindisi area[5]. From May 1915, the territory of the Italian opponent became the main target for the combat flights of the KuK Seeflugwessen crews, but occasionally the crews of the KuK Luftschifferabteilung went to the sea (on the night of February 4/5, 1915, the Flik crews. 6. bombed the port in Antivari)[6]. Similarly, as during the Serbian campaign of 1914, the Fliegerkompanien used Lohner B "Pfeilflieger" (90 HP "Daimler" engines), Lohner C "Gebirgspfeilflieger" (85 HP "Hiero" engines), Lohner C "Gebirgspfeilflieger" (140 "Hiero" engines) KM), Aviatik B.I.[7] ("Daimler" 120 HP and "Mercedes" 100 HP) and Lloyd-Doppeldecker (140 HP "Hiero" engines) supplemented over time by newer constructions such as Aviatik CI Lohner B "Pfeilflieger" and Lohner C "Gebirgspfeilflieger" "And Aviatik B.I, at a speed of 80-100 km / h, could fly for up to four hours (at an economic speed of up to 90 km / h, the average combat flight in 1914 was three ½ - four ½ hours at an altitude of 2000–2400 m)[8]. According to the realities of 1915, long-distance reconnaissance flights of the KuK Luftfahrtruppen crew led up to 500 km, the average reconnaissance flight was performed at up to 200 km; close reconnaissance flights lasted two to three hours, long reconnaissance flights took four to four ½ hours, and both types of flights were performed at an altitude of 1500–2000 m and even 3000 m[9]. Due to the small territory of the Kingdom of Serbia, the command of the Imperial-Royal Air Force assumed the maximum time needed for air operations over the enemy territory oscillating between two and a halff and three flight hours for each aircraft. The crews of Imperial-Royal Fliegerkompanien, drawing on the experience of the 1914 campaign, were to systematically identify roads, bridges, and artillery positions in Serbian territory, as well as port areas and coastal shipping on the Montenegrin coast[10]. As part of the equipment replenishment of Imperial-Royal Fliegerkompanien, the German ally gradually supplied more and more Albaros BI aircraft, which until 1916 became their basic combat equipment on the fronts of the Great War.

The Balkan Blitzkrieg, or the Genesis of the Balkan campaign of 1915–1916

In Zaječar [Serbian. Зајечар][11] *we encounter the first military echelon. The soldiers wear uniforms that they have already served*

Above: The world press in 1912 used the term puffer ... In the photo, the Lloyd C.I –from Flik 4 - next to "classic" air bombs and the so-called Veljačić bomb - Italian Front 1916 - Source: Fotoarchiv SAW-KA Wien

in two [Balkan] *wars. The coats are torn, sometimes stained with blood; it is clearly visible that these clothes were removed from the wounded and killed and handed over for reuse. Their trousers and sweatshirts were also in a deplorable condition; instead of leather boots, they usually wore bast slippers, made for domestic use, only the officers wore boots with uppers. At the same time, it was a first-class regiment supplied with uniforms! The second and third categories must take care of their clothes themselves, and they receive only weapons from the headquarters. Later, when I encountered the regiment for the second time, the army spent 42 days in the trenches and their clothing consisted of only rags. Several Serbian regiments wore Austrian uniforms that were removed from the dead and seriously wounded*[12].

The above description of the Serbian army did not fill the readers with enthusiasm, but there is also a note of war propaganda in it. A Brittan was to say the day before, First Lord of the Admiralty Winston Churchill *Serbs, a tough, militant nation, inspired by passionate patriotism stemming from centuries of struggle and oppression* [with Ottoman Turkey], *they waited fearlessly for the verdicts of fate*[13] in the campaign of 1914, starting from the clashes with K.aiserliche und Königliche - KuK - Armee -[ck] Army) "With a shield". In 1915, in the name of the proclaimed slogan "Serbien muss sterbien" after the lost campaign of 1914, now of joining the warfare of the Great War[14] on the side of the Central Powers (the Triple Alliance - Empire of Germany, Austria-Hungary, Ottoman Turkey [tur. Devlet-i Âliye-yi Osmâniyye Osmanlı İmparatorluğu]) on the 14th of the Empire of Bulgaria (Bulgarian Царство България), the last ally of the Central Powers in the Great War, pursuing the policy of being the "Piedmont of the Balkans[15] and to expand its territory at the expense of its enemy neighbor - the Kingdom of Serbia, as part of retaliation for losing the Second Balkan War[16] the field armies of Austria-Hungary, Kaiser Germany and the Bulgarian Empire broke through the resistance of the field armies of the Balkan Serbo-Montenegrin coalition in early 1916, ultimately causing their defeat. After the brilliantly carried out in 1915 campaign against the Kingdoms of Serbia and Montenegro (Serb. Краљевина Србија Kraljevina Srbij and Краљевина Црне Горе / Kraljevina Crne Gore) by the commander of the joint German-Black-Bulgarian-German Huzgar (Danzig-Danzig) forces.) Field Marshal August von Mackensen (German: Armeegruppe Mackensen), the Serbian capital of Belgrade (Serbian Београд, Hungarian

Nándorfehérvár) was seized on October 9, 1915 by the attackers, which was the beginning of the end of the Serbian monarchy[17]. On January 19, 1916, the ruler without a state, King of Serbia, Peter I Karadziordziewić, emigrated via the Duchy of Albania[18], the same was done by the King of Montenegro Nikola I Njegoš on board the Italian destroyer Regia Marina, having left his country and, like his Serbian ally, announced that he would continue the war on the side of the Entente (the territories of both countries were occupied by troops of the Imperial-Royal army until 1918)[19].

During the Great War, the Kingdom of Serbia lost over a million inhabitants, i.e., over a quarter of the population, because of the fighting and epidemics of "wartime diseases[20]- the Serbian army itself lost over 265,000 soldiers, that is 1/4 mobilized! In the fall of 1918, the wheel of "war fortune" turned away from the winners in the 1915/1916 campaign and the defeat in the Balkans affected the Central Powers, which happened with considerable help from the Western Allies (the offensive on the Macedonian [Salonika] Front of the Army of the East [Army of the Orient - fr. Armée d'Orient])[21]. However, before this happened, the 1915 campaign brought the Serbian army a "great spanking" and the name Kosowe Pole once again became part of the so-called "Serbian syndrome" in the history of Europe.

Great War campaigns in the Balkan theater of hostilities[22] including the 1915 campaign have been presented quite extensively in the literature on the subject, but most of the authors have focused on the analysis of land operations[23] or at sea operations.[24] Activities in the air during the Serbian campaign of 1915 have so far been presented comprehensively or in traces in German-language literature[25] Hungarian-speaking[26], English-speaking[27], Czech-speaking[28], Serb-speaking[29] or Polish[30].

In December 1914, because of the retreat from the territory of the Kingdom of Serbia, defeated mainly in the battles of Mount Cer (Serb. Цер, also known as the Battle of Jadar - Serb. Јадар) and the Battle of Kolubara (Serb. Kolubara) of the 5th and 6th Army Army, the Balkan-Serbian front stabilized on the Danube line (German Donau, Slovak Danube, Hungarian Duna, Croatian Dunav, Serbian and Bulgarian Дунав, Romanian Dunărea), Sava (Slovenian, Croatian, Bosnian Sava, Hungarian Száva) and Driny (Serb. Дрина). The Serbian campaign of 1914 and the war failures of the Imperial-Royal High Command of the Army (KuK Armeeoberkommando - AOK)[31] caused a serious decline in the prestige of the Habsburg monarchy in Europe because of the loss in the clash of the proverbial "David with Goliath" on the part of Imperial-Royal "Goliath", and for the Serb side it ended with practically no territorial changes, but with unprecedented human losses[32]. As a result of the failures and the policy of personnel changes, the commander-in-chief of the Imperial-Royal field armies, General Oskar Potiorek, was removed from command in the Balkans, and was replaced on October 28, 1914 by General Moritz Freiherr Auffenberg von Komarow, and finally, at the end of the year, he was appointed a new commander-in-chief at the Balkan theater of military operations of Archduke Eugeniusz "Erzi" F. Habsburg[33]. The Serbian army, despite serious deficiencies in artillery and having no advantage on land and in the air, but fighting on its territory, defended the integrity of the borders of the Kingdom of Serbia, and moreover, by December 1914, it displaced the overwhelming forces of the Imperial-Royal field armies to their starting positions from the summer of 1914[34]. As a result of the above events, AOK suspended operations against the Kingdom of Serbia, which led to an almost nine-month break in fighting[35] although the bombing of the capital of the Kingdom of Serbia by field artillery by the crews of the KuK Luftschifferabteilung aircraft and KuK Donauflotille ships continued[36]. The losses suffered significantly weakened the Serbian army[37] for which and for the Serbian people the campaign of 1914 was a great moral success (!), during which it was possible to repel several offensives of many times stronger and better armed field armies. The price paid for it, however, was exceptionally high[38] and the question arose whether in the case of the next offensive of the Imperial-Royal Army, which was only a matter of time, the strongly weakened Serbian army would be able to stop the Imperial-Royal field armies for the second time[39].

The lack of military successes of Ottoman Turkey, which had belonged to the Central Powers since 1914, at the beginning of 1915, the second year of the Great War, caused the "Serbian question" to return in the staff circles of Kaiser Germany and Austria-Hungary in March 1915. It was then that the first logistical steps were taken to prepare the next, this time "joint" offensive in the Balkans[40]. In the spring of 1915, the combined forces of the Austro-Hungarian-German armies, after the Gorlice Breakthrough on the Eastern Front (German: Ostfront) on 2-5 May 1915, gained an advantage and considerable territorial gain over the defeated field armies of the Tsarist South-Western Front Russia, although its resistance did not break, and a new front line finally stabilized. At that time, along with the gained advantage (dismissing the direct threat from the Russian army on the territory of Hungary), it was possible to resolve the above-mentioned "Serbian question"[41]. In connection with the aforementioned "humiliation" of the AOK ka and the Imperial-Royal field armies during the 1914 campaign by the weaker Serb army, the attack on Serbian territory became the main operational target of the German Supreme Command initiating the entire operation (OHL). It assumed that in 1915, the control of the Balkans as far as Istanbul (Istanbul, tur. İstanbul, gr. Constantinople) to support the Turkish ally (on August 2, 1914, Ottoman Turkey signed a military alliance with the German Second Reich, and on October 30, 1914 it joined the military operations on the side of Central Powers). A supporter of this pro-Turkish solution was the German Chief of General Staff, General E. von Falkenhayn. In the event of defeating the Kingdom of Serbia, the Central Powers, and especially the German Second Reich, would obtain a permanent railway connection with Istanbul and the Middle East (the so-called Baghdad Railway[42]), because of which it could be possible to provide military support for a militarily weak Turkish ally. Thus, the territorial contact with the allied Ottoman Turkey would be established, the railway connection would be controlled, and thanks to the help of the German ally, the Turkish army could defeat the Allied forces (ANZAC) on the Gallipoli peninsula[43]. Therefore, it is understandable that the dominant issue for OHL was to establish contact with Turkey via the Balkans, and for the AOK to knock Serbia on both shoulders, conquer Montenegro and Albania, surrender the Balkans to the domination of Austria-Hungary[44] on the entire Serbian Front (German Süd-Ostfront).

In the spring of 1915, on May 23, the Kingdom of Italy found itself on the side of the "bleeding Ententa"[45] and thus Austria-Hungary attacked from the south by a new enemy, had to repel him as part of the open seventh front of the Great War (German Süd-Westfront)[46]. From then on, the AOK's priority was to defeat the Italian army, and thus the front over Isonzo (called Soča) and in the Alps became more important than the Balkan Süd-Ostfront. The Serbian government, wanting to secure its support on the Adriatic Sea, on June 11, 1915, made a surprising move ordering unit of the Serbian army to enter the Duchy of Albania (Alb. Principata e Shqipnis) and its capital Tirana (Tiranë). Thus, the neutrality of this state was violated to prevent the "expansion of Italy" in the region, which in the late autumn of 1915 was to be beneficial for the defeated Serbian field armies retreating towards the Adriatic coast[47]. As part of "the knockdown of the Kingdom of Serbia on both shoulders, the capture of the territory of the Kingdom of Montenegro and the independent Principality of Albania, the elimination of the Kingdom of Serbia, despite the above-mentioned "Italian priority", was among the main military goals of the Danube Habsburg monarchy. The AOK intended to eliminate the threat of a one-off war on three fronts: Russian (Nord-Ostfront), Serbian (Süd-Ostfront) and Italian (Süd-Westfront). In April 1915, the task of attacking from the region of Újvidék (Croatian Novi Sad, Нови Сад, Slovak Nový Sad, German Neusatz (an der Donau) was given to command of the field army of the 5th Army, defeated in the 1914 campaign, but in connection with commencement of hostilities on the part of the Ententa countries of the Kingdom of Italy, a large part of it (six DP) had to be transferred to the Isonzo River, where from then on a new line of the Austro-Hungarian-Italian front ran (until 1917, 11 battles were fought here, which seriously exhausted both sides). Moreover, the Imperial-Royal Cavalry Group of Károl Tersztyanszky von Nádas (Armeegruppe General der Kavalerie von Teresztyanszky) was prepared for the fighting in the Balkans, strongly experienced in the fighting in the Carpathian Winterkrieg 1914/1915 soon to be transformed into Imperial- field 3rd Army and in June 1915, undertake the offensive against the Kingdom of Serbia, after earlier reconnaissance activities of the Imperial-Royal units conducted in the Dunaj-Sawa-Drina belt. In mid-August 1915, the AOK, wanting to resolve the "Serbian question", was ready to transfer to the Belgrade-Kragujevac region (Serb. Краrујевац) the entire field of the 5th Army (18th Infantry Division)[48].

In 1915, both political and military blocs, the Triple Alliance and the Three-Alliance with each other, exerted pressure on the Balkan states, which had been neutral so far and remained aloof from the politics of the great powers, mainly on the Bulgarian Empire, wanting them to have the strongest army among the Balkan countries (the mobilization capabilities of the Bulgarian army allowed it to develop from a peaceful state of 60,000 to martial law of 450,000 soldiers)[49] "make" as an ally into one of the two alliances. On the Bulgarian question, Sergei Sazonov, Minister of Foreign Affairs of Tsarist Russia, said: *We brought it back to life and it owes us its revival. Therefore, we cannot treat it as an enemy*[50]. The game of the "Bulgarian card", in which the geographical location of the country and its mobilization abilities were of great importance, by the two military-political camps engaged in military operations meant that the Bulgarian Tsar Ferdinand I Koburg, about whom the German Emperor Wilhelm II Hohenzollern was to write about the day before, that it is decorated with orders like a Christmas tree[51], territorial gains were offered in return for joining the hostilities, at the expense of the neighbors[52]. As soon as the negotiations of the Ententa countries with the government of the Kingdom of Greece came to light (Greek: Βασίλειον τῆς Ἑλλάδος), seeing that their effect could be unfavorable for the Bulgarian side, the sympathy of the Bulgarian government circles towards the Central Powers bloc quickly increased, and the price for the alliance was the promise or even the "casus belli" of incorporation into the Bulgarian territory of Macedonia ... and more[53].

Analyzing the war events of 1915 and the defeats of the field armies of the Ententa countries on the Eastern Front (Gorlice Breakthrough) and in the Black Sea (Gallipoli), the Tsar of Bulgaria on September 23, 1915 decided to sign an alliance with the Kaiser II Reich[54]. General mobilization was announced on the territory of the country on September 22, 1915[55] and thus the position of the Kingdom of Serbia became extremely bad, as it was threatened simultaneously from the west (Bosnia and Herzegovina), the north (on the Sava-Danube line) and the east (Bulgaria) by opponents with a large military advantage. The Serbian army in 1915 lacked the most modern weapons (serious shortages of equipment and ammunition occurred in artillery units) and the systematic supply of ammunition, which because of the blockade of the coast of the Kingdom of Montenegro by the Imperial-Royal Navy (KuK Kriegsmarine) and naval aviation (KuK Seeflugwessen) was getting slower and slower mainly via the Montenegrin port in Bar (Montenegro Бар, owner: Antivari). The second solution was to use the Greek port of Thessaloniki and although the Kingdom of Greece was leaning towards the Ententa camp, however, it officially held

Above: Aviatik B.II Av-3 32.15 at the front was powered by a 120 hp Austro-Daimler. Source: Aeronaut.

Below: Aviatik B.II 32.37 Av-12 flys over its airfield. Source: Aeronaut.

a position of neutrality. The King of Greece, Constantine I, did not agree with the pro-Ententa position of the Greek government, sympathizing with the Central Powers (he had studied in Germany, and his wife Sophia was the sister of the German Emperor Wilhelm II Hohenzollern)[56] and thus finally the Ententa states (British and French) decided to land their forces (Expeditionary Corps) in Thessaloniki on October 5, 1915, disregarding the position of the Greek side and thus violating its neutrality[57]. As soon as the position of the Bulgarian Empire became clear (on October 14, 1915, it entered the war on the side of the Central Powers), the governments of the United Kingdom of Great Britain and Ireland and the Republic of France declared their willingness to send military reinforcements to the Balkan peninsula[58].

Endnotes

1 *War memories ...*, op. cit., p. 31.

2 The original name of the Imperial-Royal Air Force in 1914 was KuK Luftschifferabteilung. From 8 June 1915 the name was KuK Luftfahrtruppen - see J. Rajlich, Niestěrov versus Malina. K history of prvniho taranu v dějinách letecké války, "HPM" 2009, No. 5, p. 10–11.

3 SAW-KA, Die Luftstreitkräfte Serbiens von 1914 bis Herbst 1916 von Oskar Regele, pp. 9, 11; idem - Über die Lestungstaehigkeit der oesterr.-ung. Flieger in Feldzug gegen Serbien-Montenegro im Jahre 1914 (Herzegovina and Bocche di Cattaro, August 12-December 31, 1914), op. cit. A. Nie Note (op. cit., p. 78) states quite enigmatically that in 1914 the Imperial-Royal Sea Aviation numbered 22 planes.

4 SAW-KA, Die Eroberung Serbien 1915. Tätigkeit der Fliegerverbände der österr.-ung. 3. Armee und der Serb. Luftstreitkräfte. Bearbeiter Rittm. ad Peter v. Spaits, p. 192.

5 Ibidem, p. 194. A "Quatro" class cruiser was bombed, and the crew of a Lohner LL 43 flying boat bombed the Mola-Bari-Trani railway line.

6 Ibid p. 202.

7 Montenegro im Jahre 1914. Bearbeitet von der Obst. Dr. Oskar Regele, op. cit. It is surprising that the archival list includes more names than airplanes. List of proper names of Flik planes. 2. may refer to all flying equipment used in Flik 2 in 1914.

8 SAW-KA, Die Luftstreitkräfte Serbiens von 1914 bis Herbst 1916 von Oskar Regele, p. 201.

9 SAW-KA, Die Eroberung Serbien 1915. Tätigkeit der Fliegerverbände der österr.-ung. 3. Armee und der Serb. Luftstreitkräfte. Bearbeiter Rittm. ad Peter v. Spaits, p. 237.

10 SAW-KA, Die Luftstreitkräfte Serbiens von 1914 bis Herbst 1916 von Oskar Regele, p. 14.

11 The study uses the spelling of proper names in the current wording with reference to names from the past in the wording from the era.

12 J. Piekałkiewicz, *Calendar of World War I*, Warsaw 1998, p. 235.

13 W. Churchill, The World Crisis via D. Jordan, Italy and Africa 1914–1918. From Sarajevo to Piawa and Lake Tanganyika, Poznań 2011, p. 25. A Chwalba (op. Cit., P. 211) states that the Serbian army in the 1915 campaign numbered 340,000. soldiers, of which 250 thousand. "In the field" and the Montenegrin army numbered 50 thousand. Soldiers (these forces were supported by 780 guns, including 240 heavy guns - 27 of the latest Allied shipments). During the Great War (for the Bulgarians in 1915–1918), the Bulgarian army mobilized 878,000 into its ranks. soldiers - op. cit., p. 212. Д. М. Бабац (Српска војска у великом рату, Београд 2014, p. 85) reports that the Bulgarian army during the Great War numbered 800 thousand. soldiers.

14 *History of the 20th century 1900-1934, edited by prof. Henryk Mościcki and Jan Cynarski,* Warsaw 1936, p. 233.

15 M. Радојевић, Љ. Димић (Србија у великом рату 1914-1918, Београд 2014, p. 181; M. Rauchensteiner, Der Tod ..., p. 301. After the war with Ottoman Turkey in 1877-1877, after the Congress which was organizing the status quo in Europe after its end Brliński in 1878 and after the annexation of Bosnia and Herzegovina by Austria-Hungary in 1908. The Bulgarian Tsar Ferdinand I Koburg believed *that his destiny was the rebirth of Byzantium,* which the reluctant Tsar of Russia, Nicholas II Romanov, considered a *manifestation of megalomania* - see A. Chwalba , op. cit., p. 210; *Austria-Hungary and the War*, New York 1915, pp. 4–5, 20–22, 26–27, 28–29. M. Tanta (*The Balkans in the 20th century. Political history*, Warsaw 2003, p. 129) gives the date of October 11, 1915 as the date of joining the hostilities of the Bulgarian Empire on the side of the Central Powers, and A Chwalba (op. Cit.) States that on *October 5, 1915, without declaring war, the Bulgarians moved against* [the Kingdom of] *Serbia under the slogan "liberating the brothers from Serb captivity, and on October 14* [1915] *declared war on it. Illustrated History of World War (1914–1920)* (Vol. II, ed. R. Kodźia, Warsaw 1932, p. 500) states that on October 5, 1915, the Bulgarian Empire officially joined the camp of Central Powers. On October 15–16, 1915, the United Kingdom of Great Britain and Ireland and the Republic of France declared war on the Bulgarian side - see A. Szelągowski, op. cit., vol. II, part. II, p. 779; Д. М. Бабац, op. cit., p. 91.

16 *Austria-Hungary ...*, op. cit., pp. 24, 51, 54–56; I. Kienzler, *The Great War 1914–1918,* Warsaw 2014, p. 38; *World War in the Balkans. Military and political activities in 1914–1918. Studies and materials edited by A. Krzak,* Szczecin 2014 - passim. Bulgarian "aspirations to the four seas" (Black, Aegean, Marmara, and Adriatic) were widely known in Europe. The diplomats of the Central States, wanting to drag the Bulgarian Empire to their side, promised the Bulgarian side of the part of Macedonia that belonged to the Kingdom of Serbia after the Second Balkan War, which would be recognized by the Bulgarian public opinion as its *undisputed property* and would literally become a "casus belli", and if it would

Above: Lohner Pfeilflieger AD 567 after winning the Schicht Flug speed prize. Source: Aeronaut.

Below: Lohner B.II 12.05 on its airfield. Source: Aeronaut.

join the military operations on the side of the Entente (Triple Agreement - the Republic of France, the United Kingdom of Great Britain and Ireland, the Empire of Russia, The Kingdom of Serbia and the Kingdom of Montenegro) The Kingdom of Romania (Romanian Regatul României) also southern Dobrogea (Romanian Dobrogea, Bulgarian Добруджа). Ententa diplomacy offered the Bulgarian side Greek ports and the Greek side proposed Smyrna (port of Izmir). The Bulgarian Empire demanded from the Turkish side Adrianople (tur. Edirne) and the border on the Enos-Midia line, but through the German side an agreement was signed with the "allied" Turkish government. An agreement was reached on the Marica river line and military cooperation with the German army in the Balkans - see A. Szelągowski, op. cit .; *History of the 20th century* ..., op. cit., p. 284; *Illustrated History* ..., op. cit., pp. 498-499; A. Chwalba, op. cit., pp. 209–210; M. Rauchensteiner, *Der Erste…*, op. cit., p. 479.

17 М. Радојевић, Љ. Димић, op. cit., p. 180; B. Hamann, op. cit., p. 82; The Great War 1914–1918. The Real End Belle Époque, Warsaw 2015, p. 26. The participation of Polish soldiers of the German army in the Balkan campaign of 1915 is still a "white spot" in history, while the study by R. Kaczmarek (Poles in army of the Kaiser, Krakow 2014), however, only with regard to the Romanian campaign of 1916 (pp. 404–406).

18 A. Szelągowski, op. cit., p. 780.

19 J. Pajewski, First World War 1914–1918, puB.II, Warsaw 1998, p. 304.

20 J. Friedrich, op. cit., p. 375.

21 M. Rauchensteiner, Der Tod ..., op. cit., p. 301; H. Münkler, op. cit., p. 361. The autumn offensive of the Army of the East from the Thessaloniki area, including the Serbian army actively supported since 1914 with supplies of military equipment, led to the liberation of the territory of the Kingdom of Serbia within a month. The National Council of Slovenes, Croats and Serbs (RNSChS), established in Zagreb during the fighting, as the Serbian army was approaching the borders of the collapsing Habsburg monarchy on the Danube, announced the creation of the State of Slovenes, Croats and Serbs, and on November 24, 1918, announced the unification of the South Slavic nations, Austria-Hungary under the Serbian scepter as part of the Kingdom of Serbs, Croats and Slovenes (Serb-Croat Kraljevina Srba, Hrvata and Slovenaca, Краљевина Срба, Хрвата и Словенаца). On December 1, 1918, the Kingdom of Serbs, Croats and Slovenes (SHS) was proclaimed, and then the King of Serbia, Piotr I Karadziordziewić, returned to the country to take the throne of the newly created state.

22 The dating of the years of the Great War adopted in world science for the years 1914–1918, in the case of the states of the two fighting camps that joined the coalition partners after 1914, is different. Hence, in the case of, for example, the Kingdom of Italy (Italian Regno d'Italia) and the Bulgarian Empire, the Great War is the period of 1915–1918, and in the case of the Kingdom of Romania, 1916–1918.

23 Österreich-Ungarns letzter Krieg, vol. I, Wien 1930; J. Pajewski, op. cit .; N. Thomas, D. Babac, Armies in the Balkans 1914-1918, Oxford 2001; M. Hickey, The First World War. The Mediterranean Front 1914-1923, Oxford 2002; D. Jordan, The Balkans…, op. cit .; A. Mitrović, Serbia's Great War, 1914–1918, West Lafayette 2007; IFW Beckett, The First World War 1914–1918, Warsaw 2009; M. Glenny, The Balkan: 1804–2012, London 2012; P. Hart, World War I 1914–1918. Military history, Poznań 2014; M. Ch. Ortner, Die KuK Armee und ihr letzter Krieg, Wien 2013.

24 Őesterreich-Ungarn Seekrieg 1914–1918 (Herausgegeben auf Anregung des Marine-Verbandes vom Kriegsarchiv (marinearchiv), Zürich-Lepzig-Wien (Verfast von captain Hans Hugo Sokol) 1933; T. Wywerka-Prekurat, First World War at Sea, ed. II, Warsaw 1994; K. Csonkaréti, Austro-Hungarian Navy in World War I 1914–1918, Krakow 2004; A. Samek, The Fleet That Is No Longer, Krakow 2012.

25 R. Keimel, Őesterreichische Luftfahrzeuge. Geschichte der Luftfahrt von Anfängen bis ende 1918, Graz 1981; PM Grosz, G. Haddow, P. Schiemer, *Austro-Hungarian Army Aircraft of World War One*, Mountain View-California 1993; R. Peter, *Die kuk Luftschieffer- und Fliegertruppe Österreich-Ungarns 1794-1919*, Stuttgart 1981; P. Schupita, Die kuk Seeflieger. Chronik Und Dokumentation österreichiszch-ungarischen Marineluftwaffe 1911-1918, Koblenz 1983.

26 K. Csonkar□ti, A császári □s királyi l□gierö, Debrecen 2008.

27 Ch. Chant, *Austro-Hungarian Aces of World War I*, Oxford 2002.

28 J. Rajlich, Na křidlech světove v□lky, Cheb 2014.

29 B. Ciglić, Wings of Serbia, Beograd 2009; idem - Б. Циглић, Крила Србије Ваздухопловна команда и авијатика српске војске 1912 - 1920, Београд 2009; idem - Seaplanes of Bocche, Beograd 2014; Српска aviyatika 1912-1918,Београд1993; СЈМикић, Историја угословенског ваздуховства, Београд 1933; Б.Dimitry, П. Миладиновић, М. Мицевски, КраљевскоВаздхопловство-Војно ваздухопловство Краљевине СХС / Југославије 1918-1944, Београд 2012; Н. Ђокић, Помоћ Русије Србији у стварању ваздухопловства 1901-1915, Београд, "Лет" 2000, No. 2.

30 Aviation activities in the Balkan campaign of 1914 were presented by the author of that study, including in the following studies or compact studies: The Forgotten Campaign or the activities of the Imperial-Royal Austro-Hungarian Naval Aviation on the Adriatic Sea in 1914, "*Rocznik Przemyski-Historia Wojsczości*" 2014, vol. 50, issue 1; Wings of the Habsburgs 1914. From

the Balkans and the Adriatic Sea to the Carpathians. Imperial-Royal Air Force of Austro-Hungary in the first campaign of the Great War, Oświęcim 2015.

31 Until August 1915, the AOK was divided into the Supreme Operational Command (Operierende Oberkommando - OOK) and the Supreme Stage Command (Etappenoberkommando - EOK) - see http://agad.archiva.gov.pl/pomoce/Wyw328.xml - entry from November 8, 2014.

32 H. Leidinger, V. Moritz, K. Moser, W. Dornik, *Habsburgs schmutziger Krieg. Ermittlungen zur österreichisch-ungarischen Kriegsführung 1914-1918*, Salzburg-Wien 2014, pp. 75-79; C. Falls, *The Great War*, London 1960, p. 54; B. Hamann, op. cit., p. 53. M. Rauchensteiner (Der Erste…, op. cit., p. 467) states that as of August 15, 1915, the Army consisted of 29,113 officers and 806,982 soldiers of real forces on all fronts.

33 M. Rauchensteiner, *Der Tod* ..., op. cit., p. 188; Великая борьба народовъ, Выпуск III-ій, Москва 1915, p. 59. A Chwalba (op. Cit., P. 209) states that General O. Potiorek was dismissed on December 22, 1914.

34 *The Great War 1914–1918* ..., op. cit., part I, vol. VII, Krakow 1937, p. 233.

35 M. Радојевић, Љ. Димић (op. Cit., P. 175) report that 180,000 were stationed on the Serbian border in early 1915, 240,000 in early May 1915, and 177,000 at the end of May 1915. soldiers of the Imperial-Royal Army.

36 For example, in February 1915, the shelling was harassing: on February 3, 1915, field artillery of the Imperial-Royal Army and the Air Force (KuK Luftschifferabteilung) carried out a demonstration of power from "their" bank of the Sava and fire in the Belgrade region; once again on February 10, 1915, Belgrade was shelled by artillery fire and by airplanes of the Imperial-Royal Air Force; On February 19, 1914, field artillery and KuK Donauflotille monitors continued firing on Belgrade. Once again, on April 8, 1915, another artillery bombardment of Belgrade took place - Lieutenant Великая борьба народовъ, Выпуск III-ій, Москва 1915, pp. 65-67; Великая война въ образахъ и картинахъ. Выпуск VIII-ой, Москва *1915*, p. 68.

37 J. Piekałkiewicz, op. cit., p. 245; M. Tanta, op. cit., pp. 121-122. The peace on the front prevailing from January 1915 allowed for additional mobilization of the Serbian and Montenegrin armies and for replenishing the stocks of weapons and ammunition delivered via the port of Thessaloniki via the Baghdad railway line via Ni⊠. By river transport via the Danube, the Russian ally provided the Serbian army with 120,000. rifles and 50 thousand. sheepskin coats via the port in Thessaloniki (gr.Θεσσαλονίκη, Thessaloníkē, Latin. Egnatia - Solun) and by rail via Niš (Serb. Ниш) from Egypt, a British ally delivered 500 wagons of flour. A certain amount of ammunition from the allied Republic of France was also delivered, as well as 100 doctors with medical facilities to stop the epidemic of typhus and cholera in the Kingdom of Serbia. On December 7, 1914, the Serbian government, debating after the Imperial Imperial control of Belgrade's army took over Niš, announced a declaration of the goals of the war, which included a record that the Serbs were waging a war for the *liberation and reunification of all our unbroken brothers, Serbs, Croats, and Slovenes*.

38 The Serbian Army lost 163,557 soldiers (including 2,110 officers) killed, wounded, and missing in the fighting in 1914; ck 5th and 6th field armies lost 273,804 soldiers (including 7,592 officers) killed, wounded and missing - see M. Radojeвић, Љ. Димић, Србија у великом рату *1914–1918,* Београд 2014, p. 135; M. Rauchensteiner, *Der Tod* ..., op. cit., p. 153. C. Falls (*The Great War*, London 1960, p. 54), states that the Serbian army lost 170,000 in combat. the fallen, wounded, captured and missing, while the Imperial-Royal Army lost over half of the states - i.e., approx. 215 thousand. soldiers. J. Piekałkiewicz (op. Cit., P. 115) states that the army used 462 thousand zlotys for operations in the Balkans in the 1914 campaign. soldiers, losing 274,000 in combat (killed, wounded, missing and prisoners), while the Serbian army lost 132,000. soldiers. D. Jordan (*Balkans, Italy, Africa* ..., op. cit., p. 42) states that the Imperial-Royal army lost 227,000 soldiers, H. Stratchan (op. cit., pp. 30-31) states that during four months of fighting in Galicia (German Nord-Ostfront) and the Kingdom of SernSbia (German Süd-Ostfront), the Imperial-Royal army lost a total of 957 thousand soldiers. In the winter of 1914/1915, an epidemic of typhoid fever broke out in the Kingdom of Serbia, which claimed thousands of victims (in April 1915, there were 48,000 people in hospitals, and the total losses amounted to 180,000 people - see *Great War 1914- 1918* ..., op. Cit. Part II, p. 173.). J. Friedrich (op. Cit., Pp. 362-363) reports that in 1914 the army lost 170,000. soldiers and 50 thousand. as victims of the typhus epidemic. A. Szelągowski (op.cit., P. 746) states that according to the report of the Imperial-Royal Minister of War (KuK Kriegsmininsterium) presented on September 16, 1914. at a meeting of the Council of Ministers, Imperial-Royal Army from the original state of 800-900 thousand. soldiers after the end of the 1914 campaign in Galicia and the Kingdom of Serbia fell to 130 thousand. soldiers. M. Радојевић, Љ. Димић (op. Cit., Pp. 154–155, 157) state that at the beginning of 1915 the Serbian army numbered 260,000. soldiers (men aged 18-55 were mobilized) and the army of the Kingdom of Montenegro 50,000. soldiers.

39 D. Jordan, *Italy, and Africa* ..., op. cit., p. 43.

40 J. Pajewski, op. cit., p. 303.

41 More - *Der Erste Weltkrieg 1914–1918. Der deutsche Aufmarsch*…, op. cit.; D. Jordan, *Balkans, Italy, and Africa* ..., op. cit., p. 45.

Above: Carpathian winter 1914/1915 ... A column of prisoners of war from the Russian army going into Austro-Hungarian captivity. author's collection

Above: A moment of peace on the Serbian Front ... Soldiers of the Austro-Hungarian army while resting. collection of J. Damse via P. Kazana

Above: Serbian Front 1915.... A branch of the Austro-Hungarian army during field work. collection of J. Damse via P. Kazana

Above: Front parade ... Parade of Austro-Hungarian army units on the Italian Front in 1915. collection of J. Damse via P. Kazana

Above: Goddess of the battlefield ... A sub-unit of the Austro-Hungarian army on the Serbian Front in winter 1915. collection of J. Damse via P. Kazan

Above: Officers' staff ... A group of officers of the Austro-Hungarian army in Front Serbski in 1915. collection of J. Damse via P. Kazana

Facing Page, Below: Military kitchen ... Serving a meal to the soldiers of the Austro-Hungarian army on the Serbian Front. collection of J. Damse via P. Kazan

42 The Baghdad Railway was built with German credits and about 250 km of this line ran through the territory of the Kingdom of Serbia - see P. Hart, op. cit., p. 35; *Austria-Hungary…*, op. cit., pp. 44–45. Construction of the Berlin-Constantinople railway was launched in 1888 (it entered the Turkish state near Adrianople [tur. Edirne]) when the German Bank bought a controlling stake in the railway line from Vienna to Constantinople and took over the Haidar-Pasha-Ismid section from the British side. From that moment on, one can speak of a typically German undertaking - the Berlin-Baghdad line completed in 1896 - see A. Bartnicki, *Colonial conflicts 1869-1939*, Warsaw 1971, p. 155: J. Pajewski, *Berlin-Baghdad. From the history of German politics in the Middle East in the Wilhelminian times*, "Rocznik Historyczny", R. XVIII, Poznań 1949, p. 240–241.

43 The operation in the Dardanelles, the largest landing operation in the Great War, took place between April 25, 1915 and January 9, 1916 and ended in a complete defeat of the Allied forces. It absorbed losses on their side of 87 thousand. killed (mainly soldiers of Australian and New Zealand nationality) and 165,000. the wounded - see *Illustrated History...*, op. cit., pp. 374–381; J. Piekałkiewicz, op. cit., p. 236; M. Gilbert, *The First World War*, Poznań 2003, p. 31; I. Kienzler, op. cit., p. 39. М. Радојевић, Љ. Димић (op. Cit., P. 151) report that in the fighting in the Gallipoli area, the Allied troops lost approx. 210 thousand. soldiers and the Turkish army lost 253,309 soldiers.

44 J. Pajewski, op. cit., p. 301, 303.

45 *History of the 20th century ...*, op. cit., p. 283; *Illustrated History ...*, op. cit., p. 384–387.

46 M. Gilbert, op. cit. In the history of the Great War, EO Volkmann was to write with undisguised sarcasm that *the Italian Front was the only theater of war on which Austria for a long time dealt with without significant German help* - cf. *Illustrated History ...*, op. cit., p. 387.

47 Piekałkiewicz, op. cit. In December 1912, the conference of ambassadors in London decided to create an autonomous Albanian state under the sovereignty of Ottoman Turkey and the protectorate of European powers. Initially, the capital was to be Shkodra (Alb. Shkodër or Shkodra; Serb. Скадар, owned. Skutari), which, however, was annexed in April 1913 by the army of the Kingdom of Montenegro, which led to the intervention of the Austro-Hungarian Empire trying to prevent the Serbian the neighbor's access to the sea and the seizure of the city by the Imperial-Royal army on May 14, 1913. In July 1913, the conference of ambassadors granted the Duchy of Albania independence under the supervision of superpowers, but did not demarcate the country's borders, and in March 1914, the German prince Wilhelm zu Wied became the ruler of the state. In September 1914, after the outbreak of the Great War, he resigned from the throne - see T. Czekalski, *History of Albana*, Wrocław 2009, p. 182-185, 195; H. Leidinger, V. Moritz, K. Moser, W. Dornik, op. cit., p. 182. In the name of "holy war" against the Kingdoms of Serbia and Montenegro, intelligence agents tried to eliminate Esad Pasha Toptani from political life (from February 1914, the Minister of Foreign Affairs and the Minister of War, interned for participating in a plot against Fr. Wilhelm zu Wied, who, having fled to the territory of the Kingdom of Italy in August 1914, returned to Albanian territory with Serbian help to proclaim himself commander-in-chief of the Albanian army and head of the government at the same time), who mobilized 40-50 thousand. partisan soldiers, when on May 29, 1915, 17,000 Serbian army soldiers entered the Duchy of Albania, occupying Tirana and Elbasan (Albaniopolis), supporting the power of Esad Pasha Toptani. On June 8, 1915, the Montenegrin army, having entered Albanian territory, captured Shkodra and the northern part of the country - see Д. М. Бабац, op. cit., p. 91.

48 The Serbian army forces weakened by the typhus epidemic were then estimated at 11 DP - see SAW-KA Wien, Die Eroberung Serbien 1915. Tätigkeit der Fliegerverbände der österr.-ung. 3. Armee und der Serb. Luftstreitkräfte. Bearbeiter Rittm. ad Peter v. Spaits, p. 1; D. Jordan, *Balkans, Italy and Africa ...*, op. cit., p. 48–49.

49 J. Piekałkiewicz, op. cit.

50 A. Szelągowski, op. cit., p. 779.

51 T. Aronson, *Warring Monarchs. European thrones on the eve of the disaster of 1914*, Krakow 2014, p. 18.

52 A. Szelągowski, op. cit.

53 D. Jordan, *Balkans, Italy, and Africa ...*, op. cit., p. 46.

54 A. Szelągowski, op. cit. Mobilization in the territory of the Bulgarian Empire was announced on September 22, 1915, on October 2, 1915, Bulgarian forces were massed on the Serbian border, on October 4, 1915, the Ententa states announced an ultimatum to the Bulgarian side, and on October 5-6, 1915 the main states of the Ententa (the United Kingdom of Great Britain and Ireland and the Republic of France) severed diplomatic relations with the Bulgarian Empire, tsarist Russia severed them on October 20, 1915 - cf. М. Бабац, op. cit., p. 91. According to A. Szelągowski (op. cit.), the Bulgarian-German pact was signed in Sofia on September 6, 1915, and the alliance with the Turkish side in Dimetike. The Bulgarian-German military convention obliged the Central States to join the hostilities within 30 days, and the Bulgarian Empire within 35 days in joint actions in the Balkan Peninsula against the Kingdom of Serbia (the Bulgarian side was to receive all the requested parts of Serbian territory).

55 A. Szelągowski, op. cit.

56 P. Hart, op. cit., p. 234.

57 A. Szelągowski, op. cit., p. 780; L. Höbelt, op. cit, p. 56; Д. М. Бабац, op. cit., p. 90; М. Радојевић, Љ. Димић, op. cit., pp. 171, 181. The Macedonian Front (the Salonika - the Salonika expedition) was created by the Entente states to provide military support to the Serbian army and to maintain maritime communication

with Tsarist Russia (the Niš-Salonika railway line was the main route for the supply of military equipment to the Kingdom of Greece from Parties of the Ententa countries - France and Great Britain). In May 1915, the Greek garrisons in eastern Macedonia were forbidden to interfere with the German and Bulgarian troops crossing the border, and one of the corps of the Greek army "placed itself under the protection of the Germans", remaining interned until the end of hostilities "(eastern Macedonia was soon taken over by the army). On August 29, 1915, a coup took place, as a result of which the supporters of Prime Minister Eleftherios Venizelos declared their support for the Entente, and he himself ordered the Entente countries to send an Expeditionary Corps to Thessaloniki as a condition for the accession of the Kingdom of Greece to hostilities (Tsar Nicholas II Romanov, in accordance with the policy of supporting the Serbian ally - Panslavism - see *Austria-Hungary* ..., op. Cit., Pp. 8-9, 25, 31 - agreed to participate in the fighting on this front also by the Russian division and the attack on the ports of Burgas and Varna by the ships of the Black Sea Fleet. At the time of the landing of the Allied forces in Thessaloniki, the King of Greece announced the armed neutrality of his country, and on September 4, 1915 Prime Minister E. Venizelos announced in the Greek parliament that the Kingdom of Greece would engage in a military conflict on the side of the Entanta states. On September 24, 1915, the government of E. Venizelos announced a mobilization in the Kingdom of Greece (in the event of an attack on the territory of the Kingdom of Serbia by the Bulgarian army, as part of the allied aid, the Serbian ally was supposed to support 150,000 Greek soldiers), which resulted in a sharp reaction of the King of Greece, which resulted in the government surrendering to resign. On October 8, 1915, the successor of the former prime minister, Zaimis, refused to help the Serb side and agreed to hand over the Rupel fort on Struma to the Bulgarian-German forces. In response, British and French diplomacy accused the Greek side of treason and the Allied armed forces, among others. they took the island of Corfu and the King of Greece, Constantine, received an ultimatum in which Entente diplomacy demanded the demobilization of Greek forces and the dismissal of the prime minister. The monarch agreed to meet these demands *and on September 27, 1916, Prime Minister E. Venizelos issued a proclamation calling for disobedience to the monarch*, creating a new, temporary cabinet. The Ententa countries introduced a blockade of the Greek coast to force the Greek monarch to make further concessions. On October 29, 1916, the prime minister arrived in Corfu, and on October 9, 1916, he repeated the declaration of the provisional government and called on the Greeks to join the war on the side of the Entente, which was the beginning of a national rift and the choice of a path - for most Greeks - to support the king or the prime minister. Thus, the population of the Kingdom of Greece was divided into pro-prime-pro Ententa and pro royal-anti-Ententa camps. In November 1916, there was a revolt of the Greek Fleet and on the island of Corfu and *E. Venizelos created a revolutionary government.* On November 24, 1916, Prime Minister E. Venizelos declared war on the Central Powers, and in connection with the constant anti-Allied position of King Constantine I, French troops moved to Athens, meeting the resistance of Greek army units faithful to him. As a result of repelling the French troops, the Maritime Nationale ships began shelling Athens (an eight-month blockade of the Greek coast was introduced). On December 14, 1916, the Entente issued a final ultimatum to the royal camp, after which he agreed to the Allied conditions, and the Entente on December 19, 1916 recognized the Greek provisional government in Thessaloniki as the competent government. As a result of considerable tensions on the line of King Constantine I-Entente, on June 12, 1917, the royal family left Greece in an atmosphere of pro-royal demonstrations, which were "swan songs" for the dethroned monarch. In June 1917, *in the face of the offensive of the Entente forces, the king of Greece was forced to abdicate in favor of his son Alexander*, and when the Greek prime minister was again E. Venizelos, *on July 22, 1917, the Kingdom of Greece declared war on the German Empire and the Bulgarian Empire*- por. T. Arnoson, op. cit., p. 238–239, 279, 283-284; Д. М. Бабац, op. cit., p. 90; *Illustrated History* ..., op. cit., p. 500; J. Friedrich, op. cit., pp. 358–359; *Great* ..., op. cit., p. 27; *History XX* ..., op. cit., p. 284–285; J. Friedrich, op. cit., pp. 366-367; Д. М. Бабац, op. cit., p. 90.

58 D. Jordan, *The Balkans, Italy & Africa 1914–1918: From Sarajevo to the Piave and Lake Tanganyika*, London 2008 - Polish edition - idem - D. Jordan, *Italy, and Africa 1914-1918. From Sarajevo to Piawa and Lake Tanganyika,* Poznań 2011; A. Mitrović, op. cit .; M. Glenny, op. cit .; M. Hickey, op. cit .; N. Thomas, D. Babac, op. cit. Two DPs landed on October 5, 1915 in the Thessaloniki area, and on January 16, 1915, the commander-in-chief of the Allied forces on the so-called The Salonica Abutment (Expeditionary Corps in Thessaloniki - soon the Army of the East [Army of the Orient]) became General Maurice-Paul-Emmanuel Sarrail - Lieutenant A. Szelągowski, op. cit., p. 780. *Illustrated history* ... (op. cit., p. 504) states that the Allied forces landing near Thessaloniki numbered 150,000. soldiers, while J. Friedrich (op. cit.) lists the number of 130 thousand. soldiers of the Franco-British corps supporting the 130,000 Serbian army. M. Радојевић, Љ. Димић (op. Cit., P. 181) report that the Allied forces landing near Thessaloniki numbered 120-150 thousand soldiers.

First to Recognize ... Wings over Serbia 1915

In the spring of 1915, two Fliegerkompanien - Flik 2 and Flik 4 - assigned to the field Imperial-Royal 5th Army, and soon Flik 9 joined the group of these units and as well the German-Turkish FFA to operate from the airport in the Banat area. Extended tasks for these aviation units included:
• carrying out reconnaissance flights over the Serbian border
• firing ground targets
• bombing ground targets with incendiary and destructive bombs
• photo recognition
• propaganda flights - dropping leaflets over Serbian items
• fighting enemy observation balloons
• attacking rail and road transport[1]

The crews of Imperial-Royal Flik 2 and Flik 4 initially conducted close and long air reconnaissance on the Italian Front (over Isonzo) and the crews of Flik 9 based on the airport Homokbálványos (Serb. Баваниште / Bavanište, German. Bawanischte), from the winter of 1914/1915 performed tasks of similar purpose over Serbia (from June 7, 1915, Flik 9. operated from Vrga airports in Banat and Csernahéviz [Romanian Toplec, Topleț]). Together with the German-Turkish FFA, these units worked for the command of the field command of the 3rd Army, and later the German field command of the 11th Army when the command of Armeegruppe Mackensen was in Temesvár (Romanian Timișoara, German Temeswar, Temeschwar, Temeschburg)). In addition, AOK used the services of aircraft and seaplane crews of Turkish Aviation (currently Türk Hava Kuvvetleri - volthe urethly FFA was to operate from the airport Băile Herculane[Hungarian Herkulesfürdö, German Herkulesbad])[2]. For AOK and OHL, each report about the opponent in the planning phase of the offensive against the Kingdom of Serbia was of exceptional importance (the distance between the airports of each Flik from the border line of Sava and Danube was from 12 to 50 km). Thus, the role of air intelligence in providing the latest information on the enemy's movements increased. Winter and spring reconnaissance flights were carried out not only in the border zone, but also the crews of the KuK Luftschifferabteilung went deeper into the territory of the Kingdom of Serbia. On March 17, 1915, the crews of the Imperial-Royal Air Force reported that in the railway station Mladenovac (Serb. Младеновац), large forces of Serbian infantry were transported. A similar report was received by the Imperial-Royal Command of Banat on March 18, 1915. In March 1915, the Flik crew. 4 stationed at the Csernaheviz airport, also carried out bombing raids on targets in Kraljevac (Serb. Кралевац) -Prahovo (Serb. Прахово) -Kladovo (Serb. Кладово)[3]. According to the directives of the Imperial-Royal Banat Command of April 29, 1915, the crews of Imperial-Royal Fliegerkompanien were to explore the area of Stara Pazova (Serb. Стара Пазова, Hungarian Ópázova, Slovak: Stará Pazova) -Vinča (Serb. Винча) -Grocka (Serb. Винча) -Grocka (Serb. Смедерево)- the estuary of the Jezava river - the estuary of Moravia-Kostolac (Serb. Костолац, Roman. Caştel) -Kličevac (Serb. Кљичевац) -Ram (Serb. Рам) -Kisiljevo (Serb. Кисиљево) -Veliko Gradište (Serbian. Велико Градиште) and possibly Gradina (Serb. Градина) - Tekija (Serb. Текија)[4]. In early May 1915, as part of the continuation of ground bombing of Serbian targets, the Flik crew. 4 bombed a military warehouse in Serbian territory on May 7, 1915, and on May 9, 1915 three crews dropped two incendiary bombs on the loading yard near Prahov and three carbonite bombs on the Serbian army barracks near Kraljevac[5]. The reconnaissance flights of the crew of Imperial-Royal Fliegerkompanien were carried out many times in direct contact with the Serbian Opel Thus, on May 18, 1915, after taking off from the airport in Csernahéviz at 6.40 p.m. the pilot Ada Kaleh was fired upon, as assumed by the Romanian border guards at Verciorova, in the area Orșovej (Ger. Orschowa, Hungarian Orsova, Serb. Оршава / Oršava, Bulg. Орсово) on the neutral Romanian bank of the Danube. The aviator did not violate the airspace of the neutral Kingdom of Romania, and after landing and analyzing the event, it turned out that the plane shot at a Serbian steamer sailing on the Danube just off the Romanian border[6]. On May 29, 1915, the crews of Imperial-Royal Fliegerkompanien reported from 7.00 to 9.15 recognition of the Pazov region (Serb. Пазова) -Sremčica (Serb. Сремчица) -Ralja (Serb. Раља) -Grocka, where enemy infantry units were detected, four air hangars near Sremčica, and in the direction of Sarajevo (Serbian Сарајево, tur. Saraybosna) a BP strength military camp. On the same day, aerial reconnaissance also reported the reconnaissance of the position of the Serbian army on the Sava River. Due to numerous failures of aircraft engines due to many flights, commander Flik 2 asked the command of the KuK Luftschifferabteilung to supplement his unit with two planes with more powerful engines than those used so far, so that it would be possible to fly over mountains above 1400 m. A day later May 30, 1915, aerial reconnaissance provided information about an enemy from the Kisiljevo area (Serb. Кисиљево) -Poljana (Serb. Пољана) o-Požarevac (Serb. Пожаревац, Hungarian. Pozsarevác) -Smederevo. At the railway station in Požarevac, a large group of wagons was detected, but without any troops. The region of Jarak (Serb. Јарак) -Šabac (Serb. Шабац, Hungarian. Szabács) -Zemun (Ser. Земун / Zemun, Hungarian. Zimony, German. Semlin) and Mitrovica (Sremska Mitrovica, Serb. Сремска Митр. Srijemska Mitrovica, Hungarian Szávaszentdemeter) -Zasavica (Serb. Засавица) -Jarak. On May 31, 1915, the air reconnaissance of Imperial-Royal Fliegerkompanie observed the movements of Serbian troops in the Danube-Sawa belt towards the west and over the Drina on the march towards the Bosnian border[7].

The intensity of reconnaissance flights by the crews of Imperial-Royal Fliegerkompanien on the Serbian border in

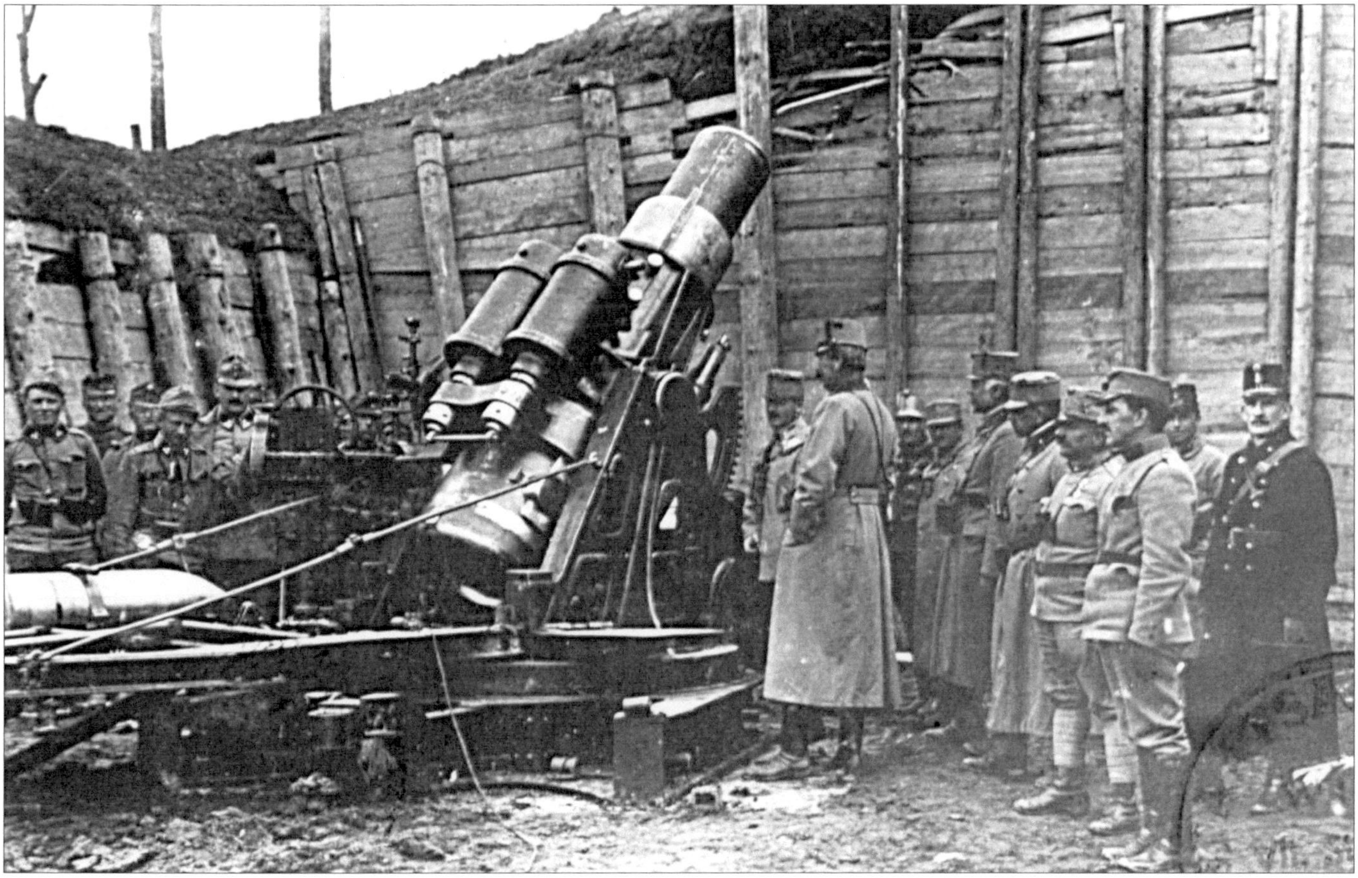

Above: Observers of tethered observation balloons were also trained to direct the fire of the Škoda mortar battery 30.5 cm M 11…In the photo, Archduke Eugeniusz Habsburg during the visit to the "Škoda" M 11. 30.5 cm mortar stand at Semlin (Zemun) on the Serbian Front in winter 1915 - source: Fotoarchiv SAW-KA Wien

June 1915 did not diminish - close air reconnaissance was carried out at 50-60 km, and long beyond this area[8]. On June 1, 1915 in Grabovac (Serbian Грабовац) -Obrenovac (Serbian Обреновац) and between Šabac and Mišar (Serbian Мисар) air reconnaissance reported the relocation of Serbian army troops[9], and on June 2, 1915, the area of Debra (Serb. Дебра) -Ub (Serb. Уб) -Obrenovac was recognized, where strong enemy military columns were detected[10]. To strengthen the air force in the Balkans,the Flik 15[11] was transferred from Nord-Ost Front (from Rzeszów) to Süd-Westfront that was in dire need of observer officers. Therefore, in connection with the forwarding of the unit to the Balkan region, commander Flik 15 asked the command that the Duke of Oblt be assigned to the unit that was waiting for short and long reconnaissance flights using radio and cameras. L. Sapieha as an observer, which was fulfilled. As of June 9, 1915 Flik 2, Flik 4 were subordinated to the field command of the 3rd Army and the German FFA 66th in Ruma (Serb. Ruma) and from 18 (21?) June 1915 Flik 15 in Újvidék (on September 2, 1915, he was assigned to the Serbian front from the Nord-Ost Front Flik. The fifth, who returned "home" himself)[12]. As part of the continued bombing of Serbian military ground targets, on June 9, 1915, four Flik crews. 9 threw 20 bombs at the arsenal in Kragujevac, causing numerous fires (the city also housed the headquarters of the Serbian Army High Command). During bombing raids also carried out against targets in the territory of the Kingdom of Montenegro, the crews of Imperial-Royal Fliegerkompanien used bombs weighing 1-25 kg[13] , and let the data on Flik 6 testify to the frequency of bombing, which by mid-June 1915 made 21 bombing flights, five times bombing the Montenegrin capital Cetinia (Montenegro Цетиње / Cetinje), dropping three or six bombs, three times bombing artillery battery positions in the Montenegrin area, in the Lovćen massif, three times bombing Podgorica (Montenegro Podgorica, Подгорица, formerly Titograd / Титоград)[14]. When attacking ground targets, airborne darts were also used (the so-called Fleszet - on June 23, 1915, incendiary darts were thrown at targets in Belgrade)[15]. The plane-artillery cooperation became stronger and stronger. In the order of June 15, 1915, the command of the fieldwork of the 3rd Army ordered an aerial reconnaissance of the Mišar-Šabac area to find Serbian artillery positions, which were then fired by the Imperial-Royal artillery in Drenovacka Island Ada (Serb. Дреновачка Ада) and Tursko Groblje (Serbian. Турско Гробље)[16]. On June 26, 1915, the Flik 15 crew, after acclimatization in the Balkan area, began combat flights. On the same day, the first large "air demonstration of power" took place by the crews of the KuK Luftschifferabteilung: four crews bombed industrial facilities in Belgrade (six bombs were dropped) and railway transports underOrșova south-west of Obrenovac (four bombs were dropped) and from the on-board weapons (km) of the Flik crew. 15 shot at Serbian targets on the island of Klenovacka Ada (Serb. Кленовачка Ада)[17]. At the end of June 1915, the bombing of targets on Serbian territory was intensified: on June 28, 1915, four crews of the Imperial-Royal Air Force attacked targets in Belgrade from the air[18]; a day later, on June 29, 1915, the Flik crew. 9 dropped one or two bombs on the Belgrade-Batajnica airport (Serbian. Батајница)[19]; until the end of the month, targets in Požarevac and near Kladov were bombed with incendiary bombs, attacking river ships and Serbian army warehouses[20]. Aerial reconnaissance of the Danube line did not weaken - on June 29, 1915, the Flik crew. 9 recognized the Danube region to the island of Ogradina (Serb. Ogradin)[21].

According to the air reconnaissance of the crews of Imperial-Royal Fliegerkompanie, the Serbian aviation (Serb.in accordance with the requirements of that time (airports were established next to the headquarters of large military units) were stationed at the Belgrade-Batajnica (weakly numbered Squadron - two planes) and Požarevac (2nd Squadron). The squadron from the Belgrade-Batajnica airport was subordinate to the command in Belgrade and the 2nd squadron to the command from Braniceva (Serb. Браничево, Braničevo). From the first airport, Serbian airmen flew to the area of Smederovo-Obrenovac-Šabac, and especially over the area of Srem and Banat in accordance with the guidelines of the Aviation Command[22]. In the 1915 campaign, the Serbian air force was reinforced with flying equipment and airmen from the allied Republic of France, where on February 28, 1915 training of Serbian pilots began. In mid-1915, a Serbian pilot school started operating in Požarevac[23] while On June 7, 1915, a group of pilots, observers and mechanics were sent to French aviation schools for three-month training in pilotage and aviation observation (this group soon returned to the country via the sea route via Thessaloniki)[24] and on June 20, 1915, six officers, six non-commissioned officers and four mechanics were sent there for a two-month flight training (this group in November 1916 was included in the Serbian air force reconstructed in the area of the city in Thessaloniki)[25]. In February 1915, at the airport in Palanka (Serbian Паланка), a large hangar and 30 tent hangars were prepared for the French squadron to operate in the Balkans, and in July 1915, 25,000 were gathered there. liters of gasoline and an order was placed for a further 75 thousand liters[26].

As part of the allied assistance, the creation and equipment of the French Escadrille MF-S 99 was carried out at the French Lyon-Bron air base, and it arrived in the Kingdom of Serbia on March 15, 1915 (12 [8?] Maurice Farman MF 11. with Renault engines equipped with on-board MGs, 99 [88?] officers, non-commissioned officers, and soldiers, including eight pilots and ten observers). Capitaine Roger Vitrat became the commander of the squadron under the command of the Serbian army, transferred to the Belgrade-Batajnica airport[27] and the area on which it was to operate stretched from Smederev on the Danube to Loznica (Serbian Лозница) on Drina (Serbian aviation crews - Serbian aviation personnel at that time consisted of 99 officers, non-commissioned officers and soldiers - while they were to operate in the Danube front zone) from Smederev to Golubač [Serbian Голубац, Romanian Golumbei, Hungarian Galambóc])[28].]). In addition to replenishment in planes, the Kingdom of Serbia received from the French ally 10 tent hangars, eight tractors, two trucks, two light cars, workshop equipment (technical facilities for Serbian aviation and the French squadron were in Palanka), two backup engines and cameras[29] (photos were taken in 9x9 or 9x12 format)[30]. In March 1915, the Serbian air force carried out reconnaissance from the Golubač-Smederevo area to the Petrovoselo line (Bačko Petrovo Selo - Serb. Бачко Петрово Село, Hungarian Péterréve, German. Batschko Petrovo Selo) -Vršac (Serb. Serbian Бела Црква / Bela Crkva, German Weißkirchen, Romanian Biserica Alba)-Golubač[31]. The first combat order in the 1915 campaign to reconnaissance the position of the Imperial-Royal Army, which was issued by the Serbian Aviation Command on April 8, 1915, ordered the French Squadron from Belgrade-Batajnica to reconnaissance the Zemun-Indjija railway line (Serbian Инђија) -Peterwardein (Serb Петроварадин, Hungarian Pétervárad, German Peterwardein) -Fruška gora (Serbian Фрушка гора, Hungarian Tarcal-hegység) -Ruma-Šabac and further along the Sava river to Zemun. It was ordered to recognize the position of the army's infantry and bomb monitors of the KuK Donauflotille. Further reconnaissance was carried out in the section Pančevo (Serbian Панчево, Hungarian Pancsova, Romanian Panciova) -Alibunar (Serbian Алибунар) -Kovin (Serbian Ковин) -Danaj[32].

As part of retaliatory raids, the Serbian air force bombed enemy targets quite actively and on April 17, 1915, Serbian airmen bombed (eight bombs) Flik airfield. 9 in Homokbálványos, on April 20, 1915 (four bombs), the same target was bombed again, and on April 25, 1915, three bombs and 500 aerial darts were dropped on the position of the observation balloon in the Belgrade-Batajnica area. The use of observation balloons on the Serbian Front was not a frequent accident and it did not cause a "balloon phobia" as, for example, on the Western or Eastern Front, where the popular "Sausages" were used en masse.

Along the front, tethered balloons were visible from afar. They were used by German divisions to observe the movements of n-pla. In my eyes, a Russian plane flew in, which caused the balloon "sausages" to drop suddenly. Like the scared snail in its shell, they wanted to hide on the ground, but not all of them made it. The aviator ignited three balloons. German observers saved themselves using parachutes. I heard one hit a stone and was killed, Oblt recalled (lieutenant) Wacław Hulewicz from the Toruń infantry regiment fighting in the ranks of the German army on the Eastern Front[33].

Arrival of French aviators to the Serbian Front and deliveries of French airplanes for Serbian and French aviation in the Balkans (as part of equipment replenishment, Serbian aviation received from a French ally two old Bleriot XI planes with 80 HP engines, as well as aerial bombs and MGs, and at the end of June 1915, the French ally delivered another 10 aircraft) enabled the Serbian aviation to perform tasks of greater importance as well as to achieve air balance over the front in the period until October 1915, which was a considerable achievement. In October 1915, with the launch of the offensive against the Kingdom of Serbia by the combined forces of the Central Powers, the situation changed[34]. However, before this happened, the April reconnaissance flights of the Imperial-Royal Fliegerkompanie crews over the hostile territory provided the Imperial-Royal Command of Banat and the AOK with a lot of information about the dislocation and the activity of Serbian aviation. On April 25, 1915, air reports of sighting enemy biplanes in Ohrid (Macedonian Охрид), towards Nikinci (Serbian Никинци) and others in Vitojevci (Serb. Витојевци) -Sawa were received by the Imperial-Royal Commander of Banat. During the flights, the crews of Imperial-Royal Fliegerkompanien saw two Farman biplane and one Bleriot monoplane in the Belgrade region and on the Kupinovo (Serb. Купиново) -Progar (Serb. Прогар) -Boljevci (Serb. Бори. Zemun (after 30 minutes to an hour, landing took place near Topčider [Serb. Топчидер] - two more Bleriot monoplane maneuvers over Belgrade). A day later, on April 26, 1915, the crew of the Imperial-Royal Air Force, after reconnaissance of the Lokva region (Serb. Локва), she saw another enemy plane in flight[35]. In April 1915, Serbian, French and mixed Serbo-French crews began to rush to the area of Dubovac (Serbian Дубовац) -Grabenac-Ulma (Serbian Уљма) -Vršac-Kakova-Oravica (Serbian Оравица) -Új Moldova[36] *while t*he crews of the Imperial-Royal Air Force on Kragujevac and other cities in the Kingdom of Serbia caused increased defense activities of the enemy air force[37]. The enemy did not behave passively and on April 24, 1915 and on the nights of April 27/28 and April 29/30, 1915, the enemy squadron with Serbo-French crews bombed KuK Donauflotille monitors operating on the Danube and Sawa[38] which was a big surprise for the Imperial Command. The activity of the crews of the French squadron increased over time and its crews, as well as the mixed French-Serb ones, carried out more and more reconnaissance flights on the section Újvidék-Subotica (Serbian Суботица / Subotica, Hungarian Szabadka) -Temesvár (Romanian Timișoara, German Temeswar, German Temeswar , Temeschburg) until Sarajevo. In May 1915, as part of the bombing raids on hostile territory, the Serbian air force dropped 1,000 air arrows on the positions of the Imperial-Royal army under Karolyfolva and Délibáb and bombed Flik 9 airfield once again in Homokbálványos[39]. On May 27, 1915, the French-Serbian crews recognized the forces of the Imperial-Royal army in the border region of the Danube and Sava, and a detailed reconnaissance was carried out in the area of Pančevo-Zrenjanin (Serbian Зрењанин, Hungarian Becskerek), Sečanj (Serbian Сечањ / Secsan, Hungarian Szécsány) -Alibunar-Kovin, between Sava and the Brčko line (Serb. Брчко)-Vinkovci (Serb. Винковци) and in the area of Deljina-Brčko-Tuzla (Serb. **Тузла**) -Zvornik (Serb. Зворник) -Šabac as well as going to the area of Zrenjanin-Újvidék and Mitrovica-Beljina (Serb. Бијељина) -Brčko[40]. The raids of the Imperial-Royal Air Force crews on Kragujevac and other cities in the Kingdom of Serbia resulted in increased retaliatory activity of the enemy air force[41] and on May 30, 1915, the crews of the French squadron bombed the targets in Kevevára (Hungarian Temeskubin, Serbian Ковин / Kovin, Roman Cuvin, Ger. Temeschkubin). In the period of May 25-31, 1915, the crews of Imperial-Royal Fliegerkompanien informed AOK about enemy planes appearing over Srem and Banat in their air reports. With the advent of June, on June 1, 1915, the Serbian air force bombed enemy targets near Orsova and the airport in Csernaheviz, and on June 9 and 28, 1915, the Flik airport was bombed again. 9. (nine bombs were dropped on them)[42]. To inflict the greatest losses on the units, on June 10, 1915, the enemy air force bombed the artillery magazines of the Petervardein Fortress with bombs, but without much effect[43]. In the period of June 11–12, 1915, in the air reports, the crews of Imperial-Royal Fliegerkompanien informed the Imperial-Royal Banat and AOK command about the flights of enemy planes detecting the railway lines and the relocation of the Imperial-Royal troops mainly in the area of Pančevo-Alibunar-Vršac-Kevevára[44] (zlotyotniska in Požarevac the pilots of Sgt. M. Tomič, Lieutenant Z. Stankovič, in the period from May to the end of August 1915, made a total of 19 combat flights over the territory of Banat, during which the positions of the Imperial-Royal army were thrown several times with darts and bombs designed by Col. artillery Miodraga Vasič - September 14, 1915 20 bombs designed by col. M. Vašić and col. Gojko Djurića weighing 2–20 kg was tested in the French squadron)[45]. French and Serbian airmen, over time, were given the task of intercepting and pursuing enemy aircraft in the air as well as air attacking the Russian and German air units located in Srem and Banat[46]. As part of the support of the Balkan ally, in November 1915, the French Ministry of War, appreciating the importance of the Balkan front, promised to provide another squadron to the Serbian Front, so that the Serbian French air force "in the field" would have 20 aircraft[47] (the base for the French squadron was to be the airport in Požarevac)[48]. Serbian aviation, despite a slightly disrespectful opinion about its capabilities on the part of the AOK and the command of the KuK Luftfahrtruppen, managed to surprise. June 1915 also brought the first air clashes with the Serbian aviation as well as losses caused by the fire of the Serbian Opel in the group of Imperial-Royal Fliegerkompanien. During one of the flights, the Flik crew. 15 (Oblt. Forst-Oblt. Hautzmayer) received a shot in the radiator of its plane and had to land forcibly with good luck not being captured. One-on-one clashes were rare in the air on the Serbian Front, but on June 7, 1915, the Flik crew. 9 fought an air skirmish with an enemy plane because of engine damage, forcibly landing in the Homokbálványos airport. Gradually "panning" in the air of the Imperial-Royal Crews aviation over Serbian territory was, however, interrupted by the Serbian air force and, for example, on June 13, 1915, by the Flik crew. 15 (Oblt. Bela Losonczy von Losoncz and its commander, pilot Oblt. Wilhelm Schreiber), while observing Serbian artillery posts near Belgrade, was attacked by a Serbian aviation plane and during the fight the observer was wounded in the shoulder (he fired five ammunition clips from the onboard MG towards enemy Farman MF 11)[49]. On June 26, 1915, the first group air clash over the Balkans in the 1915 campaign took place, in which four crews of the KuK Luftfahrtruppen encountered three Serbian airplanes in the air, which shot them from the on-board km and were also shelled by the Serbian Opel. One of the planes, damaged by an explosion of shrapnel, forcibly landed next to the Belgrade-Batajnica airport, the crew of the second, additionally damaged in a strong storm, had to forcibly land under Grabovci (Serbian Грабовци), and the other two planes were slightly damaged[50]. Three days later, on June 29, 1915, another air clash took place between two crews of the KuK Luftfahrtruppen, and a Serbian plane over the island of Huja (Serbian Ада Хуја) near Belgrade, ending with heavy damage to the Serbian enemy[51]. At the end of June 1915, there were further cases of enemy air reconnaissance aircraft exploring the territory of the Imperial-Royal Habsburg Monarchy - e.g. on June 28, 1915 at 18.15 opl. Flik 9 airport, often attacked from the air at Homokbálványos, fired on three enemy planes[52]. In the summer of 1915, as part of the increasing number of reconnaissance flights of the Fliegerkompanien KuK Luftfahrtruppen crews, some of them occasionally collaborated with the KuK Donauflotille - e.g., one of the Flik 9 crew cooperated in the Danube reconnaissance with the ship in the Flotilla[53]. At that time, the camera became a permanent part of the equipment of Imperial-Royal Fliegerkompanien planes operating over the Balkans - e.g., according to the order of 19 July 1915 of the Flik 9 and Flik 15 crews for the benefit of the Imperial-Royal Banat command to explore the area of the Kolubara-Belgrade estuary, carrying out photo reconnaissance there[54]. In July 1915, the frequency of bomb attacks against Serbian targets, especially in Belgrade, increased. 12 July 1915 in the morning hours and 23 July 1915 in the afternoon hours of the Flik 9 bombed (three bombs were dropped) military targets in Belgrade[55] (during the air raids in 1915, the crews of Imperial-Royal Fliegerkompanien used air darts in addition to air bombs)[56] and on July 23, 1915, three Flik 15 crews bombarded the capital of the Kingdom of Serbia (four carbonite bombs, weighing 20 kg each, six 13 kg bombs and about 50 incendiary bombs were dropped on targets in Belgrade)[57]. As part of the front assignment, on August 11, 1915, Flik 15 was assigned to the command of the Peterwardein fortress, in which Festungsballonabteilung 1/6 R observation balloons were also used[58] to cooperate with heavy artillery as well as with the KuK Donauflotille artillery to the east of Vezirac (Serb. Vezirac - during the autumn field offensive of the 3rd Army, another three Ballonabteilungen were assigned to the Serbian Front)[59].

After the campaign ended with Serbian Victoria in 1914, the observation balloons used by the Serbian army were damaged due to military operations and the effects of weather conditions, but after renovation they were fit for use in the winter of 1915 - e.g., on January 12, 1915, KuK Donauflotille monitors were shot by Serbian artillery, the fire of which was directed by a balloon observer (more such cases were recorded until April 13, 1915)[60]. They were most often used near Belgrade (on May 15, 1915, airmen observed the observation balloon "placed" under Slance [Serb. Slantse] east of Belgard). On June 16, 1915, the Serbian Balloon Unit returned to its home base at the Belgrade-Batajnica airfield, but due to the incoming reports of increased movements of the Imperial-Royal troops preparing to strike, on September 23, 1915, the Serbian side used an observation balloon near Smederev (its observer was observing the enemy bank of the Danube, in the direction of the march of the German 11th Army units)[61]. In connection with the "balloon irritation" on the Serbian side, they were tried to counteract balloon ups (in one of the flights the position of the Serbian observation balloon was attacked by the FFA 1 crew)[62].

In August 1915, not only the crews of the above-mentioned Imperial-Royal Fliegerkompanien conducted reconnaissance of the territory of the Kingdom of Serbia, but also the first flights were made by German and Turkish crews - e.g., on August 13, 1915, the Kraljevo (Serb. FFA from Csernahéviz airport[63]. September 1915 and with-it autumn brought about rapidly changing weather conditions, which, however, did not drastically reduce the frequency and quality of air reconnaissance on both sides, which in the case of the Imperial-Royal and German air units as the time of the army's offensive approached field Central Powers played an increasingly important role. On September 3, 1915, the crew of Flik 9 carried out from morning to afternoon the reconnaissance of targets in the Serbian territory, and on September 4, 1915 they recognized the region of the Morava valley and targets west of the Grocka-Mladenovac-Kragujevac line, going 250-300 km deep into Serbian territory[64]! The second serious demonstration of power by the KuK Luftfahrtruppen took place on September 12, 1915, mainly in the Srem and Banat regions[65]. In the second half of September 1915, the crews of Imperial-Royal Fliegerkompanien devoted a lot of attention to the reconnaissance of the Belgrade region and the Serbian

Air Force airports. On September 24, 1915 at 10:15 the command of the field of the 3rd Army reported to the AOK about the air reconnaissance of the Belgrade region (the Flik crew on the 15th reported about the cells in the wagon factory and about the movements of Serbian troops)[66]. The end of September 1915 also brought several more bombings of Imperial-Royal Air Force on Serbian targets. On September 30, 1915, there was a strong bombing of Niš (43 bombs were thrown on the seat of the Serbian press office, five people died, 17 were injured, and three bombs were dropped on targets in Lapovo [Serbian Лапово]), and on September 30, 1915, also sights in Kragujevac[67].

Endnotes

1 SAW-KA, Die Eroberung Serbien 1915. Tätigkeit der Fliegerverbände der österr.-ung. 3. Armee und der Serb. Luftstreitkräfte. Bearbeiter Rittm. ad Peter v. Spaits, p. 6.
2 Ibid, p. 8.
3 Ibid, p. 198.
4 Ibid, p. 12–13.
5 Ibid, p. 199.
6 Ibid, p. 18.
7 Ibid, p. 12-13.
8 SAW-KA, Der Einsatz der Fliegerabteilungen der Herresgruppe Mackensen im Herbstfeldzug 1915 gegen Serbien. bearbeitet von major dr. Gnamm, Kmdt. zur Luftkriegsakademie, p. 45.
9 SAW-KA, Die Eroberung Serbien 1915. Tätigkeit der Fliegerverbände der österr.-ung. 3. Armee und der Serb. Luftstreitkräfte. Bearbeiter Rittm. ad Peter v. Spaits, p. 12–13.
10 Ibid, p. 14.
11 SAW-KA, Die durchbruchsschlacht von Gorlice1915. Bearbeitet von Obst. AD Feldp. Erich Kahlen, p. 1-2. According to this document, Flik 15. In 1915 it was stationed at the following airports in Nord-Ostfront: December 20, 1914 - January 15, 1915 - Jadowniki; 15 Jan - 12 Feb 1915 - Kraków-Rakowice; 12 February - 10 May 1915 - Brzesko [Okocim-LightBrzeski]; 10–14 May 1915 - Tarnów; May 14 – June 18, 1915 - Rzeszów. From June 18, 1915, the unit was transferred to the Balkans (Obst. Emil Uzelac informed with a telegram sent from the airport in Rzeszów, the airport in Nowy Sącz [not M. Neu Sandez] and Aspern that they would be prepared to receive the planes transported to Újvidék (Serb./Ch. / Romanian Novi Sad, Нови Сад, Slovak Nový Sad, German Neusatz (an der Donau) and stationed there at the following airports; 1915 - Újvidék, October 18 – November 4, 1915 - Belgrade; 4-18 November 1915 - Aranglelovac [Serb. Аранђеловац]; November 18-December 2, 1915 - Čačak (Serb. Чачак); December 2, 1915 - January 29, 1916 - Mitrovica. During the 1916/1917 campaign, Flik 15 stationed at the following Alpine airports: January 29 - March 6, 1916 - Igalo (Serb. Игало); March 6 – August 6, 1916 - Perigne; August 6, 1916 - January 14, 1917 - Bruneck (owned by Brunico); 14 January – 16 June 1917 - Brixen (owned by Bressanone); June 16 – August 30, 1917 - Gardolo; August 30, 1917 - Dato - Bruneck - Lieutenant A. Olejko, *The Carpathian War* ..., op. cit., p. 413.
12 SAW-KA, Die Eroberung Serbien 1915. Tätigkeit der Fliegerverbände der österr.-ung. 3. Armee und der Serb. Luftstreitkräfte. Bearbeiter Rittm. ad Peter v. Spaits, p. 16.
13 Ibid, p. 209-210.
14 Ibid, p. 201.
15 Ibid, p. 209-210.
16 Ibid, p. 20.
17 Ibid, p. 201.
18 Ibid, p. 202.
19 Ibid, p. 200.
20 Ibid, p. 201.
21 Ibid, p. 27.
22 Ibidem, p. 30. December 24, 1912 by order of the Serbian Minister of Defense, Fr. Marshal R. Putnik, the Air Force Command was established (Serb. Ваздухопловна Команда), which until then was subordinate to the Serbian Supreme Command, from January 24, 1915 under General Mihajlo Živkovićovi based in Belgrade.
23 It existed however, only about month, and its participants, due to unfavorable war events in the country, were sent for further training to the Republic of France- Lt. SAW-KA, X. Mikic, Geschichte der yugoslawischen Luftwaffe. Űbersetz vom Rittm. ad P. v. Spaits, op. cit .; Б. Циглић, Крила Србије Ваздухопловна команда и авијатика српске војске 1912 - 1920, Београд 2009 - pasim; Српска aviyatika 1912-1918,Београд1993- passim; CJМикић, Историја угословенског ваздуховства, Београд 1933, passim; Б.Dimitry, П. Миладиновић, М. Мицевски, Краљевско Ваздхопловство - Војно ваздухопловство Краљевине СХС / Југославије 1918-1944, Београд 2012, passim; Н. Ђокић, Помоћ Русије Србији у стварању ваздухопловства 1901-1915, Београд, «Лет» 2000, No. 2, passim; http://sr.wikipedia.org/wiki/Ваздухопловна_команда_српске_војске_(Краљевине_Србије); http://sr.wikipedia.org/Коста_Милетић - entry from August 23, 2014; http://sr.wikipedia.org/wiki/Ваздухопловна_команда_српске_војске_(Краљевине_Србије) - entry from August 23, 2014.
24 SAW-KA, Die Eroberung Serbien 1915. Tätigkeit der Fliegerverbände der österr.-ung. 3. Armee und der Serb. Luftstreitkräfte. Bearbeiter Rittm. ad Peter v. Spaits, p. 35.
25 SAW-KA, Die Luftstreitkräfte Serbiens von 1914 bis Herbst 1916 von Oskar Regele, p. 9.
26 Ibid, p. 5.

27 Д. М. Бабац, op. cit., p. 85. On April 18, 1915, six MF 11-type airplanes flew to the Banjica airfield, out of 12 machines included in the squadron.
28 D. Babac, A. Thers, *L'Aviation Serbe 1914-1918,* Armes Militaria Magazine No. 189. p. 41–47.
29 SAW-KA, Die Eroberung Serbien 1915. Tätigkeit der Fliegerverbände der österr.-ung. 3. Armee und der Serb. Luftstreitkräfte. Bearbeiter Rittm. ad Peter v. Spaits, p. 31.
30 Ibid, p. 241.
31 The squadron from the airport in Požarevac, where the construction of two air hangars began, also used the Balloon Division (Serb. Balonska Одељење - a hydrogen generator was purchased for him in 1913) transferred here from the Belgrade-Batajnica airport. In the period from January to April 1915, this unit performed reconnaissance tasks for the Belgrade Defense Command - Lieutenant SAW-KA, X. Mikic, Geschichte der yugoslawischen Luftwaffe. Űbersetz vom Rittm. ad P. v. Spaits, pp. 30-38; Српска…, op. cit .; CJМикић, op, cit .; Б.Dimitry, П. Миладиновић, М. Мицевски, op. cit .; Н. Ђокић, op. cit .;http://sr.wikipedia.org/wiki/Ваздухопловна_команда_српске_војске_(Краљевине_Србије); http://sr.wikipedia.org/Коста_Милетић - entry from August 23, 2014; http://sr.wikipedia.org/wiki/Ваздухопловна_команда_српске_војске_(Краљевине_Србије) - entry from August 23, 2014.
32 SAW-KA, Die Eroberung Serbien 1915. Tätigkeit der Fliegerverbände der österr.-ung. 3. Armee und der Serb. Luftstreitkräfte. Bearbeiter Rittm. ad Peter v. Spaits, p. 32.
33 A. Inattentive, op. cit., p. 82.
34 D. Babac, A. Thers, op. cit.
35 SAW-KA, Die Eroberung Serbien 1915. Tätigkeit der Fliegerverbände der österr.-ung. 3. Armee und der Serb. Luftstreitkräfte. Bearbeiter Rittm. ad Peter v. Spaits, p. 17.
36 Ibid, p. 36.
37 Ibid, p. 35.
38 Ibid, p. 205.
39 Ibid, p. 36.
40 Ibid.
41 Ibid, p. 35.
42 Ibid, p. 207.
43 Ibid, p. 211.
44 Ibid, p. 34.
45 SAW-KA, Die Luftstreitkräfte Serbiens von 1914 bis Herbst 1916 von Oskar Regele, p. 12. Both Serbian designers soon created a 62 kg model of an aerial bomb - see SAW-KA, X. Mikic, Geschichte der yugoslawischen Luftwaffe. Űbersetz vom Rittm. ad P. v. Spaits, op. cit., p. 34,211; Српска…, op. cit .; CJМикић, op. cit .; Б.Dimitry, П. Миладиновић, М. Мицевски, op. cit .; Н. Ђокић, op. cit .;http://sr.wikipedia.org/wiki/Ваздухопловна_команда_српске_војске_(Краљевине_Србије)- entry from August 23, 1914. Б. Циглић, Крила Србије Ваздухопловна команда и авијатика српске војске 1912–1920, Београд 2009, p.?.
46 D. Babac, A. Thers, op. cit.
47 SAW-KA, Die Eroberung Serbien 1915. Tätigkeit der Fliegerverbände der österr.-ung. 3. Armee und der Serb. Luftstreitkräfte. Bearbeiter Rittm. ad Peter v. Spaits, p. 35.
48 SAW-KA, Die Luftstreitkräfte Serbiens von 1914 bis Herbst 1916 von Oskar Regele, p. 6.
49 SAW-KA, Die Eroberung Serbien 1915. Tätigkeit der Fliegerverbände der österr.-ung. 3. Armee und der Serb. Luftstreitkräfte. Bearbeiter Rittm. ad Peter v. Spaits, p. 219.
50 Commander of Flik 15. Oblt. W. Schreiber, after an air fight with an enemy plane and damage to his plane, was forced to land near Újvidék) - see Ibidem, p. 22.
51 Ibid, p. 219.
52 Ibid, p. 26.
53 Ibid, p. 17.
54 Ibid, p. 19.
55 Ibid, p. 201.
56 Ibid, p. 202.
57 Ibid, p. 202.
58 Ibidem, p. 202, 28. In 1898, the German army adopted as a basic observation balloon - Drachenballon - a Parseval-Sigsfeld M 98 balloon. The balloon was cylindrical and replaced the previously used spherical balloons on a tether. During the ascent, the aerostatic lift was supported by the aerodynamic buoyancy obtained by setting the balloon body at a large angle of attack with respect to the wind direction. Hence the name of the balloon "Drachen" - a kite. During the Serbian campaign of 1915, the army used, as well as during the 1914 campaign, observation balloons on the parseval-Sigsfeld Drachenballon M 98 tether and, less frequently, spherical balloons (Kugelballons) intended for free flights.
59 Ibid, p. 28.
60 Ibid, p. 40.
61 Ibid, p. 41.
62 SAW-KA, Der Einsatz der Fliegerabteilungen der Herresgruppe Mackensen im Herbstfeldzug 1915 gegen Serbien. bearbeitet von major dr. Gnamm, Kmdt. zur Luftkriegsakademie, p. 12.
63 Ibid, p. 8.
64 Ibid, p. 25.
65 Ibid, p. 24.
66 Ibid, p. 21.
67 Ibid, p. 202.

Wings over Balkans

The October date of the Austro-Hungarian-German offensive for staff officers was approaching rapidly, thus intensifying the air activities of the crews of Imperial-Royal Fliegerkompanien (from October 1, 1915, Hptm Robert Eyb from Flik became the staff officer of the Imperial-Royal Air Force in Süd-Ostfront. 15.) and allied units[1]. Before the start of the offensive, the following air force was assigned to the Mackensen Group: from September 26, 1915, the army's Flugpark 13 from Breslau (Polish: Wrocław), six German FFAs, one German-Turkish FFA (the crews were ready in Istanbul and had on a flight via Bulgarian territory to move towards the front line), four "at home" in the Balkans, Imperial-Royal Fliegerkompanien (Flik 3, 6, 9, and 15th were assigned to the 3rd Army's Field Corps) and one staff officer-airman. While the crews of the KuK Luftfahrtruppen were preparing from the winter of 1914/1915 for intensified offensive operations on the border with the Kingdom of Serbia, and mostly knew the area of future fights, the crews of six German FFAs had to be drawn from the Eastern Front (FFA 1st, 28th, 30. 59. 57. 66.[2] Only the crews of FFA 66 were briefly based in the Ruma area and made 12 reconnaissance flights over Belgrade with little "Balkan experience". The removal of the above-mentioned FFA from the Eastern-Russian war effort took place on September 18-October 6, 1915, of which the crews of the 28, 59 and 57 FFA were ready for combat flights in the Danube and Sava region on October 6, 1915![3]. The German FFA used observation B aircraft (in this version, the aircraft were equipped only with revolver pistols as personal weapons and infantry rifles) and one in the C version (in this version, the aircraft was equipped with one observer's km) Albatros B types I[4]. and LVG C.II. These planes enjoyed a good reputation among the crews, and the average flight time in the mountains was up to four hours at an altitude of 1000-2500 m[5].

Bulgarian air force (Bulgarian Вздушни Войски - Wzduszni Wojski) was to join the group of the Central Powers' air force on the Serbian Front in autumn 1915. At the outbreak of the First Balkan War in 1912, the air force of the Bulgarian Empire consisted of the Aviation Division established at the end of July 1912 (Bulgarian Аероплано Отделене - Aeroplanno Otdielienje), with three pilots and one Bleriot XI plane (a few days after the outbreak of hostilities, three Albatros F-2 planes were ordered there from Kaiser Germany to arrive in Bulgaria from Kaiser Germany). Bulgarian aviation pilots were trained in French, British, German, and Russian aviation centers, and after the outbreak of hostilities, the Bulgarian aviation was joined by volunteer pilots (by the end of October 1912, a total of 22 pilots participated in the fighting)[6]. On November 3, 1912, the 2nd Air Division was established (the commander was Старший лейтенант Dimitr Sekełarov), who was to fall under the Bulgarian field of the 1st and 3rd armies on the so-called cztaladzinski's front[7]. On November 10, 1912, the 3rd Air Division was established under the Bulgarian 4th Field Army on the Bulgarian front (the commander was Старший лейтенант Penio Popkrystew). The created aviation units were commanded by the Bulgarian aviation pioneer Major Vasil Zlatarov (he also supervised the 1st and 2nd Balloon Squads [Bulgarian Балонно Отделене - Bałonno Otdielenje] and auxiliary units). At the end of the First Balkan War, the Bulgarian air force counted 29 aircraft and 21 pilots[8]. The state of demobilization of the Bulgarian aviation was introduced after the end of the Second Balkan War on August 1, 1913[9] (at that time, the Bulgarian air force had eight training planes and eight pilots, two planes under repair and three hangars). After the outbreak of the Great War, a plan for the development of Bulgarian aviation was created (it had only five planes, of which only two were operational - the Bleriot XI type). In the summer of 1915, Col. W. Złatarow had at his disposal an Aviation Unit commanded by Cpt. R. Miłkowa (five planes - two Albatros B.I, two Bleriot XI and one Bleriot IX bis) and two Balloon Squads[10]. Just before the offensive of the field armies of the Central Powers against the Kingdom of Serbia, on September 8, 1915. Flik 5 and Flik 6 have been assigned to the field command of the 3rd Army as having officers specializing in aerial photography[11].

Photo reconnaissance was also carried out by the crews of the other two Imperial-Royal Fliegerkompanien and thus the reports of the KuK Luftfahrtruppen crews gained special importance for the command of the field armies and for the command of Armeegrupe Mackensen! Commander of Flik 15. reported on 24 September 1915 to the command of the field command of the 3rd Army, taking 12 photographs of the Serbian artillery position. In the same flight, the area south of Belgrade, in the Flik 9 operational area was explored. Flik 15 crews also recognized the area of the railway line Beli Potok (Serb. Бели Поток) -Vrčin (Serb. Врчин) and Serbian fortifications north of Boleč (Serb. Болеч) and south of the Leštane (Serb. Лештане) -Kumodraž. Кумодраж). On the same day the Flik 9 crew reported the discovery of Serbian fortifications on the section Samar-Milicevo Brdo-Lipak-Belgrade-Mokri Lug (Serb. Мокри Луг) -Erino Brdo (Serb. Ерино Брдо)[12]. Photo recognition of the crews of this Fliegerkompanie provided on September 29, 1915 valuable material from the area of Skela (Serb. Скела) and Velika Bara (Serb. Велико Бара), and on September 30, 1915, its crews recognized the Serbian fortifications in the area west of Šabac to Drina[13]. From September 18, 1915, the commander of the Mackensen Group in his headquarters in Temesvár[14] needed regular information in the area of future hostilities, and the tasks of the crews of Imperial-Royal Fliegerkompanien and the German FFA were constant:

- observation of the opponent's defense preparations
- reconnaissance of the fortifications of the Serbian army on the southern bank of the Danube
- reconnaissance of enemy artillery positions

Above: The field airfields did not provide the Fliegerkompanien crews with the comforts they had at their home airfields ... The photo shows wooden hangars at the Flik 6 field airfield on the Serbian Front in the winter of 1915 - source: Fotoarchiv SAW-KA Wien

- the location of the reserves of the Serbian army
- observation of the ground movement of units of the Serbian army
- reconnaissance of the bridges on the Danube, Morava, and Sava[15].

The number of German FFAs in Banat gradually increased. On September 19, 1915, the command in Temesvár issued an order to start air reconnaissance for the crews of FFA 69, and on September 25, 1915, the crews of FFA 1 and FFA 30 joined the above-mentioned forces and were assigned to the III Corps[16]. The scope of tasks of the German FFA crews was extended by aerial photography (25 cm "Brennweite" photo cameras were used)[17] and recognition of railway lines and airports in hostile territory[18]. The long and close reconnaissance of the flight crews allowed the command of the Mackensen Group to form an opinion about the strength of the enemy, which would soon be the target of their strike. On September 19, 1915, the command of the 3rd Army's field body ordered a long-distance reconnaissance on the Belgrade-Ralja line (Serb. Раља) -Popvic (Serb. Поповић) -Mladenovac (Serb. Младеновац) -Topola (Serb. Топола) -Kragujevac-Kraljevo and to the west of it, and to the German 11th Army of the Veliko Gradište District (Serb. Вралико Вре) Neresnica (Serb. Нересница). Photo reconnaissance for the Field Imperial-Royal 3rd Army was carried out in the Ostružnica section (Serb. Остружница)-along the Sava and Danube-Udovice (Serb. Удовице) -Kolari (Serb. Колари) -Vrčin (Serb. Врчин)

Above: In the observation plane from 1915, the main role of the observer, apart from reconnaissance, was to defend against enemy fighter planes ... In the photo from 1916, Korporal Niedermayer from Flik 4 on the Italian Front, and in the background, an observation post for a Brandenburg B aircraft equipped with a Schwarzlose machine gun - source: Fotoarchiv SAW-KA Wien

-Ostružnica for the benefit of the German 11th Army in the Udovice section - along the Danube to the mouth of the Pek River (Serb. Пек)) -Branicevo (Serb. Браничево, Braničevo) -Požarevac-Kolari[19]. By order of Field Marshal A. von Mackensen, the Flik 9 and Flik 15 crews on 15 September 1915, they were also to carry out a close reconnaissance of the Belgrade region and south of the Serbian capital. On September 24, 1915, the command of the field corps of the 3rd Army once again assessed the reports on the photo-recognition of Flik crews. 9 from the area of the northern bank of the Danube, Samar (Serb. Самар), Milicevo Brdo (Serb. Миљицево Брдо), Lipaku (Serb. Липак), Mirijeva (Serb. Миријево), Velkej Vracy (Serb. Велкаа Враца) and Flik 15 crews from the area of Banovo (Serb. Баново) and Topčcider[20].

After successful tests carried out at Nord-Ostfront in the Tarnów region in January 1915 and after the initiation of the process of introducing on-board radio stations in the KuK Luftschifferabteilung, also at Süd-Ostfront, radio-artillery cooperation on the plane-ground-ship line was initiated in the fall of 1915. Flik crews. 15 operating mainly on the Sava, performed the tasks of, among others for KuK Donauflotille monitors, and the Flik 9 crews were to be trained in this (the commander of the unit reported the need for three large transmitting radios, three receiving radios and three small ones)[21]. To disseminate this "radio news", a radio course for Flik observers was launched at the airport in Újvidék by order of 20 September 1915. 5th, 9th, and 15th, eight KuK Donauflotille officers, followed by artillery officers from the Peterwardein fortress[22]. The same, on September 20, 1915, of the Flik 15 crew recognized the area of Obrenovac, Lazarevac, Valjev andUžic (Serb. Ужице) under Lazarevac, detecting an infantry battalion, under Valjev a regiment, another regiment in Nepricava (Serb. Непричава), numerous units in Popučke (Serb. Попуцке) and nine infantry battalions near Užice. A day later, on September 21, 1915, the command of the field corps of the 3rd Army ordered aerial reconnaissance of the Drenovac region (Serb. Дреновац) -Šabac-Obrenovac-Duboko (Serb. Дубоко) -Ostruznica-Belgrade-Kostolac to detect the position of Serbian artillery[23], and on September 23, 1915, the crew of Flik 15 between hours from 2 p.m. to 5 p.m. recognized between Obrenovac and Šabac (new batteries of Serbian artillery were detected near Misary). On the same day commander Flik 15 reported the detection of the Serbian army camps near Užice, Bajin Bašta (Serb. Бајина Башта) and Valjev to the Serbian army camps, and additionally the detection of three anti-aircraft guns near Užice[24]. To strengthen the air force operating daily over Serbian territory, Flik 6 was transferred from Castelnuovo (Herceg Novi, Montenegro. Херцег Нови, Italian. Castelnuovo) to Rogatica and Sokolac (Serb. Соколац), from where its crews were to conduct reconnaissance of the Užice-Višegrad (Serb. Вишеград) area[25]. As part of the air reconnaissance, which was intensifying over the course of the days, the command of the field corps of the 3rd Army received three or four precise air reports daily[26] -for example, on September 24 and 25, 1915, it received air reports about considerable rail traffic near Belgrade. The cooperation on the plane-ship line with KuK Donauflotille monitors was also continued[27] and as part of air bomb attacks, to obtain their optimal effects, it was ordered that the raids were carried out in groups with the forces of four, six or even eight planes, i.e., the forces of full Squadrons![28]. On September 24, 1915, the Serbian airfield in Požarevac was bombed, and on September 25, 1915, by an order of an air staff officer of the German 11th Army's field staff, the following was written: *In Kragujevac, there is a Serbian ammunition factory, barracks, and a powder magazine; the raid is to be performed by F.Fl. Abt.* [FFA] *30th and 1 aircraft each from F.Fl. Abt.* [FFA] *1. and* [FFA] *66; start at 5.30 a.m.* (50 bombs weighing 20, 10 and 4.5 kg were taken on board the planes)[29]. Another bombing raids on this target took place on September 26 and 27, 1915, and on September 30, 1915, the targets in Kragujevac were bombed twice![30]

Serbian Supreme Command, trying to predict as accurately as possible both the date and the direction of the enemy strike, used subordinate aviation for reconnaissance (in the period from April 22 to September 22, 1915, the French and Serbo-French crews performed 59 sorties for five months, during approx. 140 hours flying around 14,000 km (!!!), which was a record for their efficiency), the crews of which also continued bombing targets on the "hostile" banks of the Danube and Sava. From July 1915, combat flights were performed by French and Franco-Serbian crews[31] which took its toll on the enemy side, especially in September 1915. On 8 and 9 July 1915, the railway bridge under Újvidék was bombed; On July 10, 1915, two raids were carried out on the bridges under Petervardein and Ujvidek; On July 25, 1915, the cells in Ruma were bombed with 40 kg bombs dropped on them; Flik 9 airport was extremely unlucky in Homokbálványos, which was bombed again on 27 August 1915; on 15 and 28 August 1915 the cells in Belej Cerkvi were bombed; On August 29, 1915, incendiary bombs were dropped on this city, and on September 22, 1915, it was bombed twice using air darts. The "unlucky" Flik 9 airport Homokbálványos was hit on September 9, 1915, when it was bombed again (thrown with three bombs), and on September 10, 1915, the Ruma railway station (two bombs were dropped) was bombed without causing any damage[32]. On September 11, 1915, the region of eastern and western Banat, Srem, and in Bosnia the region of Brčko-Tuzla-Zvornik and the Drina riverbed were recognized from the third dimension, and on September 12, 1915, the region of Srem-Zemun-Surčin (Serb: Сурчин) -Kupinovo-Mitrovica was recognized (Serb. Митровица) -Šid (Serb. Шид) -Ruma-Dobanovci (Serb. Добановци), area along the Sava and the estuary of Drina, Tuzla and Zvornik[33]. The commander of the Franco-Serbian Squadron was ordered on September 22, 1915 to identify the goals and fortifications of the Imperial-Royal Army north of the Palanka-Novi Sad-Vrsac line and

Belgrade-Batajnica-Ruma-Mitrovica-Šid-Palanka-Újvidék-Titel (Serb: Тител) - Zemun; in Pozarevac-Smederevo-Vršac-Bela Crkva-Moldava-Pozarevac; on the line Belgrade-Batajnica-Pancevo-Zrenjanin-Inđija (Serb. Инђија) -Zemun and the Belgrade-Batajnica-Vršac-Temesvár-Kovin region. In the territory of Srem, three, and in the territory of Banat, one Imperial Corps, were identified[34]. On September 27, 1915, a reconnaissance was carried out in Kupinovo-Zemun-Ralja, where another army unit was detected, and in the area of Pozarevac-Pančevo-Kovin, camps and the movement of the army units were detected. Nothing special was discovered in Bosnia that day, and large infantry camps were detected in Srem. The results of the reconnaissance of September 30, 1915 confirmed further enemy columns in the march, and on October 1, 1915 a large concentration of Imperial-Royal troops was detected in Belej Crkvi[35]. It "got crowded" over the Balkans, and on September 23, 1915, there was an air clash over Kupinov, during which a Serbian aviation plane shelled the KuK Luftfahrtruppen plane[36] and a week later, on September 30, 1915, an air fight between the planes of both sides took place over Kragujevac, which ended with the forced landing of a Serbian plane[37]. October brought new tasks - on October 4, 1915, aerial reconnaissance detected the concentration of the Imperial-Royal units on the bank of the Sava and materials prepared for the construction of a pontoon bridge, and on October 5, 1915 nothing special was reported from the territory of Srem[38]. The staff date of the strike was inevitably approaching and the air reconnaissance of the KuK Luftfahrtruppen and German FFA crews over the Kingdom of Serbia intensified at the beginning of October 1915 and the air bombings intensified (the crews of the 1st and FFA 66th bombed targets in Kragujevac)[39]. By order of the field command of the 3rd Army from October 1, 1915 to the Obst Group. Adler was assigned one Fliegerkompanie with the radio, and to Major Simon's group the observation balloon, which was to be "put up" in Fenki (Serbian Фенек) -Manastir (Serbian Манастир)[40]. On October 1, 1915, the FFA 1 crew recognized the Serbian airfield in Požarevac (only one or two Bleriot XI planes were detected there), the FFA 66 crew in the areaVeliko Gradištediscovered another Serbian airport on the Danube[41]. There was an increased activity of the Serbian Opel - in the report of the 3rd Army's field command it was written: "The airmen were heavily fired over Belgrade"[42]. On October 2, 1915, the command of the fieldwork of the 3rd Army ordered the reconnaissance of the Mladenovac region for the Flik 9 and Flik 15 crew's recognition of the areaRača (Serb. Рача)-Zemun and Rača-Crna Bara (perbium. Црна Бара), and the crews of Flik 5 Lajkovac region (Serb. Лајковац) and photo recognition of the Lazarevac region[43]. A day later, on October 3, 1915, the Flik crew. 15 carried out reconnaissance on the Sawa-west of the Šabac-Drina region and the Flik crews 5, 9 and 15 were ordered to actively photograph the position of the Serbian army. Each photo reconnaissance was carried out precisely, which was emphasized by the reports of the 3rd Army's field command (the Flik 15 crews were also conducting radio reconnaissance, and the Flik 9 crews were ordered to explore the border area on the Danube-Ritopek [Serb. Ritopek] -Grocka-Udovice)[44]. One of the Flik crews. 5 reported on October 4, 1915 the detection of the Serbian heavy artillery position near Kupinov[45] and due to hardware problems and the lack of regular supplies of reserves, Flik 5 had to hand over one plane to Flik 9 that day because it had only two operational planes (from Flik 15, one officer and one non-commissioned pilot had to be put at the disposal of the German command of the XXII Reserve Corps). On October 4, 1915, another air clash took place, after which the Flik crew. 5 reported the enemy plane shooting at her and "escaping from it by turns"[46]. Flik 6, whose crews joined the active crews of other Fliegerkompanien, after ferrying and deploying to the new area, could not enter the action until October 7, 1915[47]. Just before the start of the offensive, the command of the 3rd Army fieldwork ordered also increased observation of Serbian positions by balloon observers (maps of the future battlefield were operated on a scale of 1: 75,000)[48]. On October 6, 1915, the observation balloon was "erected" in Rudska-Ratari (Serb. Ратари) -Vrhovina (Serb. Врховина) in Bosnia and Herzegovina, trying to detect Serbian artillery positions, but due to strong wind in the afternoon, the balloon was pulled down. With the start of the fighting, on October 7, 1915 at 14.00 the balloon was "put up" again to observe the fire of its own artillery[49]. As part of the strengthening of own opl. because of the active activity of the Serbian air force, on October 6, 1915, in the report of the 3rd Army's field command, it was stated that "10 9 cm 75/96 cannons should be used for plot guns". They were to be installed in Cerepalja-Kačarevo (Serb. Качарево, Hungarian. Ferenchalom, German. Franzfeld) *-Pančevo-* Banatsko Novo Selo (Serb. Банатско Ново Село,Hungarian Réva-Ujfalu, Bánát-Ujfalu, Romanian Satu Nou) and Petre (Serbian Петре) next to the train station and protect the Pančevo-Alibunar and Pančevo-Antalfalva railway line (Serb. Ковачица / Kovačica, Romanian Covăcița, German Kowatschitza)[50].

"In five to 12", the last flights before the start of the offensive by the field armies of the Central Powers of the Flik crew. 15 performed on October 5, 1915, focusing their attention on the Belgrade region (reconnaissance of the Serbian airport Belgrade-Batanjica was reported); Flik crew. 5 on October 6, 1915, the traffic on the railway lines near Belgrade was reported by the Flik crew 9 reported reconnaissance of the Serbian capital area[51]. Also, the daily reconnaissance flights of the German FFA crews, which took place on the section from Petrovac to Kragujevac, were intensified just before the start of the offensive. On October 6, 1915, one plane from the FFA was shelled on the Serbian coast. The 1st was assigned to the artillery of the 11th and 105th Infantry Division to search for enemy artillery batteries[52] and in the afternoon, two FFA 1 planes were assigned to the command of the IV Reserve Corps (German crews reported no movement of river ships and artillery groups on the Serbian side)[53].

Endnotes

1 Ibid, p. 91.

2 Some of the above-mentioned FFA collaborated with the German 11th Field Army in 1915 at Nord-Ostfront. During the preparations for the Gorlice Breakthrough, the German field of the 11th Army was given three air squadrons at its disposal - the FFA. 1., 8., 21, one Ballonabteilung and the so-called Brief Abteilung Ost (BAO) operating at the headquarters - OHL. From April 26, 1915, BAO was stationed at the Czyżyny field airport near Krakow, from May 11, 1915 it was transferred to the Lipie field airport near Tarnów, in the period May 14-June 12, 1915 it was stationed at the field airport in Rzeszów, from where it was moved On June 12, 1915 to the field airport in Munin near Jarosław, from where on June 20, 1915 he moved to Wróblaczyna near Lviv, where he was stationed only for the next two days. On June 23, 1915, the BAO personnel was transferred to Berlin and the FFA 66 and 69th was created from the remaining flying equipment stationed in Rawa Ruska (the FFA 69th included a young cavalryman Manfred von Richthofen). Moreover, in May 1915, FFA 57. from the region of Brzesko (Jasień) -Biadoliny Radłowskie was assigned to the established Fliegergruppe Hptm. R. Eyba and ported via Tarnów for the German 11th Army advancing as part of the Gorlice Breakthrough - Lieutenant M. Bakun, *German Aviation on the Eastern Front 1914-1915 - an outline of the problem,* [in:] *Aviation in Gorlice Breakthrough 1915. A look after 100 years*, Rzeszów-Gorlice 2015, p. 206-207; Over the Front, 1996, vol. 12, no. 2.

3 SAW-KA, Der Einsatz der Fliegerabteilungen der Herresgruppe Mackensen im Herbstfeldzug 1915 gegen Serbien. bearbeitet von major dr. Gnamm, Kmdt. zur Luftkriegsakademie, p. 2.

4 On August 17, 1914, the Oestereichisch-Ungarische Albatros Werke factory signed a contract to produce a series of 29 Albatros BI aircraft, two were additionally ordered in March 1915, increasing the number of the series to 31 - see PM Grosz, G. Haddow, P. Schiemer, *Austro-Hungarian army aircraft of World War One*, Mountain View-California 1993, p. 86.

5 SAW-KA, Der Einsatz der Fliegerabteilungen der Herresgruppe Mackensen im Herbstfeldzug 1915 gegen Serbien. bearbeitet von major dr. Gnamm, Kmdt. zur Luftkriegsakademie, p. 4.

6 Д. Недялков, Въздушната мощ на царство България, София 2001, Vol. 1, pp. 17-19; 21-24; 26-26-49; Z. Stojczew, *Wzduszni wojski*, [in:] *Sojusznicy Luftwaffe*, Warsaw 1997, p. 3-4.

7 Д. Недялков, op. cit .; Z. Stojczew, op. cit.

8 Д. Недялков, op. cit., pp. 29-31. The author states that at the time of the end of the First Balkan War, the number of Bulgarian airplanes purchased abroad were 29. Z. Stojczew (op.cit., Pp. 4–5) states that the number of planes by the end of hostilities increased to 35 (this number should include two captured Turkish "Harlan" planes and several "private" planes belonging to airmen fighting on the Bulgarian side as volunteers). The number of planes purchased or delivered as part of the assistance to the Bulgarian Empire was 26 (four Albatros F-2 type from Kaiser Germany, three Bleriot XI type from the Republic of France, six Farman type, 11 type Bleriot XI and two type Nieuport IV - all from the Tsarist Russia. During the Second Balkan War, the Bulgarian Air Force formed the Air Company, and two Balloon Units formed the Baloon Company - the whole was commanded by Major W. Złatarew. Д. Недялков (op. Cit., P. 56) states that during the Second Balkan War, the Bulgarian air force performed 230 flights, 76 of which were combat missions. According to Z. Stojczew (op.cit.) The period of the Balkan wars ended the Bulgarian aviation with nine pilots, four mechanics, 20 soldiers and eight planes (two Albatros F-2, two Farman VII and Bleriot XXI, two Bleriot XI bis and two type Nieuport IV) at the airport near the Sofia railway station.

9 Д. Недялков, Въздушната мощ на царство България, София 2001, Vol. 2, p. 3.

10 Ibid, pp. 4-5. In July 1914, as a result of navigational errors, two Albatros B I planes from the Turkish Air Force landed by mistake on Bulgarian territory. As part of the opl. a six-gun battery of 7.5 and 8.7 cm guns was created.

11 SAW-KA, Die Eroberung Serbien 1915. Tätigkeit der Fliegerverbände der österr.-ung. 3. Armee und der Serb. Luftstreitkräfte. Bearbeiter Rittm. ad Peter v. Spaits, p. 48.

12 Ibid, p. 64.

13 Ibid, p. 65–67.

14 M. Rauchensteiner, *Der Erste…*, op. cit., p. 481.

15 SAW-KA, Der Einsatz der Fliegerabteilungen der Herresgruppe Mackensen im Herbstfeldzug 1915 gegen Serbien. bearbeitet von major dr. Gnamm, Kmdt. zur Luftkriegsakademie, p. 8; J. Pajewski, op. cit., p. 303.

16 SAW-KA, Der Einsatz der Fliegerabteilungen der Herresgruppe Mackensen im Herbstfeldzug 1915 gegen Serbien. bearbeitet von major dr. Gnamm, Kmdt. zur Luftkriegsakademie, p. 10.

17 Ibid, p. 5.

18 Ibid, p. 9.

19 SAW-KA, Die Eroberung Serbien 1915. Tätigkeit der Fliegerverbände der österr.-ung. 3. Armee und der Serb. Luftstreitkräfte. Bearbeiter Rittm. ad Peter v. Spaits, p. 49.

20 Ibid, p. 50–51.

21 Ibid, p. 53–54.

22 Ibid, p. 56.
23 Ibid, p. 60–62.
24 Ibid, p. 63.
25 Ibidem, p. 59. Composed of Flik The 6th was in Oblt at this time. Gustav Salmon from October 1915 serving as an observer there. During and after the Serbian campaign, he made 39 sorties, including 10 for photo recognition and 16 for bombing. Until his death in an air accident on October 7, 1918 (on the Albatros-Oeffag D. III from Flik 42. J) he made 157 sorties - Lieutenant J. Červenka, *Nadporučik Gustav Salmon, Zabudnutý rodák from Levoče,* "Obrana" 2014, no.7, VII 2014, p. 45
26 SAW-KA, Die Eroberung Serbien 1915. Tätigkeit der Fliegerverbände der österr.-ung. 3. Armee und der Serb. Luftstreitkräfte. Bearbeiter Rittm. ad Peter v. Spaits, p. 65–67.
27 Ibid, p. 68.
28 SAW-KA, Der Einsatz der Fliegerabteilungen der Herresgruppe Mackensen im Herbstfeldzug 1915 gegen Serbien. bearbeitet von major dr. Gnamm, Kmdt. zur Luftkriegsakademie, p. 25.
29 Ibid, p. 26.
30 SAW-KA, Die Eroberung Serbien 1915. Tätigkeit der Fliegerverbände der österr.-ung. 3. Armee und der Serb. Luftstreitkräfte. Bearbeiter Rittm. ad Peter v. Spaits, p. 201.
31 Ibid, p. 208.
32 Ibid.
33 Ibid, p. 75.
34 Ibid, p. 77.
35 Ibid, p. 78.
36 Ibid, p. 214.
37 Ibid, p. 221.
38 Ibid, p. 79–80.
39 SAW-KA, Der Einsatz der Fliegerabteilungen der Herresgruppe Mackensen im Herbstfeldzug 1915 gegen Serbien. bearbeitet von major dr. Gnamm, Kmdt. zur Luftkriegsakademie, p. 10.
40 SAW-KA, Die Eroberung Serbien 1915. Tätigkeit der Fliegerverbände der österr.-ung. 3. Armee und der Serb. Luftstreitkräfte. Bearbeiter Rittm. ad Peter v. Spaits, p. 91.
41 SAW-KA, Der Einsatz der Fliegerabteilungen der Herresgruppe Mackensen im Herbstfeldzug 1915 gegen Serbien. bearbeitet von major dr. Gnamm, Kmdt. zur Luftkriegsakademie, p. 46
42 SAW-KA, Die Eroberung Serbien 1915. Tätigkeit der Fliegerverbände der österr.-ung. 3. Armee und der Serb. Luftstreitkräfte. Bearbeiter Rittm. ad Peter v. Spaits, p. 214.
43 Ibid, p. 68.
44 Ibid, p. 69–70.
45 Ibid, p. 82.
46 Ibid, p. 214.
47 Ibid, p. 71–72.
48 Ibid, p. 89.
49 Ibid, p. 97.
50 Ibid, p. 215.
51 SAW-KA, Die Eroberung Serbien 1915. Tätigkeit der Fliegerverbände der österr.-ung. 3. Armee und der Serb. Luftstreitkräfte. Bearbeiter Rittm. ad Peter v. Spaits, p. 100.
52 SAW-KA, Der Einsatz der Fliegerabteilungen der Herresgruppe Mackensen im Herbstfeldzug 1915 gegen Serbien. bearbeitet von major dr. Gnamm, Kmdt. zur Luftkriegsakademie, p. 13.
53 Ibid, p. 14.

Fokker A.III serial A6 flown by Gottfried Banfield.

Serbian Blitzkrieg 1915

The commander-in-chief of the Central Powers forces in the planned offensive against the Kingdom of Serbia was the supporter of the frontal attack with mass artillery fire ("Feuerwalze"), Field Marshal August von Mackensen, who was one of the most outstanding, if not the most eminent commanders of the Kaiser Wilhelm II Hohenzollern's army was held for this post after the brilliant command of the Gorlice Breakthrough (May 2–5, 1915), which ended with the breakthrough of the Eastern Front. Under his orders, the combined forces of the German 11th Army (11th Infantry Division) of General Max von Gallwitz and of the 3rd Army of General Hermann Kövess von Kövesshaz formed the Mackensen Group (Armeegruppe Mackensen)[1]. In the event of his election as commander-in-chief, the decisive factor was the attitude of the new coalition partner, the Bulgarian Empire, whose government demanded that a German general took over the command of all allied forces (it was agreed to undertake military actions against his Serb neighbor only when German troops crossed the Danube)! "Black Hussar" - Field Marshal A. von Mackensen received an instruction from Emperor Wilhelm II Hohenzollern to "definitively beat Serbia, open it, and then ensure communication with Constantinople via Belgrade and Sofia"[2]. In order to strengthen the Austro-Hungarian-German strike forces and also wanting to take advantage of the possibility of an attack on the Kingdom of Serbia from the east, on August 14, 1915, the German-Austro-Hungarian side agreed with the future coalition partner on the details of the attack against the Kingdom of Serbia[3] and on September 6, 1915, a secret "treaty of friendship and alliance" was signed in Pszczyna (German: Pless-Pleß) between the German government and the government of the Bulgarian Empire[4]. The alliance treaty, the compensation treaty and the military convention were in force for a period of five years (it was also signed by the Austro-Hungarian Empire). The Bulgarian government undertook to take part in the German-Austro-Hungarian offensive in return for territorial gains (after the end of hostilities, the Bulgarian territory was to include the areas of Macedonia and the areas in the north and east of the Kingdom of Serbia)[5] officially in the warfare of the Great War, standing on the side of the Central Powers on October 14, 1915[6] (the commander-in-chief of the Bulgarian army was General Nikola Todorov Zekov). The military convention signed with the Bulgarian Empire obligated Kaiser Germany (OHL) and Imperial-Royal Austro-Hungary (AOK) to attack the Kingdom of Serbia within 30 days (each state was to deploy field armies of six German and Austro-Hungarian DP), while the party Bulgarian (five DP) within 35 days[7]. The German 11th Field Army in the Serbian section was to operate with the forces of three Corps (III, IV and X) with a total of seven DPs (approx. 72 ½ infantry battalions, 121 artillery batteries and 16 cavalry squadrons - a total of 65,000 bayonets, 1,700 sabers and 483 works), while the AOK assigned for operations the field of the 3rd Army, which was strongly affected by losses after the Carpathian period (Karpatenkrieg) 1914/1915, commanded by General Hermann Kövess von Kövesshaz[8]. Her strength consisted of 130 ½ infantry battalion, 14 ½ (15?) cavalry squadron and 127 artillery batteries (a total of 121,000 soldiers, 1,800 sabers, 501 guns and 16 mine throwers). In total, the German-Austro-Hungarian forces intended to attack against the Kingdom of Serbia numbered 392,000. bayonets, 6,000 sabers and 1,500 guns[9]. Field Marshal A. von Mackensen decided to strike Serbian territory not from Bosnia (Serbian Босна) but directly towards Belgrade, and then along the Morava Valley (Serbian Морава) to Niš (Serb.Ниш - the 3rd Army was to attack from the western side of Belgrade and the German half of the 11th Army from the eastern side). According to the above-mentioned plans, the attack by the field forces of the German 11th Army was to be carried out from the area of Srem (Syrmia, Serb. Срем, Srem, Croatian. Srijem) and Banatu (rum, Serb.Banate, part, words Banát, Hungarian Bánság or Bánát) from the Danube-Sawa line, between Kupinow [Serb. Купиново], and Belgrade, along the axis of the Belgrade-Thessaloniki railway line, the Bulgarian field of the 2nd Army was to attack through the Morava valley (Serb. Morava) to attack the enemy's rear while the field of the 3rd Army was hitting from the area of Maćwa (Serb. Мачва) as the main task having to capture Belgrade and continue the attack on Kragujevac[10]. According to the plan of the "black hussar", the Bulgarian field 1st Army was to attack in the north-east direction across the Morava valley and after capturing the Zaječar region (Serb. Зајечар) -Pirot (Serb. Пирот) attacked Niš. These three separate strike groups were to merge between Kruševac (Serb. Крушевац) and Niš. A separate strike group was the Bulgarian 2nd Army, which was to attack Skopje (Macedonian Скопје, Serbian Скопље, Bulgarian Скопие, tur. Üsküp) and Negotin (Serb. British (Army of the East - Army of the Orient) from the area of Thessaloniki[11]. Readiness for action was to be achieved at the turn of the second half of September 1915 to the beginning of October 1915, and for the units of the Bulgarian field armies the deadline was set to October 13, 1915. The opponent of the combined forces of the Central Powers, the Serbian army consisted of 11 DPs and one DK grouped in three field armies and consisted of 288 ½ infantry battalion, 40 cavalry squadrons and 678 guns - a total of 250,000 soldiers (340,000 soldiers were put into combat by the end of the campaign). The commander of the Serbian forces was Fr. regent Aleksander Karadziordziewić and the Chief of Staff was Marshal (Serb. vojvoda), Fr. Radomir Putnik. High Command of the Serbian Army exhausted by the 1914 campaign and the typhus epidemic in the winter of 1914/1915[12] expected an impact towards the Mačva plain. Serbian forces were deployed along the Sava line (the average width of the river was 300-500 m) and the Danube (the average width of the river was 1000 m). Field 1st Army (Commander-in-Chief of Gen.**Živojin** Mišić) and half the 3rd Army (Commander Gen. Pavel Juršič-Šturm) was in the

Above: Reconnaissance over the Adriatic ... 1914. In the photo, a KuK Seeflugwessen, Lohner-type flying boa,t flying over the roadstead of the KuK Kriegsmarine base in Pula - source: Fotoarchiv SAW-KA.

front line, and in the Belgrade region, which was to be defended, 20 infantry battalions were concentrated for defense (the Belgrade group was commanded by Gen. Petar Živković)[13]. On the Bulgarian border, the field forces of the 2nd Army of General Stepa Stepanović (four DPs - 59 infantry battalions, 21 cavalry squadrons and 150 guns) were concentrated, because Marshal R. Putnik initially planned a preventive attack on the territory of the Bulgarian Empire to secure the rear of the Serb's army, however, the governments of the allied Ententa countries did not agree to it, deluding themselves to the end that the Bulgarian side would remain neutral. The command of the allied Montenegrin army consisting of four field groups (Odreda) in the strength of 82 ¾ (83?) Infantry battalion and 134 guns (50 thousand soldiers) was exercised by the King of Montenegro, Nikola (Mikołaj), and the chief of staff was the Serb Col. Dušan Pešić[14]. The strength of the joint Serbo-Montenegrin forces were the brave attitude of the soldiers in the 1914 campaign and, like in 1914, the fighting on their own territory. Unfortunately, according to the maxim "nec Hercules contra plures", these advantages turned out to be too weak, in view of the numerical and technical advantage lying on the side of the field armies of the Central Powers! It is true that as part of the allied aid, on October 5, 1915 in Thessaloniki[15], in the tense political situation prevailing in the Kingdom of Greece and on the line of the Kingdom of Greece-Entente, thus violating its neutrality, landed 13,000 soldiers of the Franco-British Army of the East (Army of the Orient), to be transported up the Wardar (Macedonian Вардар - Wardar, gr. Αξιός - Aksiós) but by the end of the fighting in Serbian territory there was no real, field cooperation between them and the Serbian troops decimated by the fighting and winter[16].

For the Serbian High Command, the coming weeks were to bring combat for two, and soon for three. The front on the Sava and the Danube in the field section of the 3rd Army and the German field of the 11th Army was of a defensive character, while the front against the Bulgarian was of an offensive character. As a result of air reconnaissance reports, the Serbian High Command awaited an offensive of the Imperial-Royal and German field armies in the period 18-19 September 1915, counting on its support by strong artillery and aviation, and yet it was transferred to the Bulgarian border from the "Austro-Hungarian section" two DPs. On September 21, 1915, the order of Field Marshal A. von Mackensen, issued to the field units of the 3rd Army, ordered it to go to the attack after crossing the Danube near Belgrade and Sava west of Belgrade near Kupinov (on September 26, 1915 the units took their starting positions)[17]. The German field 11th Army, after crossing the Danube, was to attack in the Smederevo-Bazias section towards the Morava line on the Palanka-Žabari (Serb. Жабари) -Petrovac (Serb. Петровац) line. The 3rd field army was to attack south and south-east of Belgrade in the direction of the Knjaževac (Serb. Књажевац) -Avala (Serb. Авала) -Ritopek (Serb. Ритопек) line and take Belgrade. The Bulgarian Field Infantry Divisions of the 1st Army were to attack in the direction of Paraćin (Serb. Параћин) -Niš. On

October 4, 1915, from the staff of Grupearmee Mackensen, the final order to strike was handed over to subordinate units on October 6, 1915[18] from north to south on the front line running along the Danube and Sava. Along with the forces of the Imperial-Royal army stationing at the Drina, the strike was to be carried out with the forces of 143 infantry battalions and 362 guns. The enemy, the Serbian army, could deploy 202 infantry battalions and 990 guns "in the field" (more than 20 infantry battalions and 75 guns were allocated to defend Belgrade against 66 infantry battalions and 273 guns attacking, including 108 heavy ones from the 3rd field army). To defend the over 1,000-kilometer front, the Supreme Command of the Serbian Army in Kragujevac was able to deploy 420,000 "in the field". soldiers grouped in the field 1st Army on the border between Bosnia and Srem, the field 2nd Army on the Bulgarian border and the field 3rd Army facing north and defending Belgrade[19].

170 guns and heavy 420 mm mortars[20] the artillery of the Mackensen Group began on October 5, 1915 in the afternoon hours, from October 6, 1915 the shelling was cyclical and after two days of its duration, on the third day at 3.00 the first Imperial-Royal troops landed on the "Serbian coast"[21] , although on October 6, 1915, the group of General Béla Sorsich von Severin struck with the fire support of KuK Donauflotille monitors in Drenovac-Mitrovica where a bridgehead was created[22]. In terrible weather conditions, the field armies of the Central Powers attacked the territory of the Kingdom of Serbia also with the support of the air force on October 7, 1915, from the Drina line near Višegrad and Sava[23] and the corps of the German 11th Army field attacked at the above two points on the Danube towards Belgrade and Ram. During the crossing of the Sava, Imperial-Royal 19th Corps supported by KuK Donauflotille monitors[24] and Imperial-Royal VIII Corps suffered heavy losses near Belgrade. Having forced the river, the striking forces of the 3rd field army headed for the Serbian capital[25] and despite the fierce resistance of the Serbian army troops (the XXII Reserve German Corps struck through the Gypsy Island [Ada Ciganlija - Serb. Ада Цигналия] on Topčider and Banovo south-west of the city), Belgrade was fired upon by heavy artillery fire by the VIIIth Corps October 9, 1915: *When, at noon on October 8, 1915, the Allied forces decided that they were unable to continue their attack on* [stronghold] *Kalimegdan* [Kalemegdan] *and the factory suburb nearby at the desired pace, they asked for artillery support. It was provided by monitors* [KuK Donauflotille], *which were the only ones that could come close to* [the enemy's position], *so as not to endanger their own infantry on the one hand, and to keep them with the direct infantry on the other hand, close communication during the attack ... Monitors themselves exposed to the merciless fire of Serbian cannons, because they had to come quite close* [to the enemy shore] *in order to be able to accurately observe the effects of their own artillery, received data on successive targets by means of signals given by flags by the fighting troops ... At dawn on October 9* [1915] [units of the Imperial-Royal Army] *managed to break into the Belgrade citadel from the east. The entire eastern slope of* [the fortress] *Kalimegdan* [Kalemegdan], *ie the park lying to the side of the citadel, 50 meters above the river, was covered along and across deep trenches, not very visible even at close range; thick bushes grew everywhere, partially artificially planted as camouflage, until it reached the first gate of the central bastion. Here the mortar fire destroyed everything so much that in the last phase of the assault it was no longer possible to put up systematic resistance*[26].

Crossing of German troops across the Danube to the Ram (approx. 1000 m wide), behind them, who had considerable problems with the construction of bridges broken by storms for several days (Kossova - only on October 21, 1915, two bridges were completed)[27] , turned out to be easier, although in the heavy rain the fire of the German artillery was directed "blind" at times (below Smederev, two German Corps lost 42 out of 50 crossing boats). On the night of 8/9 October 1914, German troops, having the fog as an ally, flanked the positions of Serbian defense and on October 11, 1915, the city, and the fortress of Kalemegdan were seized. Thus, the defensive line of the Sava and Danube for the Serb side was lost[28]. While forcing the Drina (approx. 100-200 m wide), the Imperial-Royal units were in a difficult position, as all the pontoon bridges collapsed[29]. The attack was carried out in difficult mountain terrain conditions with strong resistance from Serbian defenders[30] and by October 16, 1915, the German field 11th Army advanced only 13 km, and the gap between the striking field troops of the 3rd Army and the German units was 16 km (allied forces joined forces on October 18, 1915). The improvement in weather conditions resulted in an intensification of offensive operations and on October 21, 1915, the German forces were on the line from the Kolubara River (Serb. Колубара) to the line of the Pek River (Serb. Пек). To achieve success and break the Serbian defense line, it was necessary to strike the Bulgarian ally[31].

According to the old rule "nec Hercules contra plures" that appeared in October 1915 in the Balkans, the Serbian army found itself in an extremely difficult position, because "the German battering ram fell on Serbia like a storm"[32]. However, with the entry into hostilities on October 14, 1915 by the Central Powers of the Bulgarian Empire (the first Bulgarian army units breached the border with their Serbian neighbor on the night of October 13-14, 1915), pursuing a policy line aimed at gaining the position of the "Piedmont of the Balkans" ("Piedmontese South Slavs"), the field armies of Austo-Hungarian Empire, Kaiser Germany and the Bulgarian Empire quickly took the Serbian army into two or even three fires[33]. The generated ratio of forces were as follows: 14 German and Austro-Hungarian Infants attacked against each other by six Divisions of the Serbian army, and in the Bulgarian section, six Divisions attacked the defense line of five Divisions of the Serbian Army; in the Montenegrin section, where the army's Imperial-Royal forces attacked on October 22, 1915 against approx. 50

thousand soldiers of the Montenegrin army supported by 135 guns were hit by 47 battalions of the Imperial-Royal army supported by 150 guns[34]. Initially not having adequate artillery support in the attack on Niš, the divisions of the Bulgarian 1st Army field, detained on the border until October 15, 1915, crossed the Serbian fortifications in the Pirot and Zaječar regions through mountain passes, crossed the Timok River (Serb.Timok) but from then on the pace of the strike slowed down quickly due to the lack of supplies as well as the staggering Serbian resistance, dominated by the long-standing anti-Bulgarian attitude of Serbian divisions. In the Macedonian direction, on Skopje (towards the southern Moravia and Vardar valleys), on October 16, 1915, the Bulgarian field army attacked the 2nd Army, with the intention of breaking the territorial connection between the Kingdom of Serbia and the Kingdom of Greece, thus commencing the so-called. Offensive in Faroe Pole (Bulgarian Овчеполска настъпателна операция)[35]. On October 18, 1915, troops of the German-Austro-Hungarian field armies entered Požarevac[36]. One day later, the units of the Bulgarian 2nd Army field attacked the Vardar valley, and when on October 21, 1915 they entered Skopje, on that day there was also an exchange of fire with units of the Allied Expeditionary Corps in the area of the Strumica railway station, located 30 km from the Bulgarian border. This event opened the 12th front of the Great War - the Macedonian (Salonic) Front[37]. Until October 22, 1915, the units of the Bulgarian field 2nd Army took over Skopje and Kumanowo (Maced. Куманово), Veles (Macedonian Велес) and Sztip (Macedonian Штип)[38] and on October 22, 1915, field marshal A. von Mackensen gave the order to "throw the main Serbian forces deep into the country and definitively defeat there"[39]. In this way, the field of the 3rd Army moved south towards Kragujevaca, 130 km south of Belgrade, the German field of the 11th Army struck on both banks of Morava, the Bulgarian field 1st Army continued the Niš, and the Bulgarian field 2nd Army in towards the Vardar valley to close the retreat to the south for the Serbian forces. The position of the Serbian army on October 22, 1915 was difficult, though not hopeless, but when Bulgarian forces joined the coalition forces of German-Austro-Hungarian near Negotin on October 26, 1915[40] , it became extremely heavy because the direction of the retreat of the Serbian army towards Saloniki via Skopje was blocked[41]. After more than a week of fighting, the Serbian army units managed to avoid being flanked on both sides, but at the end of October 1915 their position heralded defeat. On the night of October 30/31, 1915, Marshal R. Putnik ordered a general retreat to the south of the Serbian field forces of the 1st and 3rd Armies (western and southern Moravia region)[42], trying to keep them compact, although the retreat took place in a terrible rainy aura and blurred roads. Extremely difficult weather conditions also caused the attackers, and road transport began to give way to…[43] A day after the start of the organized Serbian retreat, on November 1, 1915, the field of the 3rd Imperial Army and German units were seized by Kragujevac, and on November 5, 1915, the field units of the Bulgarian 1st Army were captured by Niš and thus the left wing of the field Imperial 3rd Army and the right-wing field forces of the Bulgarian 1st Army joined forces of the German 11th Army[44]. The "sad remnants of the Erbian army" were retreating towards Priština (Alb.Prishtinë, Serb. Приштина) near the Albanian border, and on November 21, 1915, the coalition forces seized Priština and Kosowska Mitrovica (Alb. Mitrovica, Mitrovicë; Serb. Митровица, Косовска Митровица)[45]. On that day, the armed forces of the Kingdom of Serbia lost the chance to put up further effective resistance and the dramatic appeal of King Peter I to change the existing situation did not help: *Sons, you swore to defend your king and our homeland. I release you from the first part of the oath, you only must keep the second part. Anyone who does not feel like a son of this country can drop their weapons and go home*[46].

The seizure of the railway junction in Niš put the strategic Berlin-Sofia-Istanbul railway line under the control of the Central Powers[47]. On November 8, 1915, the attackers captured Kruševac (Serb. Крушевац) and Leskovac (Serb. Есковац) by pushing the Serbian forces towards Kosov (Alb. Kosova / Kosovë, Serb. Косово), the battles fought on 11-12 November 1915 with the Bulgarian 1st Infantry Division in the Leskovac region did not bring the winning result. On November 15, 1915, the Bulgarian field of the 2nd Army ended the above-mentioned Offensive in the Faroe Field (fierce fights with two Bulgarian Infantry Divisions were fought on November 11–14, 1915 in the area of the Crna River [Serb. Црна]) having a considerable advantage (3rd Balkan Infantry Infantry, 7th Rila Infantry Infantry, DK and 182 guns) over Serbian forces (weakened two or three Infantry Assaults supported by 78 guns) beaten at Kumanovo and forced to retreat along the Vardar valley towards Veles[48]. After the fighting in the Kačanika region (Alb. Kaçanik; Serb. Качаник) on 1820 November 1915, the merger of Serbian forces with the troops of the Allied Expeditionary Corps, which with the increase of forces, became the Army of the East (Army of the Orient), was unsuccessful from the south in Kosovo-Priština. By the order of Marshal R. Putnik of November 21, 1915, Serbian forces moved to the left bank of Sitnica (Serb. Sitnica) and the day before that, the Supreme Command of the Montenegrin Army agreed to the entry of the Serbian 1st Army troops into the area of Kosovska Mitrovica. With the retreat of the Serbian 1st Army, hostilities entered the valleys of the rivers Ibar (Serbian Ибар), Lim (Serbian Лим) and Tara (Serbian Тара)[49]. On November 19-24, 1915, heavily weakened by the typhus epidemic, hunger and shortage of ammunition, the Serbian army, however, maintained a compactness of forces, fought a "battle for everything and honor" in Kosovo (19-23 November 1915) overnight being in a tragic situation (on November 17, 1915, heavy snowfalls paralyzed the Serbian area), still without direct support from the allied army of the East)[50]. "Ilustrowany Kuryer Wojenny" of 25 November 1915 stated: *Accidents in Serbian territory are approaching with inexorable necessity towards this bloodiest Balkan battlefield, which as such has already passed into legend and folk song, but it has become the most vivid national memory of Serbs* [it is about battles of 1389 and 1448][51]. In the Battle of Kosovo Pole, the Serbian army, not receiving support from the allied troops withdrawn south, lost more than 2/3 of its original size[52]. As a result of the defeat, the morale of the soldiers weakened, and the compactness of the troops deteriorated significantly.

The commanders of two French DPs and soon the commander of the British DP included in the Army of the East, during the original march to north from the Thessaloniki area, received an order forbidding them to cross the border units of the Kingdom of Greece and the Bulgarian Empire, and the Allied troops moved through the Vardar valley to provide support to a weakening Serbian ally. The forces of the Bulgarian field 2nd Army (2nd Thracian Infantry Infantry, 5th Danube Infantry Infantry, 7th Rila Infantry Infantry and 11th Macedonian Infantry) replaced the French forces (57th, 156th, and part of the 122nd Infantry Division) at Krivolak on October 17, 1915. (Serb. Криволак), after a series of clashes that lasted until November 21, 1915, forced them to retreat (the French Minister of War, General Joseph Gallieni ordered the commander of the Army of the East, General Maurice-Paul-Emmanuel, on November 12, 1915 preparation of forces for the retreat from Serbian territory to the south with the task of protecting the Thessaloniki area). The disastrous weather conditions and the increasingly harsh winter meant that the hope of the Serbian Supreme Command for an attack from the south by the forces of the Allied Army of the East (Army of the Orient) to join the increasingly weaker Serbian forces turned out to be in vain[53]. The defeat of the French forces at Krivolak (until December 12, 1915, French units left the territory of the Kingdom of Serbia) was not the only defeat of the Allied forces in the clash with the Bulgarian army during the 1915 campaign. The second "beating" of the Bulgarian field troops of the 2nd Army was lowered to the Allied forces in the Battle of Kosturino, where the units of the British 10th Infantry Division directed to this area from the Gallipoli peninsula were defeated in order to support the French ally who had been defeated in the clash. The battle took place on December 6-12, 1915, and after the general attack of the Bulgarian forces on December 8, 1915, a crisis arose among the British forces and their retreat towards Salonika (on December 12, 1915, the German OHL prohibited Bulgarian units from entering the territory of the Kingdom of Greece with British-French units). After these two defeats, the situation on the Macedonian (Salonitian) Front stabilized, changing the nature of hostilities from offensive to positional. Thus, the Serbian troops were completely on their own, as the Allied Army of the East, having found themselves in the region of Thessaloniki, took a defensive position, which resulted in tensions between the Entente and King of Greece, Constantine I[54].

The front line, due to the continuous retreat of Serbian forces to the south, along with the above events, was rapidly approaching the borders of the Principality of Albania and the Kingdom of Montenegro. On November 21, 1915, the Supreme Command of the Serbian Army gave its troops another order to retreat, and on November 24, 1915, due to the difficult situation, the Serbian Supreme Command ordered the march of strongly weakened forces (approx. 220 thousand soldiers) towards the coast of the Adriatic Sea through the territory of Montenegrin and Albanian[55] called "Serbian Golgotha". Strongly weakened by the march, hunger and weather conditions, the Serbian army, and crowds of Serb refugees, including intellectuals, headed on foot towards the Adriatic coast with the intention of reaching the section from Shkodra (alb. Shkodër / Shkodra, Serb. Скадар, owned by Scutari) after Dürres (Alb.Durrës, owned by Durazzo). On November 28, 1915, the Serbian government arrived in Shkodra, and on December 6, 1915, the Supreme Command of the Serbian Army arrived there[56] and on December 2, 1915, the Serbian army retreated to Albanian[57]. The general retreat of the decimated Serbian field armies, through the territory of the Kingdom of Montenegro towards the Principality of Albania, literally became a "death march" in the nightmarish winter weather conditions prevailing in the Albanian mountains. It became such not only for the units of the Serbian army, but also for the masses of civilians leaving the Kingdom of Serbia[58]. As Winston Churchill recalled, *the tenacity of the Bulgarian pursuit exposed the retreating Serb soldiers and civilians to the worst horrors of war and winter. Tens of thousands of defenseless people died*[59]. A 150-kilometer march through the Albanian mountains, lasting for three weeks, was costly (20,000 civilians died, many of them murdered in retaliation by Albanian tribes who suffered a lot from Serbs in 1912–1913)[60]. *People participating in the retreat told inconsistently* [spelled in accordance with the original] *about frost, hunger, wonderful scenery, roads covered with carcasses, about people dying in roadside ditches and about Albanian ambushes. We have heard of Russian and British ministers sleeping on straw with the Serbian Foreign Minister, his wife and son, while Italian and French ministers, secretaries, consuls, servants, and God knows who else slept in the next room. Braided about the king* [Peter I] *lying on a cart pulled by four oxen and sharing the hardships of the road with ordinary privates*[61]. To prevent the remains of the dead Serbian rulers from contaminating them with the invaders' hands, the *Serbs took coffins and sarcophagi from the monastic crypts and now transported them on two-wheeled carts, and if the roads did not allow it, the monks accompanying them carried them on barges. This poignant and, given the circumstances, tragic concern was not limited to the dead monarchs. King Piotr* [I], *too old and too twisted with rheumatism to move on his own, and his son Aleksander, who had to undergo appendectomy in a roadside cottage, rested on a platform pulled by donkeys or on a stretcher*[62]. On January 19, 1916, the King of Serbia, Peter I, left for exile via the Duchy of Albania[63]. As the German Chief of General Staff Gen. E. von Falkenhayn recalled, assessing the Serbian retreat, only a few unfortunates escaped to the Albanian mountains, losing all artillery and everything they could not take with them. The Serbian army ceased to exist, losing around 50,000 soldiers during the 1915 campaign[64]. On December 7, 1915, Marshal R. Putnik was seriously ill, carried by soldiers on a stretcher, resigned and was succeeded by general Živojin Mišić, who was appointed voivode[65]. 1/3 of the Serbian army reached the destination of retreating through the Albanian mountains, which was the coast of the Adriatic Sea, from the ports in Durrës andVlorë be evacuated by sea to the Greek island of Corfu (Kerkyra, Italian Corfu) occupied by the troops of the Ententa[66]. The retreating Serbian troops took 24,000 with them to the Albanian mountains. POWs from the Imperial-Royal Army and when they reached their destination, those who did not die of typhus and cholera were interned in camps in the Kingdom of Italy and Sardinia (Italian: Sardegna, Sardigna)[67]. The Railways of the Great War ...

An important, strategic railway connection on the Berlin-Istanbul line was reopened on January 15, 1916, and on January 18, 1916 in Niš, Emperor Wilhelm II Hohenzollern appointed the Bulgarian Tsar as field marshal of the German army[68] and the same, during the meeting of the two rulers, offered the following toast to his German ally: *Hail, commander, emperor, and king. From ancient Naissy, all the peoples of the East greet you, the savior who brings prosperity and salvation to the oppressed*[69].

Endnotes

1 *Illustrated History* ..., op. cit., pp. 500-501; History XX..., op. cit., p. 285.

2 J. Pajewski, op. cit., p. 303. A. Chwalba (op. cit., p. 211) reports *that Mackensen's Army Group had over 300,000 soldiers and 1,200 guns.* М. Радојевић, Љ. Димић (op. Cit., P. 176) state that the combined multinational forces of the Central Powers dedicated to fighting the Serbian army numbered around 800,000 soldiers.

3 A. Chwalba, op. cit., p. 210.

4 L. Höbelt, *Stehen oder Fallen? Österreichische Politik im Ersten Weltkrieg,* Wien-Köln-Weimar 2015, p. 54; H. Münkler, op. cit., p. 360; B. Hamann, op. cit., p. 82; Wielka..., op. cit., p. 26; Д. М. Бабац, op. cit., p. 90; M. Rauchensteiner (*Der Tod* ..., op. Cit., P. 301) incorrectly gives the date of the agreement on November 6, 1915 - surprisingly, this error also appears in the extended and supplemented new study by the same author (*Der Erste* ..., op. cit., p. 487).

5 J. Pajewski, op. cit .; M. Tanta, op. cit., p. 128; D. Jordan, *Balkans, Italy and Africa* ..., op. cit .; J. Piekałkiewicz, op. cit .; Z. Stojczew, op. cit., p. 5. On the German side the agreement was signed by General E. von Falkenhayn, on the Austro-Hungarian side by General F. Conrad von Hötzendorf, on the Bulgarian side by Col. Ganczew. The Bulgarian side expressed its readiness to enter Macedonia with the forces of one DP. In addition, it also demanded that part of the territory of Turkish Thrace be handed over to him from Edirne (Bulgarian Adrianople) and the railway line leading to Dedegacz (Bulgarian Alexandropol). As a result of the above-mentioned agreement, it was also supposed to receive in return the area on the Maritsa (Bulgarian Maritsa), at the expense of the territory of the Ottoman Empire (the Turkish-Bulgarian agreement provided the Bulgarian side with the railway line with a part of Thrace on the right bank of the Maritsa) - see *Great War 1914–1918* ..., op. cit., pp. 174–175.

6 A. Szelągowski, op. cit.

7 *The Great War 1914–1918* ..., op. cit., pp. 175–176. According to the allied arrangements, the Bulgarian 1st Army of General Kliment Boyajiev was to cooperate with the main German-Austro-Hungarian forces against the Serbian army, while the Bulgarian field Army of General Georgi Stojanov Todorov was to operate independently in the Macedonian section. In the literature on the subject, there is also information that the Bulgarian armies were to operate with the forces of four DPs and that the Bulgarian army forces directed to offensive actions against the Kingdom of Serbia consisted of 11 reinforced DPs (240 [350?] Infantry battalions, 44 cavalry squadrons and 198 artillery batteries - a total of 240 thousand soldiers, 5,700 sabers and 934 [1,400?] works.)

8 After the Gorlice Breakthrough on the Eastern Front in May, On June 10, 1915, the field of the 3rd Army was disbanded and the Army Command headed the Army Group of General Karl Kirchbach's Cavalry (Armeegruppe General der Kavalerie Kirchbach), transformed into the field 1st Army. As part of the planning of the offensive against the Kingdom of Serbia, at the beginning of September 1915, the existing Army Group of General Karl Teresztyanszky was strengthened and transformed into the new field 3rd Army - seehttp://agad.archiwa.gov.pl/pomoce/Wyw328.xml - entry from November 8, 2014.

9 *The Great War 1914–1918* ..., op. cit. Д. М. Бабац, op. cit., pp. 93-94; D. Jordan, Balkans, *Italy, and Africa* ..., op. cit., p. 51. J. Pajewski (op. cit., p. 304) gives the number of 330 thousand. German, Austro-Hungarian and Bulgarian soldiers, J. Friedrich (op. cit., p. 366) lists the number of 350,000. soldiers grouped in three field armies, and M. Rauchensteiner (Der Tod ..., op. cit., p. 300) gives the number of 500,000. troops striking against 250,000 soldiers of the Serbian army. The German 62nd Infantry Division was to strike from the Višegrad area.

10 Д. М. Бабац, op. cit., p. 93-94; *The Great War 1914-1918* ..., op. cit.

11 *Illustrated History* ..., op. cit., pp. 500-501; Д. М. Бабац,

op. cit., p. 94.

12 М. Радојевић, Љ. Димић (op. Cit., P. 162) indicate that because of the epidemic, around 400,000 people died at the turn of 1914/1915. Serbian population, 35 thousand. soldiers and 30 thousand. captives.

13 Д. М. Бабац, op. cit., p. 85. The Supreme Command of the Serbian Army, on the request of the Russian Supreme Command (Stawka) of April 18, 1915, regarding the readiness to support the Russian offensive on the Eastern Front, replied that the Serbian army would be ready for allied operations in the south in 20 the day it lasted. Allied aid to the Serbian army reached not only by sea via the Adriatic Sea (port of Bar) and via the Aegean Sea (port of Thessaloniki), but also by river ships of the Danube. The artillery defense of Belgrade included three 138 mm M 1893 heavy guns from the French ally, four batteries of 120 mm Mark III naval guns supplied by a British ally (the British headquarters of the support forces was in Smederevo [Serb. Смедерево, Hungarian. Szendrö , Vég-Szendrö, Semendria] and was subordinate to Rear Admiral Troubridge) and the Russian side provided two old 152 mm / 45 M 1877 naval guns. М. Радојевић, Љ. Димић (op. Cit., P. 162) state that Serbian forces intended to defend Serbian territory in 1915 numbered approx. 300 thousand soldiers.

14 SAW-KA, Die Eroberung Serbien 1915. Tätigkeit der Fliegerverbände der österr.-ung. 3. Armee und der Serb. Luftstreitkräfte. Bearbeiter Rittm. ad Peter v. Spaits, pp. 2-3; J. Piekałkiewicz, op. cit., p. 236; J. Pajewski, op. cit.; Д. М. Бабац, op. cit., p. 91. In 1915, the Montenegrin army mobilized 2/3 of the manpower. At the end of September 1915, after the end of the concentration of forces, on the northern and north-western border of the Kingdom of Serbia, a total of 14 German and Austro-Hungarian DPs were grouped against the five Divisions of the Serbian Army, and on the Serb-Bulgarian border against the Bulgarian Divisions of the six Divisions were grouped.

15 P. Hart, op. cit., p. 234.

16 Д. М. Бабац, op. Cit., P. 103; D. Jordan, *Balkans, Italy, and Africa* ..., op. cit., p. 60; J. Piekałkiewicz, op. cit., p. 239; M. Gilbert, op. cit., p. 220; *The Great War 1914-1918* ..., op. cit., p. 183. When on October 14, 1915, the Bulgarian Empire began hostilities on the side of the Central Powers, French troops from the Army of the East (Army of the Orient) reached the Greek-Serbian border, and a further 18,000 landed in Thessaloniki. soldiers. At the height of the 1915 campaign and the troubles of the Serbian army on November 10, 1915, the Allied forces grouped in the Thessaloniki region numbered over 80,000. soldiers, by the end of November 1915, strengthened by four British Infantry Officers, to reach the number of nearly 150,000. soldiers! A supporter of seeking military success in this part of the "old continent" was British Finance Minister David Lloyd George, "obsessed with dreams about achieving a great breakthrough in the Balkans." The command of the British army sent the 22nd, 26th, 27th, 28th and temporarily the 60th Infantry to the Salonite Front, while the command of the French army sent a total of nine Infantry to it. From April 1916, these forces began to be joined by Serbian DP, delivered from the island of Corfu, evacuated there from the continent by sea on board Allied ships and Serbian ships, as well as one Italian DP and one Russian BP. The most dangerous opponent of these forces, however, was ... endemic malaria, which was growing massively in this region. Only British units recorded 162 thousand. cases of it - cf. P. Hart, op. cit., p. 235–236.

17 SAW-KA, Die Eroberung Serbien 1915. Tätigkeit der Fliegerverbände der österr. ung. 3. Armee und der Serb.

18 Rauchensteiner, op. cit., p. 301; *Illustrated History* ..., op. cit., p. 501. This date was established as the commencement of hostilities the day after the declaration of war on the Serb side by the Bulgarian Empire.

19 Д. М. Бабац, op. cit., p. 85.

20 According to Д. М. Бабаца (op. Cit., p. 94–95), about 400 guns and 30 thousand guns were intended for shelling Belgrade. artillery shells, including the highest caliber of 305 and 420 mm.

21 SAW-KA, Die Eroberung Serbien 1915. Tätigkeit der Fliegerverbände der österr.-ung. 3. Armee und der Serb. Luftstreitkräfte. Bearbeiter Rittm. ad Peter v. Spaits, p. 44; M. Gilbert, op. cit., p. 220. *Great* ... (op. cit., p. 26) states that on October 6, 1915, 300,000 people joined the offensive actions against the Serbian army soldiers.

22 SAW-KA, Die Eroberung Serbien 1915. Tätigkeit der Fliegerverbände der österr.-ung. 3. Armee und der Serb. Luftstreitkräfte. Bearbeiter Rittm. ad Peter v. Spaits, p. 79-80; M. Rauchensteiner, *Der Erste*..., op. cit., p. 483.

23 SAW-KA, Die Eroberung Serbien 1915. Tätigkeit der Fliegerverbände der österr.-ung. 3. Armee und der Serb. Luftstreitkräfte. Bearbeiter Rittm. ad Peter v. Spaits, p. 79-80.

24 Ibid, p. 45.

25 Ibid, pp. 79-80; M. Pöhlmann, H. Potempa, T. Vogel, op. cit., pp. 91-92. A. Szelągowski (op. Cit.) States that on 6-9 October 1915, units of the combined German and Austro-Hungarian armies crossed the Danube. Belgrade was attacked by Imperials of the 57th and 59th Infantry Corps which crossed the Danube from the direction of Zemunu, Imperials 85 and 86. The reserve BP of the 43rd Infantry Division and Imperials 87 and 88. BP from the reserve of the 44th Infantry Division of the German XXII Corps forced Sawa at the height of Ada Island. The artillery support of the 8th Corps was provided by 70 heavy and 90 light guns and five M 11 "Škoda" mortar batteries, 30.5 cm. The defense of

Belgrade was provided by approx. 16 thousand. soldiers (20 infantry battalions) and 77 guns - Lieutenant Д. М. Бабац, op. cit., p. 94–95.

26 J. Piekałkiewicz, op. cit., p. 241. During the fights for Belgrade, the VIIIth Corps lost seven thousand. soldiers, almost twice as many as the troops of the German 11th Army forcing the Danube. A. Szelągowski (op. Cit.) States that Belgrade was seized on October 8, 1915 by units of the Imperial-Royal army. *Illustrated history* ... (op. Cit., P. 501) states that Belgrade was seized on 9 October 1915 by German-Austrian troops and M. Rauchensteiner (Der Tod ..., op. Cit., P. 300; idem - *Der Erste Weltkrieg und das Ende der Habsburgmonarchie 1914-1918,* Wien-Köln-Weimer 2013, p. 483) gives the date 11 October 1915 as the day of the capture of the Belgrade region. After the seizure of the Serbian capital, the front line moved some 30 km south of the city. On the night of 13/14 October 1915, the Serbian troops defending Belgrade, retreating south, reached the line of the Awala Mountains (Serb. Авала) where they were attacked a day later by advancing enemy forces. The fierce fighting in the Belgrade region is evidenced by the fact that the Serbian 1st Army withheld the enemy forces striking from the Srem direction for a period of 10 days, and the Chief of Staff, Armeegruppe Mackensen, was to characterize him with the words: "Very strong resistance! Serbs are fighting bravely! At that time, the Montenegrin army repulsed the attack of the 62nd Infantry Division on the border of the Kingdom of Montenegro.

27 J. Pajewski, op. cit.

28 Д. М. Бабац, op. cit., p. 96; *The Great War 1914-1918 ...*, op. cit., p. 180; J. Friedrich, op. cit., p. 363; M. Pöhlmann, H. Potempa, T. Vogel, op. cit.

29 D. Jordan, *Balkans, Italy, and Africa ...,* op. cit .; J. Pajewski, op. cit.

30 J. Pajewski, op. cit.

31 D. Jordan, *Balkans, Italy, and Africa* ..., op. cit.

32 *History XX*…, op. cit., p. 285.

33 Д. М. Бабац, op. cit., p. 97.

34 Ibidem, p. 98. The striking forces on the border of the Kingdom of Montenegro were supported from the air by about 36 aircraft. At that time, the Serbian air force operated with a force of 18 planes grouped in two squadrons, one of which operated on the section of the Bulgarian offensive. Serbian opl. consisted of field guns type Schneider M 1907 cal. 75 mm and Krupp M 1903 cal. 75 mm. The High Command of the Serbian Army, after being transferred from Belgrade to Kragujevaća, ordered the creation of the first anti-aircraft battery on 24 August 1915 to protect its berth.

35 Д. М. Бабац, op. cit., p. 98.

36 Србија у великом рату *1914-1918*, Београд 2014, p. 26

37 M. Gilbert, op. cit., p. 220.

38 Д. М. Бабац, op. cit .; A. Szelągowski, op. cit. German 62nd DP led an attack in Vi⊠egrad on the Drina River.

39 J. Pajewski, op. cit.

40 A. Szelągowski, op. cit.

41 Д. М. Бабац, op. cit., p. 100. On October 29, 1915, the High Command of the Serbian Army established contact with the command of the Allied Army of the East, trying to establish a potential area of cooperation on the Bitola (mac. Битола) -Drač (Serb. Драч).

42 Д. М. Бабац, op. cit., p. 101.

43 H. Münkler, op. cit., p. 361; J. Pajewski, op. cit.

44 *Illustrated History* ..., op. cit., p. 501.

45 Ibid, A. Szelągowski, op. cit.; H. Münkler, op. cit.; M. Радојевић, Љ. Димић, op. cit., p. 182. On October 27, 1915, Varna was shelled by the ships of the Russian Black Sea Fleet.

46 A. Chwalba, op. cit., p. 211–212.

47 J. Piekałkiewicz, op. cit., p. 242; D. Jordan, *Balkans, Italy and Africa* ..., op. cit., p. 57; J. Pajewski, op. cit., p. 305. On January 17, 1916, an expressway from Berlin, decorated with the signs of victory, entered the train station in Istanbul along the route of the Baghdad Railway - A. Chwalba, op. cit., p. 213.

48 J. Friedrich, op. cit., p. 370, 373; Българската армия в Световната война 1915–1918, София 1938, Vol. III, p. 5.

49 Д. М. Бабац, op. cit., p. 102–103.

50 H. Stratchan, op. cit., pp. 154–155; J. Friedrich, op. cit., p. 373; A. Chwalba, op. cit., p. 212.

51 "Ilustrowany Kuryer Wojenny" no. 40, from 25 November 1915, Bytom, p. 2.

52 J. Friedrich, op. cit., pp. 370, 373. M. Бабац (op. Cit., P. 110) states that during the battle in Kosovo Pole the Serbian army numbered approx. 300 thousand soldiers.

53 Д. М. Бабац, op. cit., pp. 104, 110; J. Piekałkiewicz, op. cit., p. 245.

54 J. Piekałkiewicz, op. cit., pp. 244-245; D. Jordan, *Balkans, Italy and Africa* ..., op. cit., p. 60; A. Chwalba, op. cit., p. 214.

55 Д. М. Бабац, op. cit., pp. 104, 110. According to the reports of the Serbian army, after a month of fighting, there were about 250,000 soldiers. In the literature on the subject, you can also find data that 140,000 Serbian soldiers moved towards the Adriatic coast.

56 Ibid, p. 106. M. Радојевић, Љ. Димић (op. Cit., P. 183) report that the Serbian High Command and the government reached the Albanian coast on November 25, 1915.

57 A. Szelągowski, op. cit. As a result of the retreat of Serbian forces, the German troops reached the border of the Kingdom of Greece, and the field troops of the 3rd Army entered the Duchy of Albania through the territory of the Kingdom of Montenegro.

58 M. Радојевић, Љ. Димић, op. cit., pp. 184–185.

59 D. Jordan, *Balkans, Italy and Africa* ..., op. cit., p. 58.

60 What the neighborly relations were like in 1912, on

the threshold of the emerging Albanian statehood on the Albanians (Arnauts) -Serbians-Montenegrins line, can be proved by the fact that then approx. irregular Albanian tribal forces of irregular forces, known for their bravery and ruthlessness, could fight the Serbs and Montenegrins in Sanjak Novopazarski and the Greeks in southern Albania, northern Epirus (Al. Chamuria) and Macedonia on the Turkish side. As volunteer troops (tur. Gönüllü müstahfiz), Albanians constituted a significant armed force, which the Ottoman command took seriously when rearming it (Albanian leader from Kosovo, Isa Bojentin was to receive as many as 60,000 rifles and rifles for rearmament and equipment from the Turks) - see D. Wydranowski, *The Battle of Kumanów (23-24 October 1912). Armed forces of the parties, operational activities, military, and political consequences, [in:] "Balcanica Posnaniensia"*, Acta et studies XX 2013, *ed. I.* Czamańska, W. Szulc, p. 154.

61 M. Gilbert, op. cit., p. 224.

62 T. Aronson, op. cit., p. 222. In one of the accounts, it was written: Suddenly, someone banged hard on the door. A hoarse voice called out in commanding Serbian, "Open up!" My host obeyed the order. Through the open door, the riders could be seen in the darkness of the night. Two of them had already dismounted and with great care lifted the third one, wrapped in a large cloak, from the saddle. They took him carefully into the room. They were Serbian officers. I considered the third one to be their wounded colleague. "Light the fire" they ordered the owner of the hut. The smoldering logs shot flames and in their light, I recognized the third, by his dry gypsy face, tousled wisps of gray hair and hanging white mustache, the king of Serbia, Piotr [I] - see J. Piekałkiewicz, op. cit.

63 D. Jordan, *Balkans, Italy, and Africa ...,* op. cit., p. 50. The Austro-Hungarian diplomacy considered the idea of uniting the Kingdoms of Serbia and Montenegro under the rule of King Nikola I of Montenegro and providing Montenegrins with an advantage in this country, which was not implemented.

64 A. C. Недок, Повлачење српске војске ка албанском приморју и њена евакуација на Крф 1915-1916, Београд 2014, pp. haughty. 94 (95?) Thous. killed, approx. 174 (150, 175?) thousand soldiers in captivity and 800 guns.

65 A. Chwalba, op. cit., p. 212.

66 *Illustrated History ...*, op. cit., p. 501. A. Szelągowski, op. cit. *The History of XX ...* (op. cit., p. 285) states that the Allied ships transported 50-60 thousand to the island of Corfu. soldiers of the Serbian army. B. Hamann (op. Cit.) Reports that 140,000 people were evacuated from the Albanian coast to the island of Corfu. soldiers of the Serbian army, while M. Gilbert (op. cit., p. 224) reports that 260,000 people reached the Adriatic coast. Serbian soldiers. In fact, according to Serbian sources, which are the most reliable in this matter, allied ships and ships carried 77,475 soldiers and 77,278 wounded from the Albanian coast to Corfu - see A. C. Недок, op. cit., p. 122. However, there is no consensus among Serbian authors on this point, as Д. М. Бабац (op. Cit., P. 110) states that until February 21, 1916, 43 Allied ships and ships evacuated to the island of Corfu, to Bizerta in North Africa and to French and Italian over 169,000 soldiers of the Serbian army. As of 27 February 1916, the Serbian army consisted of the 1st field army, col. Mikhail Wasić (Moravian and Vardar DP), field 2nd Army of General S. Stepanović (Šumadijsk and Timock DP) and field 3rd Army of General Paweł Jurisić-Szturm (Danube and Drinsk DP). Konica DP, General Borća Borćević, existed as an independent one, while the entire Serbian army was commanded by Fr. Regent Alexander - Lieutenant Д. М. Бабац, op. cit., pp. 115-116. М. Радојевић, Љ. Димић (op.cit., p. 188–189, 192) report that between 12 December 1915 and 22 February 1916, 11,650 wounded were evacuated from the Albanian coast to Brindisi, Lipari, Marseilles and Bizerta, 130,840 soldiers to the island of Corfu and 4100 to African Bizerta using 28 Italian ones.

67 J. Pajewski, op. cit., p. 305; *The Great War 1914–1918 ...*, op. cit., p. 184. B. Hamann (op. cit.) reports that 20,000 people were evacuated from the Albanian coast to the island of Corfu. POWs from the Imperial-Royal Army.

68 M. Gilbert, op. cit.

69 A. Szelągowski, op. cit., p. 780; J. Friedrich, op. cit., p. 380.

Habsburg Wings Over the Balkans 1915

The commander of the field aviation of the 3rd Army was Baron Hptm. A. von Mandelsloh, to whom Flik 9 was subordinate at the airport Ruma and "his" Flik 15 at the airport in Újvidék. From the first day of the fighting, the crews of both Imperial-Royal Fliegerkompanien (during the crossing of the Sawa River on October 7, 1915, the crews of the KuK Luftfahrtruppen provided airborne bomb support to the army's Imperial-Royal units)[1] were tasked with recognizing the movements of enemy forces, especially the position of the Serbian artillery, which, despite being weaker in numbers, "bit off" the attackers very effectively with its fire. The commanders of the field staff of the 3rd Army, who expected the most precise information from the airmen, put special emphasis on photo recognition, although Flik 15 was composed of only two photographers and in Flik 9 there was none[2]! During the fights in the field section of the 3rd Army, "its" air force actively supported the attacking its own troops and the divisions of the German 11th Army[3] (German divisions supported the crews of Flik 9, Flik 15 and the German-Turkish FFA - many flights were made to photograph the Kostolac area)[4]. On October 7, 1915, the Flik crew. 15 also carried out radio-cooperation flights with the Imperial-Royal artillery batteries near Belgrade, and the "balloon support" was also used to precisely fire from the Imperial-Royal artillery (Festungballonabteilung 1/6. R was assigned to the 19th Imperial Corps with the task of photographing the Zabrez area [Serb. Забрез])[5]. Despite the equipment problems, the crews performed combat flights in terrible weather conditions on "whatever could" – for example: Flik 6 operated from the airport in Rogatica on October 7, 1915 for the benefit of Imperial-Royal 62nd DP, originally having only two planes in stock, and after replenishment of the stock, four aircraft armed with MG's. The enemy's air force operated sporadically on that day (in the order of the 3rd Army's field command it was written - its own troops were bombed by the enemy's plane near Surčin [Serb. Сурчин]")[6].

On the next day of the offensive, the crews of the KuK Luftfahrtruppen recognized the state of defense of the enemy. On October 8, 1915, the Flik crew. 15 reported heavy shelling over Belgrade[7] and heavy traffic on enemy railroads, and in the afternoon and evening three German FFA crews recognized the area behind the front, where near Pozarevac a column of infantry in the march with the strength of approx. regiment was recognized[8] (FFA 66. was assigned on that day to the command of the German X Reserve Corps)[9]. The following days brought increased aviation activity. On October 9, 1915, the Flik crew. 5 and Flik 9 as part of radio-cooperation, performed flights for the artillery group supporting Imperial-Royal 53rd Infantry Division north of the Zivaca region (Serb. Зиваца) -Bara (Serb. Бара)[10]. The same, on October 9, 1915, of the Flik crew. 5 despite the changing weather conditions, they recognized the movement of the Serbian army troops retreating along roads from Belgrade to the south and the area of Skela-Debec (Serbian Дебец) -Banjani (Serbian Бањани) -Piroman (Serbian Пироман) -Stepojevac (Serbian вацое) -Obrenovac[11]. In the attack section of the XIX Corps, a photo-exploration of the Jarak-Drenovac-Šabac-Misar-Kupinovo-Zbrec [Serb. Збрец] -Smederevo-Kostolac [Serb. Костолац])[12]. On the next day, October 10, 1915, the crews of Flik 5 during the flights, detected a strong retreat of Serbian army units, and the crews of Flik 9 made four flights detecting the Serbian fortifications near Drenovac and bombing them (three bombs weighing 10 kg each were dropped). In the afternoon, a large Serbian military camp near Kumodraž and a column of troops in motion on the section Železnik (Serbian Железник) -Sremčica (Serbian Сремчица) were detected, targets in Drenovac, positions of five infantry battalions and two artillery batteries in the area between Velylki batteries were bombed again Sopot (Serb. Велки Сопот) and Djurinac[13]. Same on October 10, 1915, the crew of Flik 15 and Flik 5 also conducted active photo reconnaissance (50 photos of enemy targets were taken). Weather conditions on October 11, 1915 did not improve and the day turned out to be flightless. October 12, 1915 the crew of Flik 9 only made 2 flights, and Flik 5 - 1 reconnaissance flight[14] and the reconnaissance continued the next day. On October 13, 1915, the Flik 5 detected an enemy camp near Kumodraž, two enemy columns on the march and a camp of Serbian troops near Zvetkov[15]. The crews of only two Flik 5 and 6 performed by: Flik 5 - 1, 24, and 26 October 1915, 15 long-distance reconnaissance flights, two radio flights - cooperation with artillery and the Flik 6 made only three long distance reconnaissance flights[16].

The crews of the German FFA on October 9-13, 1915, conducted a long-distance reconnaissance, and despite the fog in the afternoon of October 9, 1915, three crews from FFA 1 and one from FFA 69 recognized the base of the Serbian army[17] as well as "zipped up" to the region of Palanka-Lapovo and the Morava Valley. On October 10, 1915, the crews of the FFA 1. recognizing the Palanka-Perečin and Perečin-Zaječar railway lines discovered a large train depot near Požarevac, a lot of traffic on the retreat routes of the Serbian army, a field airfield and nine artillery positions. A day later, on October 11, 1915, the FFA 1st crews recognized the Palanka-Smederevo railway line, on October 12, 1915 the Smederevo-Markovac railway line (Serb. Марковац) and Velika Plana (Serb. Велика Плана) -Krana (Serb. Крана), and on October 13, 1915, aerial reconnaissance reported heavy fighting in the Požarevac region and in the Morava Valley[18]. After a few days, on October 19, 1915, the crews of the German FFA delivered three more reports proving the retreat to the south of the Serbian army[19] and from the 13 flights of German crews made on October 25, 1915, as many as seven reports spoke of a wide retreat of Serbian troops (the report of the FFA 57 crew for the German command of the XXII Reserve Corps spoke of the mass retreat of the enemy to the south-

Above: Wacht am Schwarze Meer ... In 1918, the troops of the Imperial-Royal army before Finis Austriae reached the Black Sea, occupying the territory of the present Republic of Ukraine. The photo shows a patrol of the soldiers of the Imperial-Royal army in the Odessa region - source: "*Das interessante Blatt*" 1918

west along the Kragujevac-Topola-Arangjelovac route)[20]. As the field offensive activities of the Central Powers' armies developed, the number of Imperial Balloon units on the Serbian front was increased, and on September 21, 1915, Ballonabteilung 1/5. B was assigned to the German 11th Army, a 12 October 1915 Ballonabteilung 2/4. R and 1/3. B went to the KuK Krigsmarine base in Pula (owned by Pola, German by Pulei)[21] and on October 18, 1915 they were assigned to the command of the field command of the Imperial-Royal 3rd Army[22]. During this phase of the fight, observation balloons were "placed" in height - spherical balloons M 96. 1900-2000 m, observation balloons M. 98. 1800-1900 m. observation and flight crew from Flik 6[23] which was rare.

The crews of the enemy air force in the period preceding the Central Powers offensive, operating mainly from the Belgrade-Batajnica airport, received thanks from the Serbian Supreme Command for the reconnaissance flights (on September 27, 1915, the Franco-Serbian squadron was divided into two sub-units - one was assigned to the Danube section of the front the second on the section on the Sava - this division lasted until the Bulgarian Empire entered the war on October 14, 1915)[24]. At the time of the start of the fighting in Serbian territory, with the support of own troops, on October 9, 1915, the reconnaissance of the Imperial-Royal forces was carried out in the Pančevo-

Homokbálványos-Kubin-Ram area, and on October 10, 1915 in the Smederevo-Kubin (Serb. Ram-Bela Cerkva-Mladenovac and Kupinovo-Ašanja (Serbian Ашања) -Torvanik (Serbian Товарник) -Nikinci (Serbian Никинци) -Jarak. On October 11, 1915, due to terrible weather conditions, no combat flights were performed, and on October 12, 1915, reconnaissance of the left bank of the Danube was carried out, detecting the concentration of heavy forces near Kubin. In the direction of the main attack by the field armies of the Central Powers, after the fall of Belgrade, on October 13, 1915, the Belgrade-Kupinovo-Ašanja region was recognized[25] and a day later, on October 14, 1915, bridges under construction and a pontoon bridge in the Gypsy Island, near Belgrade on the left bank of the Sava were observed near Kubin[26]. At the end of October 1915, the pace of flights of the crews of Imperial-Royal Fliegerkompanien did not slow down, although the weather conditions were deteriorating day by day (on October 23, 1915, combat flights were carried out even though the clouds were 300 m above the ground)[27]. The crews of Flik 6 from the airport in Rogatica, until October 23, 1915, recognized the Drina area, even though only one plane was operational (three were in rail transport, two were damaged during forced landings in western Bosnia during ferrying)[28].)[1]. As part of the aerial harassment of the retreating troops of the Serbian army, the Flik 15 On October 24, 1915, dropped 16 bombs on a detected train[29] and the Serbian infantry camp near Topola and Natalinci (Serb. Наталинци) and reconnaissance in the area of Palanka-Ratari-Arangjelovac-Bukulja (Serb. Букуља) -Orlivica (Serb. Орливица) -Vic (Serb. Виц) on October 25, 1915, the Topola-Kragujevac route was explored[30]. The command of the field corps of the 3rd Army ordered the crews of Flik 6 the bombing of the Serbian army camp near Vardište (Serb. Вардиште - its crews recognized the Rogatica-Užice area)[31]. As a result of a change during the front line due to the retreat of the Serbian army to the south, on October 27, 1915. Flik 5 and Flik 15 changed field airports and transferred them to Arangjelovac[32]. The German crews of the 28th, 1st, 3rd and 69th FFA made two flights to the area of Kragujevac-Tribunovo (Serb. Tribunovo) on October 26, 1915, and due to the deterioration of weather conditions, for the next four days, until October 30, 1915. the crews of the German FFA did not perform the sorties[33].

During the retreat of the Serbian army, its air force constantly provided information from the "third dimension" about the movements of the enemy forces. On October 25, 1915, the Serbian French crew recognized the region of Jagodina (Serb. Јагодина) -Topola-Jagodnia (three hours spent in the air did not bring anything special), another airman recognized the region of Niš-Weternica (Serb. Ветерница) -Vranje (Serb. Врање) -Kumanovo (Macedonian. Куманово, Alb. Kumanova, Kumanovë) detected about 800 horse-drawn vehicles on the way. Three further flights were made on the line Niš-Leskovac (Serb. Лесковац) -Western Morava-Stubla (Serb. Стубла), another two to the region of Niš-Knjaževac (Serb. Књажевац) -Zaječar (Serbian Зајечар (Serb.) -Niš where a large infantry camp was discovered near Rakovica and Kula (Serb. Kula, Hungarian. Kúla) and a large artillery park near Kul[34]. With the entry into hostilities on 14 October 1915 of the Bulgarian field armies, the attention of the Serbo-French crews gradually shifted to the newly formed Bulgarian front. During the retreat of the Serbian 3rd Army, they were identified in Svilajnac (Serb. Свилајнац) -Cuprija (Serb. Ћуприја) -Varvarin (Serb. Варварин) -Kruševac (Serb. Крушевац). airport grounds[35].

At the time of the accession of the Bulgarian Empire to the Great War, the Bulgarian aviation had only six pilots on the side of the Central Powers[36] while the flying equipment consisted of only five planes (three of the Bleriot XI type and two of the Albatros B.I type). On the day of the beginning of the fighting on October 14, 1915, the Bulgarian High Command planned five flights with all forces in the Pirot area, which, however, was thwarted by bad weather conditions (Capt. Penio Popkrystew damaged the Bleriot XI bis plane during the landing). On October 18, 1915, the "Solun" observation balloon was used in the fighting to correct artillery fire despite enemy fire (the Balloon Division was transferred to Niš, where it was stationed until 1916 after the end of the fighting)[37]. A few days after the start of the fighting, the German ally delivered (Bulgarian Божурище) near Sofia (Bulgarian София) FFA 69 with six LVG C. II planes[38] (six more aircraft on order were expected). They were the first armed planes in the Bulgarian aviation, and Bulgarian flying personnel were trained on new flying equipment under the supervision of German pilots (pilot Franke trained the first seven pilots and nine observers). In connection with the improvement of weather conditions, in the period of 21–24 October 1915, the crews of the Aviation Division made the first combat flights for the command of the Bulgarian 1st Army field in the fortified region of Niš, after which they made combat flights for the Bulgarian 2nd Army on the Front Macedonian (Thessaloniki - a combat airfield of four crews and LVG C.II airplanes was the landing area located next to the army headquarters in Belica [Maced. Белица])[39]. During the fights, changes in the organization of Bulgarian aviation took place - an Aviation Team was created, consisting of: 1st Aviation Division at Belica airport (84 ground and flying personnel, including six pilots, seven observers and six LVG C. II aircraft), 2nd Division Aviation at the Udovo airport (Bulgarian Удово - 56 flying and ground personnel, including six pilots and five observers, three LVG C. II and three Otto C. I), the Balloon Department deployed to Ruse (Bulgarian Русе - one observation balloon and 154 ground and flying personnel, including two observer officers), command of the airship base in Jambol (Bulgarian Ямбол - 32 ground personnel) where soon the German 18th Airship Division and the SL 10 airship were to appear, the Aviation School at the Bozuriszte airport, Aviation Workshops and the Balloon Base[40]. From mid-October 1915, for the field of the Bulgarian 2nd Army, two crews from the field airfield in Toplet made reconnaissance flights of the Timok River valley (Serb., Bulgaria, Bulgaria). Timok) and the Sofia region. During the field fighting of the Bulgarian 1st Army, there was no air support, and the German crews throughout November 1915 monitored the retreat of the Serbian army to the south from the air[41].

November 1915 brought a strong weather breakdown, because of which there was a decrease in the pace of air operations on the Serbian Front, but still the tasks of the air crews of the Imperial-Royal and German FFAs were dominated by the reconnaissance and destruction of detected targets from the air. On a monthly basis, the Flik 9 made 10 reconnaissance flights, due to bad weather conditions seven flights were unsuccessful: Flik 15 made seven flights without success, and the Flik 6 -11 flights without success[42]! On November 1, 1915, the crews of Imperial-Royal Fliegerkompanien explored the area from the border of the Imperial-Royal Monarchy to the Morava Valley[43] - Flik 9 near Čačak, observed traffic on the bridges on the retreat roads of the Serbian army, which were photographed and then destroyed by bombs by the crews of Flik 15[44]. November 2, 1915 was a day of considerable aviation activity. 8 km north of Požega (Serbian Пожега), a 3 km long transport column of the Serbian army was recognized in the march; on the same day a large rolling stock was detected near Travnik in a retreating march to the south-east; between Belosava (Serbian Белосава) and Witanowa (Serbian Витнова) about 750 horse-drawn vehicles were sighted on the route Kragujevac-Kraljevo; no movement was detected from Kraljeva south-west towards the mountains. A large group of Serbian horse-drawn vehicles were detected near Belušić (Serb. Белушић) and Čedić, and no enemy field fortifications were reported south of the Morava line[45]. The retreat of the Serbian army continued and was constantly watched from the third dimension. On November 4, 1915, the FFA 57 crew detected strong columns of the Serbian 1st Army field in the march on the Kraljevo-Konarevo (Serb. Konarevo) section, and until November 6, 1915, the Flik 9, Flik 15 and FFA. 57 provided systematic reconnaissance reports of the main directions of the Serbian retreat[46]. As a result of the day-to-day worsening weather conditions, combat flights were carried out in the high mountains, constituting a real challenge for the crews of the KuK Luftfahrtruppen. On 8 November 1915, the crews of Flik 15 reported the detection of 30 to 200 horse-drawn vehicles on the Kraljewo-Raška route, while on another flight, in the Užice-Sjenica region (Serb. Сјеница) -Novi Pazar (Serb. Нови Пазар, bosn. Novi Pazar) small units of the Serbian army on the move[47]. November 9, 1915 Oblt. Löw of Flik 9 reported the detection of enemy forces in the march near Sjenica, and the report of the commander Flik 15 of November 11, 1915, spoke about the results of photo recognition of the Ivanjica-Sjenica-Novi Pazar-Ivanjica area. Despite the worsening weather conditions, combat flights continued, and the fog and low clouds became the "daily bread" of the airmen[48]. On November 12, 1915, the crew of Flik 9 made a combat flight in bad weather conditions near Sienica, by detecting units of the Serbian army and performing photoreconnaissance (around 2 p.m. between Razdeginja [Serb. Раздегиня], and Budevo [Serb. Будево] a transport column approx. 1 km long was detected with a mixed set of vehicles)[49]. The bombing raids carried out by the crews of the KuK Luftfahrtruppen constantly took their toll on the Serbian troops: on November 7, 1915, the Flik 6 bombed a Serbian army camp near Goražde (Serbian Горажде), On November 17, 1915, Mitrovica was bombed, and on November 21 and 22, 1915, near Mitrovica, the FFA 57 crew discovered the encampment of the Serbian army, which was then bombed by it and by the Flik 9 (each crew dropped seven [two?] Bombs)[50]. Wanting to know about the movements that were in a permanent retreat, but still a strong opponent in the Serbian-Albanian-Macedonian-Greek border, on November 15, 1915, on the orders of Field Marshal A. von Mackensen, the field air force of the 3rd Army and the German 11th Army had recognized the area to the line Užice-Mitrovica (Kosovska - Serb. Косовска Митровица, Kosovska Mitrovica, Alb. Mitrovica-Mitrovicë) -Pristina (Alb.Prishtinë, Serb. Приштина)-Skopje and the Ipek (Serb. Ипек) -Prizern (Serb. Призрен, Alb./ Bosn. Prizren). A day later, on November 16, 1915, in an air report by the FFA 57 crew the detection of large enemy infantry forces in the area north of Raška, on both sides of Brvenica (Macedonian), was reported to the field command of the 3rd Army. Brevenica)[51] and a day later November 17, 1915, a heavy snowfall caused combat flights on the entire Balkan front to be suspended! Reconnaissance flights, often performed in terrible weather conditions, were carried out on routes of 320-340 km[52]. On November 18, 1915, despite the dense fog, which made daily combat flights a "gamble" for air crews, reports of the reconnaissance of the FFA 57 crews assigned to the German XXII Reserve Corps command arrived from the front, in which it was reported that a long column of enemy infantry had been detected opposite Mitrovica, and near Kosutica hostile rail transport[53]. As part of another change during the front line and the need to shorten the distance to the battle line, Flik 6 from November 17, 1915, was gradually transferred from Rogatica to the region of Čačak, in order to cover the entire area of the Serbo-Albanian-Macedonian-Greek border with the same range. On November 19, 1915, aerial reconnaissance was carried out in Priština, where the retreat of Serbian forces towards Mitrovica was observed, on November 21, 1915, the FFA 57 crew detected the camp of Serbian troops near Mitrovica[54], and on November 22, 1915, the crew of Flik 9 recognized the route Raška-Mitrovica-Priština and Novi Pazar-Ipek (near Sinaja [Serbian Синая] and Veliki Istok, camps of Serbian infantry in the strength of three battalions were detected and a battalion of Serbian infantry on a march to the north under Cerefulj)[55]. The hostilities quickly spread across the Serbo-Montenegrin border, and intensified also on the Albanian-Montenegrin border, where the crews of Imperial-Royal Fliegerkompanien and the FFA of 57 were increasingly venturing. On November 23, 1915, the commander of the field c. K. 3rd Army demanded the reconnaissance of the Bar area (Бар, Montenegro, Antivari) -Virpazar (Вирпазар, Montenegro) -Podgorica-Andrijevica (Андријевица, Montenegro) -Ipek-Mitrovica, and on November 24, 1915, air reconnaissance was ordered in the area of the Kingdom of Montenegro. On that day, two crews were sent to the area of Čačak-Mitrovica-Ipek-Andrijevica-Podgorica-Rijeka Crnojevica-Virpazar-Stari Bar (Стари Бар, Montenegro). During the retreat of the enemy forces, more and more urgently, the command of the field Imperial-Royal The 3rd Army needed data on the enemy from the Lovćen massif, and from the route Danilovgrad (Даниловград, Montenegro) - Nikšić (Montenegro Никшић) -Šavnik (авник, Montenegro) -Plevlja (Пљевава, Serb. Čačak. Report from the commander of Flik 15 on November 25, 1915, reported that one of the crews had discovered a small Serbian force in the Plevlja region, and on November 28, 1915, the crew of Flik 15 performed a close reconnaissance, but due to engine failure it had to forcibly land. On November 29, 1915, the forces of about 1-1½ of the Serbian Infantry Brigade (BP) were detected, as well as railway transport opposite Ipek, and as part of the Flik ferrying. 15. moved from the field airport in Čačak to the next field airport in Novi Pazar[56].

Along with the retreat of the Serbian army to the south, the Serbian air force was also re-positioned there (within three weeks the front line retreated approx. 250 km to the south)[57]. During the retreat of the Serbian army, the crews of the Franco-Serbian squadron provided information on the movement and movements of Imperial and German troops, which proved particularly valuable at a time when the Serbian government was considering the evacuation of the army and large numbers of civilians towards the Adriatic coast through the mountains Principalities of Albania and the Kingdom of Montenegro. On these extremely difficult November days, the crews recognized the directions of the attack of the Imperial-Royal and German troops, being the only source of the latest information about the enemy for the Serbian High Command! On November 2, 1915, the crews recognized the area of Raška (Serbian Рашка) -Čačak-Gornji Milanovac (Serbian Горњи Милановац) -Arangjelovac-Kraljevo, where enemy forces were detected on the march, and on November 3, 1915 the Raška-Čačak-Kragujevac-Kraljevo area was explored, where the forces of the 19th and 8th Corps were recognized for nearly four and a half hours. The reconnaissance flights continued on November 5 and 6, 1915 - on the first day in the Raška-Kraljevo regionGuča (Serb. Гуча)-Čačak-Raška, and on the second day the French crews recognized the Raška-Ivanjica region (Serb. Ивањица) -Arilje (Serb. Ариље) -Čačak[58]. As a result of the failure, the number of operational airplanes gradually decreased. On November 8, 1915, the region of Usce-Kraljevo-Guča-Arilje-Ivanjica-Raška was identified - in the second flight, the region of Trjicevo-Samolica-Gnjilane (Alb. Gjilan, ser. Гњилане) -Mitrovica was identified. As a result of the retreat towards the borders of the Duchy of Albania, the squadron had only two operational planes, the crews of which were supported by Serbian troops during the battle for Kačanik, and after this battle only one plane remained operational and thus the recognition of the enemy's "third dimension" forces practically decreased to zero[59]! As a result of the actions of December 4, 1915, the 62nd DP reported the shooting down of an enemy plane near the road Plevlje-San Sehir-Kadic (Serb. Кадиц) -Pribiljevic[60]. Until November 15, 1915, Raška was the last combat field airfield in the Kingdom of Serbia, which French airmen called the "Eagle's Nest", after which the unit was transferred to Mitrovica, and on December 7, 1915, all its personnel were in Shkodra. As soon as the Serbian army marched into the Albanian mountains, the equipment that could not be taken with you was burned, and the planes flew to Shkodra[61] (acc five planes were delivered there by the Serbian command)[62] from where two or three combat flights a week were carried out in combat conditions more and more resembling "polar", during which the crews had no cover - MGs, and on airplanes unsuitable for high-altitude flights, tasks were carried out over high and dangerous mountains. Airplanes were used to transport mail and communication between individual points controlled by Serbian forces. "Hazardous" flights were performed mainly on the routes Shkoder-Durrës-Elbasan (formerly Scampa) and Shkoder-Durrës-Vlorë, but the lack of fuel and oil (the weekly needs of the Squadron were 500 liters of gasoline and 50 liters of oil) resulted in a reduction of flights to two or three a week. Moreover, the unit was troubled by failures (on 27 November and 10 December 1915, there were two aircraft failures in Shkodra, where the planes were immediately repaired)[63]. During the tragic retreat of the Serbian army, in the middle of winter, French and Serbian pilots were the first in the world to transport the wounded from Pristina to Shkoder by air. During its retreat through the territory of the Principality of Albania and the Kingdom of Montenegro, Sgt. M. Tomič maintained communication by plane between the staffs of the retreating Serbian field armies and the Serbian High Command, transporting senior officers, mail, and money by flight[64]. The squadron left Shkodra at the end of December 1915[65] (last operational planes flew to the Greek island of Corfu)[66], where a new squadron of Serbian aviation was created, headed by Major Kosta Miletič[67].

Aviation activities in the Balkan region were gradually fading away. In the period of October 5th-December 2nd, 1915, the weather conditions were exceptionally good for only 13 days, the weather conditions were terrible for 23 days, and 11 days were days with mixed weather conditions[68]. During the Serbian campaign of 1915, the crews of four Imperial-Royal Fliegerkompanien worked in combat for almost 2 months, making combat flights over the Balkans and during that time flying around 22,000 people. km with Lohner B "Pfeilflieger", Lohner C "Gebirgspfeilflieger", Lohner D "Pfeilflieger" (B.III), Albatros B.I and Aviatik B.I and B.II[69]. The reasons for the losses incurred during combat flights and the places of forced landings were different. Engine failures and shots down by Serbian Opel cars were dominant. In the run-up to the offensive of the Mackensen Group, on June 28, 1915, a forced landing of a plane from Flik 15 took place due to a storm at Zemun; On 7 July 1915, the Flik 15 plane landed forcibly at Homokbálványos; On August 28, 1915, a German-Turkish crew forcibly landed near Csernaheviz due to an engine failure; On September 20, 1915, planes from Flik 9 and Flik 15, forcibly landed at Tiszakalmanfalva (Serb. Будисава / Budisava); On September 21, 1915, the Flik 15 plane landed by force at Loznica. During the offensive operations, on October 11, 1915, during a combat flight in the region of Kragujevac, the Serbian Opl. She shot down a heavy-duty aircraft of Aviatik B near Plevlje[70]. On November 2, 1915, the Flik 6 plane forcibly landed near Užice; On November 14, 1915, the Flik 6 plane forcibly landed near Pribojem (Serb. Прибој); On November 20, 1915, a report about a missing plane from Flik 6 arrived two weeks earlier under Zaborak (Serb. Заборак) and its crew in captivity in Montenegro[71]. The losses of the German FFA during the Balkan campaign of 1915 were as follows: because of the Serbian air force - 0, because of bad weather conditions - 2 planes crashed; 1 plane caught fire during a bombing raid on Kragujevac - cause unknown[72].

The Balkan campaign of 1915 was not the end of the history of Serbian aviation in the Great War. It was revived thanks to the training and equipment support of the Allies on the island of Corfu, where Serbian pilots found themselves at the French Levkimos airport, and then in the Thessaloniki area. In March 1916, the Allied Conference in Paris granted the Serbian army material support and as of March 14, 1916, the Serbian air force consisted of the command, Detached Squadron (12 pilots and 4 mechanics) in Thessaloniki and the Balloon Division (132 people)[73]. On April 8, 1916, 10 officers and 10 non-commissioned officers were sent to a pilot training to Chatoreaux in France, and to an observation course in Sezes near Thessaloniki, 20 officers, and then on August 20, 1916, they were assigned to the Serbian squadron (until autumn 1916 in Serbian aviation served 32 new pilots, 31 observers and 14 mechanics)[74]. In early May 1916, a Serbian aviation training center was established near Thessaloniki, where 5 mixed Serbo-French Squads were soon established[75] (On May 9, 1916, a group of 8 officers, 3 mechanics and 85 soldiers of the Serbian army were transferred from Corfu to Thessaloniki to supplement the units created there). On June 4, 1916, the Serbian government paid 30,000 gold dinars for training another group of pilots in the Republic of France[76], and on June 12, 1916, a decision was made to create the above-mentioned 5 mixed Serbo-French Squadrons: N 87 (Nieuport) for the disposal of the Serbian Supreme Command; N 84 (Nieuport); MF 82 (Maurice-Farman), N 82 (Nieuport) for the field 2nd Army, MF 98 (Maurice-Farman) for the field 1st Army and MF 99 (Maurice-Farman) for the field 3rd Army. The 5 Squadrons, equipped with Maurice-Farman and Nieuport planes, were commanded by a French pilot, Maj. Rode Toulouse, and their airports were at Vertekop and Gorgopi. When these Squadrons entered combat, they were used for combat flights on the Macedonian front, and then another French aviation officer, Major Roger Vitrat, took command of these forces[77]. Due to the constant shortages in the flying personnel, another 20 students of the art of flying were sent to the flight school in France on June 14, 1916, who soon strengthened the group of Serbian aviation pilots. On June 20, 1916, another 10 junior officers assigned to the Serbian air force were sent there, and at the beginning of October 1916, 11 officers were sent to the observers' school in Sedes[78].

As of August 18, 1916, the MF 82, N 87 and MF 99. squadrons were stationed at the airport in Vertekop, the N 84 Squadron was stationed at the airport in Šamlija, and the MF 98 Squadron at the airport in Gorgop. September 1916, Squadron N 87 was transferred to the Sorovič airfield and the MF 98. Squadron to the Ostrovo airfield[79]. Until 1918, the crews of these Squadrons fought on the Macedonian front, and in the fall of 1918, they became the nucleus of the air force of the reborn state.

Air operations in the Balkans in 1916-1918 took place in the operational region of the Albania Group[80] (Armeegruppe Albanien) with considerable intensity but did not achieve the same pace as on the Italian front over Isonzo. The region includes the school unit Fliegerersatzkompanie (Flek.) In Raijlpvac, the 14th "Balkan" Flik 6 was stationed at the Shkodra airport in the period from January 30, 1916 to the end of June 1917, and then operated from the Kavaja (Alb. Kavajë) airport near Tirana until September 1918. From April 1918, Flik was transferred to the Duchy of Albania. 64./F (Fernaufklärungs-Fliegerkompanie - F-Flik) to the airport in Fjeri (2 pilots, 2 observers and 62 people of ground staff - from December 1917 stationed at the airport in Igalo in the Bay of Kotor). These units used 8 planes, including six observation planes (types Lohner B. VII, Brandenburg C. I, Lloyd C. V) and 2 fighter planes (in Flik 6./F, Albatros D. III and Aviatik D types).. I). On September 27, 1918, the units of the KuK Luftfahrtruppen were ordered to retreat from the Macedonian front. This way Flik 6./F went to the airport in Podgorica, and in mid-October 1918 Flik 64 / F went to the airport in Igalo. October 12, 1918 Flik 34./F and 64./F ended up at the airport in Zemun, and in the Bay of Kotor region was Flik 1. J (Jagd-Fliegerkompanie - J-Flik). One of the Allied airmen who made combat flights over the Balkans in the fall of 1918 was capitaine Perrin de Brichembaut, a French pilot belonging to the aeronautical section of Wardar. In his memoirs, he wrote: *Today on my faithful slope* [XIII C 1?] *From 1,800 meters, in the same area, I observed numerous military movements, large groups, new batteries and a tethered balloon, brought from a distance ... Flying over the peaks at a height of about 2000 meters, I saw a German artillery observer. Then, through my binoculars, I saw some two types giving me friendly signs with their hands. I make a flat turn so as not to show my colors to them, I take the direction to the north and come back. The entire crew of the observatory stepped outside - six or seven Swabians. They stand together, beckoning me in awe of Fritz so low flying in the mountains. I target them and shoot without stopping. Two of them remained on the rock. Previously recognized by battery artillerymen and fired sharply, I flew maybe one meter above the observatory's binocular scissors and 1,800 meters above sea level*[81].

Endnotes

1 SAW-KA, Die Eroberung Serbien 1915. Tätigkeit der Fliegerverbände der österr. ung. 3. Armee und der Serb. Luftstreitkräfte. Bearbeiter Rittm. ad Peter v. Spaits, p. 203.

2 Ibid, p. 45.

3 Ibid, p. 203.

4 Ibid, p. 107.

5 Ibid, p. 109–110.

6 Ibid, pp. 209–210.

7 Ibid, p. 217; D. Jordan, *Balkans, Italy and Africa* ..., op. cit., p. 53.

8 SAW-KA, Der Einsatz der Fliegerabteilungen der Herresgruppe Mackensen im Herbstfeldzug 1915 gegen Serbien. bearbeitet von major dr. Gnamm, Kmdt. zur Luftkriegsakademie, p. 17.

9 Ibid, p. 18.

10 SAW-KA, Die Eroberung Serbien 1915. Tätigkeit der Fliegerverbände der österr.-ung. 3. Armee und der Serb. Luftstreitkräfte. Bearbeiter Rittm. ad Peter v. Spaits, p. 98.

11 Ibid, p. 102.

12 Ibid, p. 104.

13 Ibid, p. 113.

14 Ibid, p. 115.

15 Ibid, p. 119–121

16 Ibid, p. 238.

17 SAW-KA, Der Einsatz der Fliegerabteilungen der Herresgruppe Mackensen im Herbstfeldzug 1915 gegen Serbien. bearbeitet von major dr. Gnamm, Kmdt. zur Luftkriegsakademie, p. 18.

18 Ibid, p. 19.

19 Ibid, p. 29.

20 Ibid, p. 31.

21 SAW-KA, Die Eroberung Serbien 1915. Tätigkeit der Fliegerverbände der österr.-ung. 3. Armee und der Serb. Luftstreitkräfte. Bearbeiter Rittm. ad Peter v. Spaits, p. 259.

22 Ibid, p. 135, 259. Ballonabteilung 1/5. B On October 12, 1915, he left Fishamend by train and on October 14, 1915 in Zemun he was assigned to the command of the 8th Corps, and on October 24, 1915 he was transferred to Belgrade. Ballonabteilung 1/6. On October 20, 1915, he was prepared for action at Zemun, and on November 6, 1915 he was near Belgrade with his assignment to the 19th Corps (the balloons of this unit were used to work with 24 cm mortar batteries near Valjevo). On November 9, 1915, near Belgrade, Ballonabteilungen 1/5 "worked" for the Third Army's fieldwork. B, 2/4. R and 1/3. B, 1/6. R and 1/5. R. The M. 98 observation balloons used during the 1915 campaign had a capacity of 1200 m3 and 750 m3, and the M 96 Kugelballons, intended for free flights, had a capacity of 600 m^3, 1300 m^3, and 1000 m^3.

23 Ibid, p. 260.

24 Ibid, p. 119-121.

25 Ibid, p. 125.

26 Ibid, p. 126.

27 Ibid, p. 254.

28 Ibid, p. 137.

29 Ibid, p. 203

30 Ibid, p. 139.

31 Ibid, p. 203.

32 Ibid, p. 141.

33 SAW-KA, Der Einsatz der Fliegerabteilungen der Herresgruppe Mackensen im Herbstfeldzug 1915 gegen Serbien. bearbeitet von major dr. Gnamm, Kmdt. zur Luftkriegsakademie, p. 32.

34 SAW-KA, Die Eroberung Serbien 1915. Tätigkeit der Fliegerverbände der österr.-ung. 3. Armee und der Serb. Luftstreitkräfte. Bearbeiter Rittm. ad Peter v. Spaits, p. 145.

35 Ibid, p. 146.

36 Ibid.

37 Д. Недялков, Въздушната..., op. cit., p. 7–8.

38 SAW-KA, Der Einsatz der Fliegerabteilungen der Herresgruppe Mackensen im Herbstfeldzug 1915 gegen Serbien. bearbeitet von major dr. Gnamm, Kmdt. zur Luftkriegsakademie, p. 28; Д. Недялков, Въздушната..., op. cit., pp. 6–7. Z. Stojczew (op. Cit., P. 5) states that the Bulgarian Aviation Division commanded by Capt. At the front, R. Miłków had six pilots, eight observers, and at the end of the year, the German ally delivered six LVG C I and one Otto CI aircraft to the Balkans.

39 Д. Недялков, Въздушната..., op. cit., pp. 13-16. During the fighting in the Aegean Sea, Struma and Maritsa, the command of the Bulgarian 10th Infantry Division received a separate Aviation Unit (two pilots, three observers and three LVG C. II planes), 13 Otto C planes were also used in the fighting. And two observation balloons.

40 Ibid, p. 19.

41 SAW-KA, Der Einsatz der Fliegerabteilungen der Herresgruppe Mackensen im Herbstfeldzug 1915 gegen Serbien. bearbeitet von major dr. Gnamm, Kmdt. zur Luftkriegsakademie, p. 36.

42 SAW-KA, Die Eroberung Serbien 1915. Tätigkeit der Fliegerverbände der österr.-ung. 3. Armee und der Serb. Luftstreitkräfte. Bearbeiter Rittm. ad Peter v. Spaits, p. 238.
43 Ibid, p. 150.
44 Ibid, p. 153.
45 Ibid, p. 151.
46 Ibid, p. 156.
47 Ibid, p. 165.
48 Ibid, p. 166–169.
49 Ibid, p. 170.
50 Ibid, p. 203.
51 Ibid, p. 171.
52 Ibid, p. 177–178.
53 Ibid, p. 172.
54 Ibid, p. 174.
55 Ibid, p. 176.
56 Ibid, p. 180–182.
57 Ibid, p. 185.
58 Ibid, p. 158.
59 Ibid, p. 186.
60 Ibid, p. 221.
61 Ibid, p. 187.
62 Ibid, p. 189.
63 Ibid, p. 190.
64 SAW-KA, X. Mikic, Geschichte der yugoslawischen Luftwaffe. Űbersetz vom Rittm. ad P. v. Spaits, p. 21–30.
65 D. Babac, A. Thers, *L'Aviation Serbe 1914-1918,* Armes Militaria Magazine No. 189. p. 41–47.
66 SAW-KA, Die Luftstreitkräfte Serbiens von 1914 bis Herbst 1916 von Oskar Regele, p. 6.
67 SAW-KA, X. Mikic, Geschichte der yugoslawischen Luftwaffe. Űbersetz vom Rittm. ad P. v. Spaits, p. 21-30; Српска…, op. cit .; CJМикић, op, cit .; Б.Dimitry, П. Миладиновић, М. Мицевски, op. cit .; Н. Ђокић, op. cit .;http://sr.wikipedia.org/wiki/Ваздухопловна команда српске војске (Краљевине Србије); http://sr.wikipedia.org/Коста_Милетић - entry from August 23, 2014; http://sr.wikipedia.org/wiki/Ваздухопловна команда српске војске (Краљевине Србије)- entry from August 23, 1914. Б. Циглић, Крила Србије Ваздухопловна команда и авијатика српске војске *1912–1920,* Београд 2009 - pasim
68 SAW-KA, Die Eroberung Serbien 1915. Tätigkeit der Fliegerverbände der österr.-ung. 3. Armee und der Serb. Luftstreitkräfte. Bearbeiter Rittm. ad Peter v. Spaits, p. 239.
69 Ibid, p. 240.
70 SAW-KA, Die Luftstreitkräfte Serbiens von 1914 bis Herbst 1916 von Oskar Regele, p. 14.
71 SAW-KA, Die Eroberung Serbien 1915. Tätigkeit der Fliegerverbände der österr.-ung. 3. Armee und der Serb. Luftstreitkräfte. Bearbeiter Rittm. ad Peter v. Spaits, p. 246.
72 SAW-KA, Der Einsatz der Fliegerabteilungen der Herresgruppe Mackensen im Herbstfeldzug 1915 gegen Serbien. bearbeitet von major dr. Gnamm, Kmdt. zur Luftkriegsakademie, p. 52.
73 SAW-KA, X. Mikic, Geschichte der yugoslawischen Luftwaffe. Űbersetz vom Rittm. ad P. v. Spaits, p. 21-30; Српска…, op. cit .; CJМикић, op, cit .; Б.Dimitry, П. Миладиновић, М. Мицевски, op. cit .; Н. Ђокић, op. cit .;http://sr.wikipedia.org/wiki/Ваздухопловна команда српске војске (Краљевине Србије); http://sr.wikipedia.org/Коста_Милетић - entry from August 23, 2014; http://sr.wikipedia.org/wiki/Ваздухопловна команда српске војске (Краљевине Србије) - entry from August 23, 1914.
74 SAW-KA, Die Luftstreitkräfte Serbiens von 1914 bis Herbst 1916 von Oskar Regele, p. 10.
75 SAW-KA, X. Mikic, Geschichte der yugoslawischen Luftwaffe. Űbersetz vom Rittm. ad P. v. Spaits, pp. 21-30; Српска…, op. cit .; CJМикић, op, cit .; Б.Dimitry, П. Миладиновић, М. Мицевски, op. cit .; Н. Ђокић, op. cit .;http://sr.wikipedia.org/wiki/Ваздухопловна команда српске војске (Краљевине Србије); http://sr.wikipedia.org/Коста_Милетић - entry from August 23, 2014; http://sr.wikipedia.org/wiki/Ваздухопловна команда српске војске (Краљевине Србије) - entry from August 23, 1914.
76 SAW-KA, Die Luftstreitkräfte Serbiens von 1914 bis Herbst 1916 von Oskar Regele, p. 13.
77 Ibid, p. 6.
78 SAW-KA, X. Mikic, Geschichte der yugoslawischen Luftwaffe. Űbersetz vom Rittm. ad P. v. Spaits, pp. 21-30; Српска…, op. cit .; CJМикић, op, cit .; Б.Dimitry, П. Миладиновић, М. Мицевски, op. cit .; Н. Ђокић, op. cit .;http://sr.wikipedia.org/wiki/Ваздухопловна команда српске војске (Краљевине Србије); http://sr.wikipedia.org/Коста_Милетић - entry from August 23, 2014; http://sr.wikipedia.org/wiki/Ваздухопловна команда српске војске (Краљевине Србије) - entry from August 23, 1914.
79 SAW-KA, Die Luftstreitkräfte Serbiens von 1914 bis Herbst 1916 von Oskar Regele, p. 7, 15. In the history of air operations over the Balkans, on the night of 17/18 May 1916, the Zeppelin LZ 85 airship bombed Saloniki (at the mouth of the Vardar it was shot at by enemy artillery). On the night of September 30, 1916, Sofia was bombed (on October 25, 1916, the crew of one plane bombed targets in the rear of the Bulgarian army during the fighting in the Thessaloniki area).
80 After the defeat of the Kingdom of Serbia and the occupation of the Principality of Albania in 1916, the 19th Corps remained in Albania - the later Albania Army Group - http://agad.archiwa.gov.pl/pomoce/Wyw328.xml - infrom 8 November 2014.
81 R. Umiastowski, *People of the Sky*, Warsaw 1930, p. 20-21.

Above: Lohner B.I *Bote* Av-3 32.15 at the front was powered by a 90 hp Austro-Daimler. Source: Aeronaut.

Below: Lohner B.VII 17.61 in flight. Source: Aeronaut.

Above: Lohner Type C "C 6". Source: Aeronaut.

Above: Lohner B.III 14.57 after a bad landing. Source: Aeronaut.

This Page: On August 15, 2019 in Czereśnie near Lubaczów, today in the Republic of Poland, and in 1914 in Galicia, on the initiative of the author, Dr. Gerhard Artl, Gerald Penz, and the mayor of Cieszanów, Zdzisław Zadworny, a grave renewed after many years was unveiled on Polish Soldier's Day. Hptm. Oskar Rosman, at the outbreak of the Great War in 1914, was the commander of field aviation of the Imperial-Royal 4th Army (Flugreferent) and commanded the first Luftschifferabteilung Fliegergruppe created in K.u.K. composed of Flik 8 and 10, who died in an air accident on August 29, 1914.

Made in the USA
Las Vegas, NV
10 November 2026

57846472R00155